PAN-SPECIES LISTING

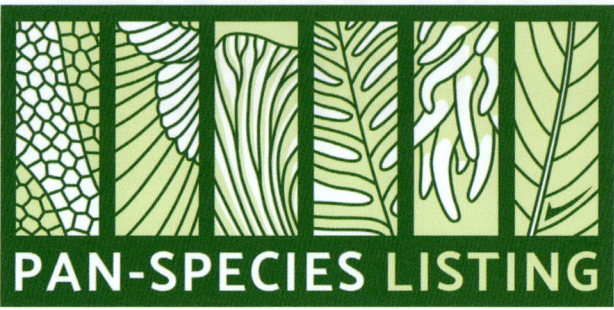
© Mark Lawlor

PAN-SPECIES LISTING

How to Become a Super-Naturalist

GRAEME LYONS

PELAGIC PUBLISHING

First published in 2026 by
Pelagic Publishing
20–22 Wenlock Road
London N1 7GU, UK

www.pelagicpublishing.com

Pan-Species Listing: How to Become a Super-Naturalist

Copyright © 2026 Graeme Lyons
Photographs and figures © the author unless otherwise stated in the caption

The right of the above to be identified as the author of this work has been asserted by him in accordance with the UK Copyright, Design and Patents Act 1988.

All rights reserved. Apart from short excerpts for use in research or for reviews, no part of this document may be printed or reproduced, stored in a retrieval system, or transmitted in any form or by any means, electronic, mechanical, photocopying, recording or otherwise, now known or hereafter invented, without prior permission from the publisher.

https://doi.org/10.53061/SDTY1821

Without limiting the exclusive rights of any author, contributor or the publisher, any unauthorised use of the contents of this publication to train generative artificial intelligence (AI) is expressly prohibited. Pelagic Publishing also exercises its rights under Article 4(3) of the Digital Single Market Directive 2019/790 and reserves the entirety of this publication from the text and data mining exception.

A CIP record for this book is available from the British Library

EU Authorised Representative: Easy Access System Europe –
Mustamäe tee 50, 10621 Tallinn, Estonia, gpsr.requests@easproject.com

ISBN 978-1-78427-517-4 Pbk
ISBN 978-1-78427-518-1 ePub
ISBN 978-1-78427-519-8 PDF

Cover artwork by Rachel Hudson

Designed and typeset by BBR Design, UK

5 4 3 2 1

Printed in the Czech Republic by Finidr

This book is dedicated to Ewart Gardner, to Steve Cooper and to the memory of my mother, Irene

Without Ewart's knowledge, passion for the subject and kindness I might never have been able to write the book, let alone live the kind of life I live now. The birding and twitching were exciting and unforgettable, but encouraging me to get into botany and giving me my first book on wildflowers changed everything for me. Steve got me hooked on moths and butterflies while I was at secondary school. I'll never forget the first moth trap I opened. It was autumn 1990 and all the moths looked like dead leaves: Angle Shades, Canary-shouldered Thorn and Setaceous Hebrew Character. So much incredible and bizarre variety with such ludicrous and wonderful names, all right on my doorstep. I've never lost that sense of childlike wonder at our natural history…

My late mother, Irene, was buried the day before I started work on this book. Without her, it never would have been written. Thanks, Mum, for giving me a long leash and for putting up with the moths flying around the kitchen every morning.

Here she stopped and, closing her eyes, took a deep breath of the flower-scented air of the broad expanse around her. It was dearer to her than her kin, better than a lover, wiser than a book. For a moment she rediscovered the purpose of her life. She was here on earth to grasp the meaning of its wild enchantment and to call each thing by its right name, or, if this were not within her power, to give birth out of love for life to successors who would do it in her place.

Boris Pasternak, Dr Zhivago

Contents

	Foreword by Chris Packham	x
	Acknowledgements	xi
1.	A brief history of (ti)me	1
2.	My approach to this book	8
3.	What is pan-species listing (PSL) and what are its benefits?	9
4.	From inception to where we are today: a PSL timeline	22
5.	The PSL website and how to take part	28
6.	Accessing the different taxonomic groups	51
	6.1 Cyanobacteria (blue-green algae)	56
	6.2 Algae (photosynthetic eukaryotes other than plants)	57
	6.3 Eumycetozoa (slime moulds)	61
	6.4 Protists other than algae and slime moulds (a polyphyletic group)	63
	6.5 Lichens (lichenised fungi)	68
	6.6 Fungi other than lichens (a polyphyletic group)	70
	6.7 Bryophytes (mosses, hornworts and liverworts)	77
	6.8 Vascular plants (clubmosses, horsetails, ferns, naked-seeded plants and flowering plants)	81
	6.9 Porifera (sponges)	87
	6.10 Ctenophorans (comb jellies)	90
	6.11 Cnidarians (jellyfish, sea anemones, corals and hydras, etc.)	92
	6.12 Molluscs (slugs, snails, limpets, bivalves, squids, chitons and sea slugs, etc.)	99
	6.13 Bryozoans (sea mats, hornwracks and lace corals, etc.)	108
	6.14 Annelids (earthworms, marine worms and leeches, etc.)	110
	6.15 Platyhelminths (flatworms, tapeworms and flukes, etc.)	115
	6.16 Pycnogonids (sea spiders)	118

	6.17	Arachnids (spiders, harvestmen, scorpions, pseudoscorpions, ticks and mites, etc.)	119
	6.18	Myriapods (millipedes, centipedes, pauropods and symphylans)	129
	6.19	Crustaceans (woodlice, amphipods, crabs, lobsters, crayfish, barnacles, shrimps and copepods, etc.)	133
	6.20	Entognatha (springtails, proturans and two-tailed bristletails)	141
	6.21	Insects: Archaeognatha and Zygentoma (three-tailed bristletails and silverfish)	144
	6.22	Insects: Odonata (dragonflies and damselflies)	146
	6.23	Insects: Orthopteroids (grasshoppers, bush-crickets, crickets, groundhoppers, stick-insects, cockroaches, earwigs and mantids)	149
	6.24	Insects: Hemipteroids (true bugs, leafhoppers, aphids, whiteflies, scale insects, psyllids, psocids, thrips and lice, etc.)	154
	6.25	Insects: Hymenoptera (bees, ants, wasps and sawflies, etc.)	164
	6.26	Insects: Coleoptera (beetles)	173
	6.27	Insects: Diptera (true flies)	184
	6.28	Insects: Lepidoptera: butterflies	196
	6.29	Insects: Lepidoptera: moths	199
	6.30	Insects: Remaining small orders (mayflies, stoneflies, caddisflies, lacewings, scorpionflies, snakeflies, alderflies, stylops, web-spinners, fleas)	209
	6.31	Echinoderms (sea urchins, sea stars and sea cucumbers, etc.)	215
	6.32	Tunicates (sea squirts and salps)	218
	6.33	Fish (a paraphyletic group)	220
	6.34	Reptiles (a paraphyletic group)	226
	6.35	Amphibians	228
	6.36	Birds	230
	6.37	Mammals	234
	6.38	Other animals (nematodes, tardigrades, rotifers, other worms, hemichordates, etc.)	239
7.		Pan-species listing in unusual habitats and specific situations	244
8.		How to become a super-naturalist: hints and tips	259
		The correct way to write common names of species	289
9.		Can pan-species listing change your life?	301
10.		Pan-species listing and collaborative competition	315
11.		Pan-species approaches to surveying and monitoring	327

12.	Pan-species listing of sites	338
13.	Public engagement and PSL field meetings	343
14.	Representation and demographics in pan-species listing	356
15.	Neurodivergence, natural history and pan-species listing	363
16.	Threats to pan-species listing	371
17.	Pan-species listing in other countries	377
18.	Lifetime strategies for pan-species listing	381
19.	The future of pan-species listing	385
	Index	387

Foreword

'Ligules', he said, 'ligules are hard. They all look the same, grasses are next to impossible.' I crouched down and plucked a stem, peeled back a leaf and peered into my lens at the papery frill, a fragile little fragment of tissue that I'd never considered before. I then picked a second stem and teased out its little ligule. It was entirely different. You see for me, the beautiful, playful, naughty devil has always been in the detail. The devil brings me happiness, the devil lights my darkness, the devil is the sparkling star that fires my interest which then burns like a fabulous yellow Roman candle. And then I have to tell you all about it.

I love life. All life. And as mine will always be too short, so I want to suck it all up while I can. I want to see it, hear it, smell, touch and taste it all. And then make a list of it all. I love lists. I have lists of lists. I rarely forget anything on those lists, but I make them, update them and sometimes revise them. This can be annoying… taxonomists like changing names, I don't, but I grudgingly do.

I can't call myself a Pan-species Lister. Because if I'm not into something I just can't be interested in it at all. So forgive me, I'm a specific species lister… Nevertheless, this clever and immensely resourceful book is an overdue gem, not just a repository of repositories, a glorious catalogue of all the catalogues, the ultimate list of all the joyous lists, a key to the keys to find all the beautiful things that join life together but a story told by a brilliant naturalist.

Enjoy, and never ever be afraid of ligules, or spikelets or racemes!

Chris Packham

Acknowledgements

Many thanks are due to Sarah Whild, Andy Musgrove, Seth Gibson, Finley Hutchinson and Mark Gurney for agreeing to read the manuscript and offering invaluable early guidance. Thanks also to Karen McDermott for enduring this project for three years and for reading the whole book twice. I am hugely grateful to Mike Prince for doing most of the work on the new website, to Brian Eversham for contributing valuable text on microscopic organisms, to Gerald Legg for his help with all of the marine sections, and to Chris Raper for his help with the UK Species Inventory and his expertise in the Tachinidae.

I also want to thank the following people for their help with the different taxonomic groups: Howard Matcham and Chris Carter (freshwater algae and cyanobacteria); Stephen Plummer (slime moulds); Nicola Bacciu (lichens); Dick Alder, Claire Blencowe and Stephen Plummer (fungi); Fay Newbery (aquatic hyphomycetes); Paul Bowyer (bryophytes); John Poland (vascular plants); Martin Willing (molluscs); Seth Gibson and Brian Boag (platyhelminths); Tylan Berry, Richard Gallon and Helen Smith (arachnids); Christian Owen and Steve Gregory (myriapods and isopods); Duerden Cormack (myriapods and springtails); Warren Maguire (marine crustaceans); Peter Sutton (orthopteroids). Jim Flanagan (hemipteroids); Sally Luker (aphids, psyllids and scale insects); Derek Binns (thrips); Mike Edwards, Gavin Broad, Derek Binns, Simon van Toller and Steven Falk (Hymenoptera); Mark Telfer, Martin Fowlie and Simon Van Toller (Coleoptera); Mark Gurney (weevils and soldier beetles); Chris Bentley, Andy Musgrove, Alice Parfitt and Steven Falk (Diptera); Barry Warrington (Agromyzidae); Tony Davis (Lepidoptera); Nicolas Jouault (some of the marine groups); Tom Brereton (marine mammals); Laurie Jackson (terrestrial mammals).

I am also grateful to Bob Foreman for helping me with several maps, and to Mick Crawley for insightful discussions about the extreme end of biological recording.

I am hugely grateful to Andy Darling, my therapist, for helping me overcome the impostor syndrome that came with writing this book, guiding me through the world of neurodivergence and generally helping me through some pretty rough times.

Finally, I wish to thank Sally Luker, Sarah Patton, Nicola Bacciu, Libby Morris, Jen Fellows, Laurie Jackson and Anna Maka for discussions about representation at the more involved end of natural history and pan-species listing, and Esmond Brown for talking to me about representation of people of colour in natural history. And to Joe Myers for his input on access to wild places for those with mobility issues. Thanks to John Boback for his thoughts on PSL in the USA.

Our 'one-hour bioblitzes' at Global Birdfair are a great way to pass on knowledge and encourage new people – young or old or anywhere in between – to get into biological recording via pan-species listing. Starting early really does pay off in terms of developing incredible natural history skills, and you won't go wrong with the PSL approach. From left to right: Andy Musgrove, Graeme Lyons, Billy-Joe Beatwell, Clare Beatwell, Deborah Sacks, Harrison Knight, Michael Knight and Sullivan Knight. (Emily Knight)

Chapter 1

A brief history of (ti)me

This book aims to be a guide to – and history of – pan-species listing (PSL), as well as a record of my own personal experience of studying and engaging with natural history, documenting some of the challenges I've experienced along the way. I hope it will inspire others and show them that anyone can do this, even those who have been dealt a difficult hand in life. I shall start by giving you a brief history of me and how I got to see 9,637 species in the British Isles by the time I was 47.

I was born in 1978 in the former mining town of Rugeley in Staffordshire. My father, who had worked first as a miner and then as a gamekeeper after the mines closed, left when I was five. Mum brought up my two older sisters and me for about five years, in a little council house about a mile from Cannock Chase, where we were well below the poverty line, but I have very happy memories of this part of my childhood.

Then my stepdad came on the scene, and soon afterwards my younger sister was born. From then on my home life became very difficult, there was neglect and emotional abuse, and I was quite badly bullied at school, all of which affected me deeply. Around this time other mental health problems developed, all of which went unrecognised by my family.

At around the same time, one of my teachers, Ewart Gardner, started to take me out birding with the local Young Ornithologists Club (YOC). I loved birding, and once we had managed to see all of Staffordshire's commoner birds, we started to travel a little further afield to look for rare breeding birds, such as Black Grouse *Lyrurus tetrix* at Swallow Moss in north Staffordshire. This soon developed into full-on twitching. My best twitching day ever was in autumn 1992, when we visited Spurn, Flamborough Head and other sites nearby. In a single day we saw Hume's Leaf Warbler *Phylloscopus humei*, Pallas's Warbler *Phylloscopus proregulus*, Asian Desert Warbler *Curruca nana*, Richard's Pipit *Anthus richardi*, Isabelline (as it was called then) Shrike *Lanius isabellinus*, Great Grey Shrike *Lanius excubitor*, Shore Lark *Eremophila alpestris* and more. As I write this it is hard to believe, but it really happened – I was there.

Every Saturday morning when I wasn't out with the YOC I would go down to the farmland around San's Brook and work my way up to Cannock Chase, walking miles patch listing on my own, and savouring the solitude. Back then there were breeding Willow Tit *Poecile montanus*, Turtle Dove *Streptopelia turtur* and Grey Partridge *Perdix perdix* as well as a significant wintering flock of Tree Sparrow *Passer montanus*. It was just farmland, but it was mine. It was how I taught myself the intimacies of bird song and calls. It was mindful, and it distracted me from a noisy and claustrophobic home life. I learned to enjoy my own company and, most importantly, I learned that I could figure things out on my own.

When I was a little older, Ewart gave me a copy of *The Wild Flowers of Britain and Northern Europe,* by Richard Fitter, Alastair Fitter and Marjorie Blamey. This book opened up a whole new world of botany, which would later become a solid foundation for everything that came afterwards. During my third year at secondary school, a new teacher, Steve Cooper, sparked my interest in moths by letting me help him ID them from an actinic moth trap that he had set up on the school roof. With his help, I also managed to build my own trap in craft, design and technology (CDT) classes. I ran this trap continuously for five years in my garden, as well as my first year at university. I always kept a copy of Bernard Skinner's *Colour Identification Guide to Moths of the British Isles* nearby. My head was always in a book.

My world fell apart the day I finished my GCSEs. Mum took me to the park in a panic, after a passing comment from a neighbour. She bought me a cake and told me that my father wasn't my biological father. "When can I meet him then?", I asked. "He died when you were seven", she replied. I'll never fathom why she took until I was 16 to tell me, but it wasn't in my best interests. My world pivoted. Who was I now? Where did I really come from? This initiated a period of self-medicating with alcohol, and more, that went on for about 20 years. So much pain and devastation had been unleashed on me and my siblings, and we are still paying the price today. Yet, despite all of this, I feel that I got off lightly compared with other family members.

During my A level studies I didn't get on with the biology teachers (Steve Cooper was not teaching A levels in those two years), who seemed completely uninterested in my passion for natural history. In the whole two years we had just one field trip – to

'Do you like my owl?'

the limestone Peak District, an area where I had spent a great deal of time botanising with Ewart. We were tasked with recording plants in some quadrats, and when I was told that I had identified the Common Rock-rose *Helianthemum nummularium* incorrectly and that it was in fact a buttercup, I decided that a biology degree was not for me. There was no support from my family to encourage me to reconsider this decision, and I probably wouldn't have listened to them if they had tried!

Ewart was a big influence on me, as he had been to university, and at that point no one in my family had done so. Having Ewart around had a much more motivating effect than I realised at the time, as it sparked the attitude that 'if they can do it, I can do it'. This has become a mantra by which I live my life, as well as a core tenet at the heart of PSL. I had great physics teachers, loved science fiction and was enjoying reading Stephen Hawking's *A Brief History of Time*. I have also never forgotten Ewart's throwaway remark that conservation jobs involved 'getting paid peanuts to count Puffin eggs on a freezing cold island'. So I studied astrophysics for four years at Sussex University.

I picked the university that was both as far away as possible from Rugeley, whilst also being situated in the most promising-looking landscape for nature. Academically I did very well in my first year, but just scraped a degree in my final year. Apart from running a moth trap in the first year, most of my spare time at university was spent drinking and chasing girls but I soon picked up natural history again after finishing my degree. After a couple of years without making a single record, I got back into birds in the autumn of 2000, a year after the publication of *Collins Bird Guide* – the kind of book I had always dreamt of! I went to Beachy Head by bus in the hope of seeing some migrants, and found my first ever Osprey *Pandion haliaetus*.

I didn't know what to do next. By this time I was covered in tattoos, had 15 piercings and a mass of long dreadlocks. I called in at the Job Centre to see an advert for residential volunteering at RSPB Dungeness. I knew Dungeness well, having run a moth trap there while on holiday when I was a teenager. I went for the interview, was successful, but then had to wait six months for a place to come up. In the meantime I worked in a piercing and tattoo shop, and before I knew it I was at Dungeness, under those big skies and a long way from all my friends back in Brighton. I loved it, though. In fact, I often think I was never happier until I went completely freelance 18 years later. I worried that everyone else would be way ahead of me after spending four years studying natural history at university, but I was wrong about this. Knowing my birds, plants and moths since my teens actually put me at a huge advantage over most biology graduates and the other volunteers I met (except of course for my good friend Chris Bentley, another volunteer who had also put in a huge amount of time learning natural history in his own time, *before* his placement).

However, it soon became clear that the world of conservation was very white and very middle class. I recognise that, as a white male in the UK, I am already in a position of some privilege (I will discuss this in more detail in Chapter 14), but volunteering was so much easier for those from a more affluent background. Most had already learned to drive; I tried when I was at home doing my A levels but when I asked Dad for help, all I got was "what's the point of learning to drive if you can't afford a car?" I had to learn that skill while volunteering, which resulted in me having driving lessons in Staffordshire (an emergency stop for a flock of Waxwing

Bombycilla garrulus during my second lesson didn't go down well), Kent, Norfolk and eventually Anglesey – where I finally passed on my fifth attempt. Life was made harder by not being able to drive, and by having to sign on every two weeks and explain why I couldn't apply for a casual job. Because of a lack of support when I was younger, it took a lot of work on my part to overcome the feeling that this kind of life was not for the likes of me.

I volunteered for four and a half months at Dungeness, and then got my first paid job in conservation, as a Reserve Assistant at RSPB Titchwell Marsh. I didn't mind about the low pay – on my walk to work on my first day I saw a Gull-billed Tern *Gelochelidon nilotica*, and I found an Iceland Gull *Larus glaucoides* on a WeBS count, my very first task. Then short-term contracts followed each other in quick succession – from Titchwell Marsh to a Black Grouse survey in North Wales, to a tern colony on Rhosneigr, back to volunteering at South Stack and then to the Woodland Bird Survey in Shropshire and Wales. In summer 2003 I got my first big break, working for Malcolm Ausden in reserves ecology – a dream post in which I stayed until 2008. I really believe that when I blurted out 'Green Sandpiper!' in the middle of the interview, as I heard one call, it helped to get me the job.

Working for Malcolm was great. He taught me my grasses, sedges and rushes, which I myself went on to teach for 15 years. I learned so much about ecology, habitat management, conservation grazing and identification. I also learned about National Vegetation Classification (NVC) communities, how to map them and use GIS, and about experimental design and statistical analyses – all skills that I use to this day. Best of all, though, I got to work on around 200 nature reserves across England, Wales, Scotland and Northern Ireland, in the company of some great naturalists, such as Mark Gurney, Mark Telfer and Matt Self. There were five unforgettable winters spent monitoring fish by electrofishing with Matt. At one point towards the end of those five years, surrounded by so many people with doctorates, I nearly decided to do a PhD myself when I heard that someone had dropped out at short notice of one that would be supervised by the RSPB. However, in the end I decided that, at this stage of my life, three or more years of what I was doing already would be more useful experience than focusing narrowly on one specific area (controlling Marsh Ragwort *Jacobaea aquatica* on the Somerset Levels). It turned out to be the right call.

After seven years working for the RSPB, I longed to be back in Brighton, and one day I saw an advert for an ecologist at Sussex Wildlife Trust. I applied, got the job and stayed in that role for nearly 12 years. It was a fantastic experience, involving the design and implementation of a monitoring strategy for 32 nature reserves. We were a very good team, and everything worked seamlessly for a number of years. James Power was an excellent manager and strategic thinker, and we rarely disagreed. He was calm, considered and selfless, keen to facilitate the work of his staff. I also met Mike Edwards, a dedicated entomologist and ecologist with a brilliant mind who has helped me a great deal over the years, especially in freelance entomology. We still work closely together today, and I shall always be grateful to James and Mike for their support and friendship.

Soon after I started working for Sussex Wildlife Trust, I began doing occasional freelance work, and in my spare time I started a blog (at that time called The Lyons Den) and drifted into the world of listing everything (this activity didn't have a

Electrofishing is the most fun you can have with 240 volts and water. (Matt Self)

name then). Little did I know that I was inadvertently training myself to become an entomologist. When the credit crunch hit in 2008 and my monitoring budget to pay external specialists dried up soon after, I was in a position to start surveying the reserves as an in-house entomological surveyor. This wasn't a perfect arrangement, but it was a start. After 11 years I made the switch to part-time work at the Trust in early 2019, and finally left and became a full-time freelancer in December of that year, just before the pandemic.

Freelancing has been very good for me. I have written standardised methods for monitoring invertebrates and plants on rewilding projects, and applied these at various leading rewilding projects throughout the UK. I have developed methods for monitoring biodiversity on farms across south-east England, and I am currently booked up two years in advance. I manage a database of well over 350,000 records. I have become a specialist in spiders, probably seeing more than anyone in the British Isles (559 species). I am a leading advocate of the pan-species listing movement and have seen over 9,637 species on these shores. I have a side project about a little fella that does natural history that has over 26,000 followers and over 1.5 million hits on my blog. I sit on various committees in Sussex and now have more influence than I did when I worked for the nature conservation charities.

And here comes the plot twist that even I have not yet come to terms with. I have just been diagnosed with autism and attention deficit hyperactivity disorder (ADHD) – a combination known as AuDHD. After a number of people mentioned that it might be worth getting an assessment, I started reading around the subject and bit the bullet. Being AuDHD is a hugely positive factor for my work, and is probably

Me surveying invertebrates at West Beach, 2025. (Vicki Richardson)

the reason I am doing what I am doing – it's like a superpower on that front. I know I have an unusually retentive memory, which helps a great deal. It is the other aspects of AuDHD that have caused me problems – how I interact, my anxiety around other people, why I can't walk away from a fight, why I often appear blunt and rude, and why I won't do anything I'm not interested in (but can't stop doing the things I am interested in). It is also why I spent 20 years trying to drink myself into the shape of an extrovert – a behaviour that is known as masking. The way that I have approached the writing of this book would give most people nightmares. I am remarkably chaotic and messy internally, but the silhouette I cast just about passes for 'normal', so I can get by in a world designed for people who are not like me. I decided to get the assessment mainly to inform this book, and indeed the diagnoses have significantly changed some of the content. Notably, a chapter on neurodivergence that wasn't remotely on my radar before I did this research.

I overcame a difficult childhood, the associated mental health problems and undiagnosed AuDHD, and then I managed to build a successful career in conservation as an influential entomologist, arachnologist, botanist and freelancer – despite the odds being stacked against me and some poor decisions being made early on in life. This book sets out a framework of natural history within which to operate – a framework that I helped to build and influence, and that has become an ever-present rock in the maelstrom of noise that is my life. It is now also used and enjoyed by an ever-growing number of naturalists. Natural history has been my lifelong, guiding light in the darkness, and PSL has provided me with structure in a world of bewildering confusion. I see it as a road map to completing in a lifetime as much of the infinite jigsaw puzzle that is the wildlife of the British Isles as I can. I know that I will never finish it; however, it's not the finishing that matters, but the journey itself. *That is what PSL is*, and my journey proves that PSL is for anyone and everyone, regardless of their background or personal challenges.

Chapter 2

My approach to this book

If I had concentrated solely on the practical aspects of pan-species listing (PSL) this book would have been much shorter, much less informative and probably less interesting, concerned only with the listing process, the website and field meetings. However, I have also described all of the conservation and biological recording benefits of PSL – what some might call 'pan-species recording'. These are arguably the most important aspects of PSL. Indeed, for me, the listing and recording of species are one and the same thing, so in this book, pan-species listing can be assumed to include pan-species recording – one can't exist without the other.

I have adopted a very personal approach, using my own experiences, history, anecdotes and humour to add colour to the text. For me and many others, PSL is a movement – changing British natural history at the coalface. I see being a pan-species lister a bit like being a Jedi: both require training, dedication, a set of skills that few people have, and a desire to use them to do good, with the help of some guidance and structure. Others might not agree, simply regarding this activity as a bit of fun, which it certainly is – pan-species listers rarely take themselves too seriously, me included. Of course, both views are true, but if some of us hadn't shifted the focus more towards recording, the benefits to wildlife might not have been as great.

I don't believe that PSL is the only way to become a 'super-naturalist' or just a competent all-round naturalist, but it is the route I have taken. Nor am I suggesting that this method is better than your current methods if you are an experienced naturalist – rather it is something you can do in addition to what you are already doing. Because PSL works for me I believe that it could work for anyone, and by describing my own personal experiences I hope to enable other people to benefit in the ways I have done, so that ultimately wildlife will benefit, too.

This book is also a snapshot in time of PSL about 15 years after its inception. For that reason, the numbers of species listed will become out of date almost immediately, but they are there to demonstrate how dedicated, obsessed or driven (depending on your viewpoint) we are both as a community and as individuals. Lists and numbers mean a lot to pan-species listers, so I wanted to acknowledge their importance by including them in the book.

I live and breathe natural history, and have been dedicating most of my time to it for decades. However, I'm aware that very few people will be able to invest this much time and energy in PSL. That doesn't matter – it's a unique journey for each person, and the aim of this book is to inspire: don't let the big totals at the top of the rankings put you off starting, be inspired by them instead!

Throughout this book, species are indicated by their common name (if a widely used common name exists) followed by their scientific name – for instance, Sprawler *Asteroscopus sphinx*.

Chapter 3

What is pan-species listing (PSL) and what are its benefits?

Most birders keep a 'life list' – a list of all the birds they have seen in the UK or the British Isles since they developed an interest in the subject. A smaller number of individuals (often a subset of the birders) will keep a list of, for example, moths, butterflies, dragonflies or orchids. Some botanists keep a note of exactly how many species of vascular plant they have recorded, too. Some people might list several or all of these groups. But suppose you were to keep a list of every living species you have seen – not just every bird, butterfly, moth and flowering plant, but also every species you have seen from the less well-known groups, such as spiders, beetles, mosses, lichens, fish, molluscs, pseudoscorpions, caddisflies, starfish, anemones, bats, fleas, flies, cetaceans and crustaceans. What would that look like?

I wrote the following paragraph over ten years ago, but it's just as relevant now as it was then:

> A pan-species list is a list of all the animals, plants, fungi and protists you have seen in Britain, Ireland and the Channel Islands. Whether a Daisy or a Death's-head Hawk-moth, a Killer Whale or a Killer Shrimp, all species count as equal on your pan-species list. Although this may seem like the trivialisation of natural history to the accumulation of a big list, it's what is behind the list – how you get there – that makes this approach to natural history so powerful. Add a healthy dose of competitiveness in the form of the rankings page and, thanks to Mark Telfer, pan-species listing was born. Will this bring about a 'renaissance of the all-round naturalist'?

The goals of PSL

Today the main aims of the pan-species lister and the PSL movement as a whole are:

- to provide a standardised and transparent mechanism for all-round naturalists to maintain their life lists across all taxa, using the most up-to-date species inventory in a gently competitive manner through gamification
- to do so in an enjoyable, accepting and engaging way through the provision of structure and community, while simultaneously becoming a formidable source of knowledge when we are considered collectively
- to bring about a renaissance of the 'all-round naturalist' in a world where there is a growing tendency to be drawn into specialisms

- to create a framework that helps you to become as competent as possible at identification across many different taxa, and to break down the barriers that deter people from tackling difficult groups
- to pass on the baton of species identification and recording to the next generation, in a world where identification skills are at risk of being lost
- to generate millions of high-quality biological records of under-recorded groups by promoting taxonomy, field identification, microscopy and, most importantly, the amazing wildlife of the British Isles.

The benefits of PSL

Obviously there are plenty of other ways to achieve these benefits and, as I mentioned in Chapter 2, PSL isn't the only way to become a super-naturalist or an all-round naturalist, but it really has worked for me, so here's why it could work for you, too.

Each species counts as one

One of the unexpected conservation benefits of PSL is that each species counts as one. A Death's-head Hawk-moth *Acherontia atropos* counts as one, but so does Annual Meadow-grass *Poa annua*. You do not get more points on your list for the Alchymist *Catephia alchymista* than you do for a Large Yellow Underwing *Noctua pronuba*. Himalayan Balsam *Impatiens glandulifera* will get the same score as Snowdon Lily *Gagea serotina* (not that we don't value our native species above our non-natives from a conservation perspective, but we certainly value the importance of recording non-natives, even the invasive ones). This equivalence changes the very nature of how you operate as a naturalist.

Of course, in many respects this is not very different to birding. However, it doesn't take long to see most of the common breeding and regular wintering birds in the British Isles, whereas a lifetime is not long enough to see all of the common species in every other taxonomic group.

PSL has changed the way I think. I have always been more focused on guilds (species that share the same resources) when providing habitat management advice for a site. In that context, thinking in terms of entire species lists might seem impossible, but in fact it isn't, and 'Embrace the complexity' has become a motto of mine. It might seem daunting at first, but as you gain experience you can learn to sift your way through a list of thousands of species in order to come up with the best way to manage a site (within the resources available) so as to cater for as many of those species as possible. Many people when confronted with such bewilderingly large numbers tend to focus on just one or two species, sometimes at the expense of all the others, which can potentially result in failure. However, if we consider all of the taxonomic groups present at a site, it can change the way that we approach the conservation of such a huge diversity of species.

PSL improves your species ID skills, by pushing you outside your comfort zone

By far the greatest benefit of starting PSL is that it will push you outside your comfort zone into ever more strange new worlds of natural history. Fifteen years ago I would never have believed that I would develop such a deep interest in spiders and the marine environment, and derive so much enjoyment from them, but these are now the areas that I spend most time on outside of work.

PSL improves your taxonomic knowledge

You can't work your way through all of these species groups without learning a thing or two about taxonomy – something I missed out on at university. I would certainly not admit to having total mastery of this area, but I know more than I once did, and I can hold my own when talking to most other naturalists on the subject. I have also learned a lot about taxonomy while working on this book!

PSL teaches you how to handle large data sets

As I write, I am preparing to give a talk on the biodiversity of the Pevensey Levels, including a 'gap analysis' (that is, to highlight what taxonomic groups are poorly recorded) of the entire species list for the levels. I asked Bob Foreman, Biodiversity Data Lead at the Sussex Biodiversity Record Centre, for all the data they hold on the designated area, consisting of around 70,000 records. At one time I would not have known where to begin, but because I've become used to processing such large amounts of data regularly, this analysis shouldn't take me much longer than an hour or two. With just 2,615 species recorded on such a large and heavily designated area (almost 1% of the area of Sussex returned just 0.5% of Sussex's records held in the SxBRC), I was able to show which taxa were under-recorded, suggest possible reasons for this and offer practical solutions to try to rectify the situation. There is now a pan-species list for the Pevensey Levels for others to maintain and update.

PSL forces you to look at the common species, not just the rarities

If you want to do well in PSL you really do have to work your way through entire species groups. This means that you'll end up looking at a lot of common species if you have taken on a large number of different groups. Casual recording in natural history can be a little more superficial, focusing on the rarities or the big showy species. However, the PSL approach is all about trying to master entire groups, by trying to 'complete the set'. I have found that this approach to listing and recording has significantly influenced the recording methods I use as a freelance ecologist. Now, within the taxa I am surveying, I always try to record everything I can. Although I love seeing rare species and I still do a small amount of 'twitching', rarity chasing alone isn't going to pay the bills.

PSL promotes long-term thinking

In an age of instant gratification (just think of the way smartphones reduce our attention span), where long-term thinking seems to be on the decline, having a potentially lifelong hobby is particularly rewarding and worthwhile.

Planning weeks, months, years and decades ahead (often all at the same time) is complicated, but it's a great skill to develop. In recent years, my long-distance excursions for the purpose of PSL have been limited to holidays, which generally only happen in late March and late September, fortuitously coinciding with the best tides of the year! Building your knowledge of species that you have not yet seen is an integral part of mastering any group, and for the pan-species lister this is magnified across multiple taxa. You need to learn about a lot of species of which you have no experience, so that you are prepared when you do finally encounter them. Books are a great way to acquire this knowledge, as is looking through your social media feeds from the more active members of the PSL community. I recently asked a friend which species of sea slug she had seen, so that I could learn more about the ones I am likely to see on the shoreline. I then spent half an hour researching the four species whose names I didn't recognise. Not all of the information stayed with me, but I did remember Dusky Doris *Onchidoris bilamellata* enough that when I finally saw one in September 2025, I recognised it.

If only I could turn back the clock and revisit all those sites I visited with the RSPB, especially the Scottish reserves, but from a PSL perspective. I have seen 559 species of spider in the British Isles, but I have not recorded a single spider in Scotland. Therefore my advice to all is: whatever stage in life you start PSL, you should never squander an opportunity.

Back in October 2013 I joined in a PSL twitch to Cornwall with Mark Telfer, Seth Gibson and Neil Fletcher. We were part of the original 20 or so listers, and the target was a Hermit Thrush *Catharus guttatus* at Land's End. It was great to see the thrush, but it turned out not to be the highlight. In fact, the most visually impressive species of the day was the Blue-rayed Limpet *Patella pellucida*. While the most unexpected find of the day was a spider, an unusual species of *Enoplognatha* that was found in a wall by the car park at Long Rock Beach. It was immature and most probably something very rare or new to Britain. Nine years later, I sent the photo and the details of the record from my database to the County Recorder for Cornwall, who

Immature female *Enoplognatha mandibularis*. Long Rock Beach, Cornwall, October 2013.

reported that he was almost certain that it was *Enoplognatha mandibularis*, which would be a spider new to the UK.

In 2023, I went back to Cornwall to interview some of the more active pan-species listers there. I finally got to meet Sally Luker, one of the original pan-species listers and one of the few women in the top 100 (I shall be discussing representation in Chapter 14). I also had a wonderful few days with Louis Parkerson and Finley Hutchinson, both very active pan-species listers and students at the Penryn Campus of Exeter University. On my last morning I visited Long Rock Beach car park, exactly ten years since my previous visit and encounter with the unidentified spider. I got out my suction sampler (an electric leaf blower modified to suck invertebrates into a bag) and ran it between the rocks in front of the bewildered-looking occupants of some parked cars. I found an adult male and an immature of this spider in the very first sample! I jumped in the air, whooping with excitement, as it was indeed *Enoplognatha mandibularis*, although it felt like my third first for the British Isles, it was actually my first (because it took a decade to conclude). I really do like to play the long game, and what longer game could there be than pan-species listing?

I plan to write a sequel to this book in roughly 25 years' time when I'm in my 70s, perhaps I will call it "Pan-species Listing – I once claimed I was a super-naturalist!". I wonder how high the species totals will be then, and how different the rankings will look. Will I have reached the mythical 20,000 species mark? Is that even possible in a lifetime? We shall have to wait and see.

PSL generates lots of good-quality, usable records, especially of under-recorded groups

As PSL has been going for around 15 years now, we really are starting to see its positive effects on biological recording – the two are intrinsically linked. This book is probably more about biological recording than it is about PSL, and I've pushed hard to make sure that the website is more aligned with biological recording, too.

I started keeping a Recorder 6 database to keep track of my PSL efforts in 2012, and it currently holds 353,253 records, this repository of data is how and where I keep track of the huge number of records I collect through my freelance work. According to Bob Foreman at Sussex Biodiversity Record Centre, I have recorded more species in Sussex than anyone else. That's at least 7,689 species (80% of everything I have ever seen), and would put me in eleventh place on the national rankings at the time of writing. I'm also now the most prolific individual recorder in Sussex ever. To date, the Sussex Biodiversity Record Centre holds 225,184 of my records, while the next highest recorder has not yet reached the 110,000 mark. This means that roughly 1 in 50 of all the records in my local records centre are mine, and that proportion is rising.

PSL creates competent all-rounders who also develop specialisms

One of the most important things to remember about PSL is that you should avoid spreading yourself too thinly at any one time – you have a lifetime to master all of these groups. Do the things that really interest you and then the whole endeavour will be self-sustaining. Yes, this is about being an all-rounder, but we all end up

Nine sea slugs, representing nine of the most exciting wildlife encounters I have ever had. From top left to bottom right: *Facelina auriculata*, *Polycera quadrilineata*, *Facelina bostoniensis*; *Archidoris pseudoargus* (Sea Lemon), *Pelagella castanea*, *Doris verrucosa*; *Okenia nodosa*, *Edmundsella pedata*, *Aeolidia filomenae*.

specialising in one or more areas at any one time. For me it's currently entomology, arachnology and the marine environment, but especially spiders. However, by the time you read this book it might be sea slugs, as I seem to have a growing need to get to the coast more often to search for these wonderful little blobs of colour.

There is no point trying to force yourself to be interested in a particular group. I have tried hard with lichens but I just find them really difficult, and – I hate to admit it – somewhat dull. I would love to look at them in more detail, to ignite that spark of interest, but I simply don't have the time right now. Perhaps when I slow down one day, they'll become more appealing.

PSL enables you to make a huge contribution in a relatively short space of time

There are enough groups and subgroups for everyone to carve out a niche for themselves. As you become really knowledgeable in one area, you can then start to draw other people in by sparking their interest too. I have seen this happen with spiders. At the time of writing, there is a huge increase in recording, which is even changing the statuses of our spiders! My advice is to have lots of fingers in lots of pies, but also to plonk yourself firmly in the middle of one or two big pies and take those to the next level.

Within less than a decade, Andy Musgrove has become a national specialist on sawflies, and has written a status review on them for Natural England. Meanwhile

James McCulloch, who is a PhD student, is now County Recorder for springtails in Surrey, has recently become the national recorder and is the top lister for them on PSL. He has also written a book on the springtails of Surrey that is soon to be published.

The unexpected find is often better than what you are looking for

In late 2022, while on holiday in Pembrokeshire with my partner, Karen, I had just one target for the trip, Scaly Cricket *Pseudomogoplistes vicentae*. I also had a site for this species, Marloes Sands. I spent hours looking for it without success, only to spot (with minutes to spare as the tide was coming in) a rather striking spider covered in golden hairs run across the shale in front of me. It was unmistakably the exceptionally rare *Callilepis nocturna*. This stunningly beautiful spider is classified as Vulnerable and Nationally Rare, and is known from only three locations in the UK. I knew of the Isle of Wight population and the one in Devon, but I didn't know that this species was at Marloe Sands, too. This perfectly demonstrates the exciting unpredictability of PSL, and also why it is the ideal hobby for someone with ADHD (see Chapter 15), because there is always a plan B, or a plan C... "Ooh look, a beetle!"

All of my efforts to find my original target had been unsuccessful, but instead I found something far better, in a completely different taxonomic group. A failure became an epic win, and for just a brief moment I also believed that I had found a spider new to Wales. In fact I returned from that holiday with four lifers, demonstrating a very 'pan' spread across different taxonomic groups, none of which were targets – *Callilepis nocturna*, Boat Bug *Enoplops scapha* (found as a nymph by Karen),

The immature *Callilepis nocturna* that appeared in front of me while I was looking for Scaly Cricket *Pseudomogoplistes vicentae*.

Horse Mackerel *Trachurus trachurus* and, best and most overdue of all, Otter *Lutra lutra*!

Twitching isn't quite the same for a pan-species lister as it is for a birder. A day out will often be driven by the search for one or more *resident* target species, where there's usually a population present rather than just one individual. The stakes therefore never feel quite so high – a lone specimen of Monkey Orchid *Orchis simia* is never going to suddenly fly off! These target species are often just the catalyst to get you out of the door and into a new area, though (like twitching) they can be frustrating and disappointing if you don't see a bird or other species on a twitch or targeted trip. This experience is known as 'dipping out', and the higher the stakes, the bigger the dip. It's usually the unexpected species that you find, often while dipping out on the target species, which are the real source of joy. Such finds are arguably a better result for conservation, too, as they are often new site records (rather than the same species being recorded on the same site by many different people each year). These unexpected moments are perhaps the best thing about a day out pan-species listing.

Literally anything can happen at any moment!

Not only could you find the unexpected while out searching for a specific species, but also you just never know when you might strike gold. I particularly enjoy coming across lifers when I'm 'off duty'. I was once backstage at the Green Man Festival in Wales and found a beetle on the ceiling in a toilet. It was *Oedemera femoralis*, a large and impressive species associated with dead wood; this was only my second encounter with it at the time. Super Furry Animals were playing that day, and when one of the band members walked in as I was balancing on the urinal to get the beetle down off the ceiling he became intrigued, so I showed him the beetle – that was a special moment. Another time I was walking home from the gym and found a big weevil that I didn't recognise. I had several miles to walk, so I used an empty Tic Tac container as a weevil transportation unit. The weevil turned out to be just *Brachypera zoilus*, but at the time I had never even heard of it.

Tic Tac Tick Tactics – it's always worth having a container on your person for when you spot something you've never seen before. Like this *Brachypera zoilus* (other weevils are available).

PSL is a hobby that you can take with you wherever you go. At any moment a random organism can come into your field of view and bring a little joy to your day, and by recording it you can add to our understanding of the distribution and ecology of our species. Even though, as a pan-species lister, you are maximising the number of groups that you cover, you might still be surprised how often these chance encounters occur. And there are a bewildering number of new species crossing the Channel due to climate change, as well as accidental introductions, further increasing the frequency of encounters with new species.

If you put the hours in, PSL pays off

I typically make over 50,000 records of some 4,000 to 5,000 species a year. Amount of time spent in the field, therefore, is the most significant factor for me to be able to achieve this – resolute determination will not get you very far if you don't put in the hours. The best natural history find of my career happened one weekend in June 2016 when, instead of taking part in a PSL recording event in Norfolk as I had hoped to do, I was trying to complete some territory mapping of farm birds, as these surveys had been delayed by recent bad weather. I was on an arable farm near Bishopstone in East Sussex, and had found a few plants of Field Gromwell *Buglossoides arvensis*, which was a lifer, but despite this I was thinking wistfully about all the lifers I was missing on the field trip in Norfolk. Then I saw on the ground at the edge of a field of wheat what I at first assumed to be a Rose Chafer *Cetonia aurata*, not particularly scarce on the Sussex Downs. However, this species was behaving oddly, wriggling around on the ground, and I soon realised that it was the massive, metallic, rainbow-coloured carabid, *Calosoma sycophanta*! Overcome with excitement I immediately

Mythically rare and painfully beautiful, behold the mighty *Calosoma sycophanta*!

phoned Mark Telfer, who was at the Norfolk field event. He promised to drive straight down if I found another one, but alas that did not happen. It had been just a chance encounter with a very rare migrant. At the time this was the first UK record in 19 years, the first record of a living animal in 23 years, and the first Sussex record since the nineteenth century. I have never been as happy to miss a field meeting as I was that day. And it just goes to show how, if you put in the hours, no matter where you are, eventually you will find something amazing.

PSL is for people of all ages

The two youngest pan-species listers on the rankings are five and seven years old, while the oldest is 82! I've recently recruited the well-known Sussex naturalist and coleopterist, Peter Hodge. On his first night he stayed up until nearly 4 am ticking off beetles on the website! PSL can be as exciting and addictive for a lifelong naturalist as it is for someone who is completely new to it.

You can get lifers every day... for life!

When I think of my twitching days, I remember the feeling of nervous anticipation as we got within a few miles of the bird, as I worried that it would fly off before we could get to it. Then there is the excitement of actually seeing a new species. PSL can give you all that every day if you put the time in. These days, I get a huge thrill not just from seeing a new bird, but also from seeing a new spider, bug, beetle, mammal, moth or vascular plant – and particularly a new sea slug!

The box-fresh Golden Twin-spot that arrived on a stormy night in Brighton.

I generated my first pan-species list on 1 August 2010, about 15 years before the publication of this book. I had recorded 2,748 species back then, whereas at the time of writing I am on 9,637 species. That is equivalent to 1.24 lifers a day, every single day. You don't get that with birding! In 2024, I kept track of how many lifers I got each day, too. By 6 February 2024 I already had 47 lifers, which is an average of 1.27 lifers a day, slightly up on my 15-year average. By 14 November I had 393 lifers, which is 1.23 lifers a day, almost the same as my 15-year average – a remarkable consistency across these different time scales.

Even after 35 years of mothing, I am still getting lifers in my garden actinic, such as this Golden Twin-spot *Chrysodeixis chalcites*. Most of my lifers these days are from specimens collected over the summer during work, but I still get out regularly over the winter, even when I'm writing. This really is a lifelong hobby that can generate little packets of joy and dopamine with an incredible regularity and consistency.

Are there any downsides to PSL?

Personally I don't believe that there are any drawbacks, but there have been some criticisms of the concept, though these have either ignored or dismissed its conservation benefits. I shall attempt to discuss the reasons for this here.

PSL and competitiveness

Many people are put off by the perceived competitiveness of PSL. Yes, at its heart there is an element of competitiveness, but this is a movement based on cooperation, collaboration and inclusiveness. The competitive aspects can both inspire people by showing them what can be achieved higher up the rankings, and, most importantly of all, facilitate a framework that enables us all to be continuously competing *with ourselves*. I find it really heartening when I see someone, who I know puts in the leg work, start to shoot up the rankings. I am less inspired by those who are time-and-time-again, simply ticking species off a list after they have been shown them by an expert, as this doesn't even generate any new records. However, such behaviour is rare in PSL, and generally speaking we all help each other out.

Sometimes people will overtake you on the rankings. At the lower end, movement is turbulent. The convection currents of new recruits bubble upwards, as young stars rise fast, displacing some as they progress. However, the drop-off rate can also be high at the bottom end of the rankings. Mark Telfer coined the term 'base camp' for the plateau where a number of us sit under the dizzying heights of the top lister, Jonty Denton, a naturalist who has seen well over 13,000 species. Up at base camp, our lists grow more slowly and are more widely separated, here movement in the rankings occurs at a glacial pace. Many of us have edged ever closer to the arbitrary but seemingly unreachable threshold of 10,000 species, which at the time of writing is still around 500 species away for us at base camp.

What exactly is twitching and why does it get so much bad press?

Twitching does get a lot of bad press, and indeed there are always a few people whose behaviour ruins it for everyone else. It can also incur a large carbon footprint, but you can limit this by car sharing, using public transport or reducing the distance that you travel to more local twitches. Serious twitchers are often extremely good birders and competent naturalists, and many find their way into PSL, or make significant contributions to conservation in other areas of their lives, through work, surveying or volunteering, for example. I used to twitch much more than I do now, and I strongly believe that some twitching is part of a healthy, balanced diet of natural history.

Twitching is fairly well defined in relation to birding. It usually involves travelling (often, but not exclusively, long distances) to see an individual rare bird from faraway lands, which someone else has found and that you might not have another chance to see on these islands for a very long time, or possibly ever again. The bird could potentially fly off at any moment and, most important of all, it will usually be a species that you have not seen.

PSL is about so much more than a list – it is about the unbridled joy and enthusiasm that we experience when interacting with the natural world, and in that respect the situation is no different for twitchers. PSL is also undoubtably about a way to get regular dopamine rewards, especially for those who are neurodivergent, though this reward drug is released in the brain of anyone who experiences something new.

This kind of structuring of our lives is often particularly important for neurodivergent people. So if you are the kind of naturalist who turns their nose up at the idea of listing or twitching, perhaps it's worth taking a step back to consider why people are doing it. They might be autistic, and this activity might be really important to them, and if they're not neurodivergent, it's still likely to be really important to them, possibly for all of the same reasons.

One of the last twitches I went on that had this effect was not to see a bird, but the Walrus *Odobenus rosmarus* in Hampshire. That tusked tonne of blubber was an unexpected wonder to behold and moved me to tears. Yet some naturalists would look down on me for even going to see this, as it falls into the remit of twitching. However, they are wrong to do so.

Would going to see a plant or insect that someone else has found be considered twitching? I believe the answer is yes, especially if it was not normally present in that particular area. A couple of years ago I was lucky enough to see the first Monkey Orchid *Orchis simia* to appear in Sussex. Previously I had many times pencilled in the exact week when I would need to drive to Kent to see these plants at their known site, but my summer fieldwork commitments meant that I always ended up cancelling the dates in my diary as they approached.

If you have to drive to a specific site to see a species that someone else has discovered, such as a new population of Praying Mantis *Mantis religiosa* on the Isle of Wight, then I'd consider that twitching, with or without large crowds being present. Fast forward a few years. If that species now occurs on the site annually, apparently without the possibility of suddenly disappearing or not being there the following year (with any crowds now completely absent and the associated stress minimised), then this is starting to feel some way from what most people would term 'twitching'. There's clearly a blurred line between what exactly is and isn't twitching. Whatever you call it, all such excursions still have a carbon footprint, the potential to have an ecological impact if not carried out sensitively and potentially rather limited benefits if only one species is targeted (and hopefully recorded) at a well-visited site. I imagine that a lot of naturalists who don't consider themselves twitchers engage in plenty of this kind of activity.

Yet no PSL twitch happens in a vacuum. While I was in that Sussex valley twitching the Monkey Orchid I made quite a few other records, including Dingy Skipper *Erynnis tages*, Spotted Flycatcher *Muscicapa striata* and Firecrest *Regulus ignicapilla*. I bumped into a naturalist whom I had known on social media for some time but had not met in person before. I advised on the management of the orchid for the estate, we discussed its provenance, and I saved myself a much longer journey to Kent. This certainly felt like twitching to me. So much negativity about twitching has also been directed at PSL over the years. This includes the perception that twitchers and/or listers don't care about the sites or the welfare of the animals and plants that they go to observe and then 'tick off' or list. Yet for me and all the people I know personally who are involved in PSL, this activity is much more about conservation than it is about 'collecting'. Many of the pan-species listers I have met are lifelong, passionate naturalists and nature conservationists, who are simply using the PSL approach as a framework for seeing, enjoying, recording and sharing as much of the UK's wildlife as they possibly can, in the time that they have left to them.

If you drive any distance to see a rare migrant or a rare resident, there is clearly a cost to the environment in terms of carbon emissions. On the other hand, there is a benefit to the environment in terms of what you learn from the experience. Seeing new counties, regions and sites helps to make you a better naturalist and ecologist, especially if you make a whole day of it and record as much as you can while you are there, or stop off at other sites on the way. Sharing your records on iRecord or with your local records centre is also of huge value, and sharing your knowledge and experiences on social media helps others. Alternatively, you can just enjoy the drive and then indulge yourself in some observations of nature, in the same way that you would drive to a football match or to visit friends or family. Naturalists need to be less harsh on twitchers. And maybe birders who twitch should all get into PSL, so that they can see how many more ticks they could get right on their doorstep, and reduce their carbon footprint at the same time.

Chapter 4

From inception to where we are today: a PSL timeline

How did PSL begin?

Around 2010, a number of UK naturalists had (independently of each other) started to come up with this idea, but it was Mark Telfer who really put PSL on the map. He hosted the then rankings on a small section of his excellent website (sadly no longer active) by putting his list up as a single numerical total, along with the totals of around five other people. These six totals were ranked in order of size, with the highest number at the top, and in that moment the idea of 'rankings' was born, along with the embryonic concept of PSL.

For the first time, not only could the sum total of a person's natural history identification and recording exploits be represented as a single figure, but also it could be measured against the corresponding figures for everyone else who had compiled their list. This is a double-edged sword, as the rankings approach draws in some more competitive people and puts others off. Yet, as I explained in the previous chapter, if you have focused on the competitive element of PSL before you have learned anything else about it, you have totally missed the point, and I hope this book will demonstrate the relative insignificance of the competitive aspect.

By 2011 there were still less than ten people on the rankings list, but after that the number started to grow significantly, and by 2013 there were close to 40 of us. By that time, too, a thriving community had developed and we held our first in-person field meeting (described in more detail on page 346).

The old website

Mark needed a new approach, as it was becoming virtually impossible for him to keep up with the almost daily updates from 40 people. He and I met up in autumn 2013 and discussed our options. We felt that some kind of purpose-built website was needed, but we knew that we couldn't do this ourselves, as we lacked the necessary IT skills and didn't have the time to learn them either.

That December, during a Sussex Wildlife Trust (SWT) work team meeting, Charles Roper, who was working at the Sussex Biodiversity Record Centre, suggested that I should contact David Roy of the Biological Records Centre (BRC). Charles thought that the potential benefits that PSL could have for biological recording might well convince the BRC to fund the building and hosting of a website, including the all-important rankings. I put together a proposal and David Roy had the foresight to see the long-term benefits that PSL could have in enabling large numbers of high-quality records of under-recorded groups to be created.

It wasn't just the BRC that saw the benefits: SWT and the Sussex Biodiversity Record Centre were able to offer several days of staff time to help in building the website. The BRC funded John Van Breda to create the site using Indicia (the same software that was used to develop iRecord). I designed the structure of the website, and Charles Roper and Bob Foreman helped to build it with John over two days in April 2014.

It soon became clear that we had an opportunity to produce something more sophisticated than the simple rankings that we had been using up until now. We wanted rankings for different taxa, not just for our overall species totals, but this would involve allocating all of the species in the UK to a manageable number of groups. Mark Telfer had the relevant expertise and immediately offered to do that for us, and this is how the 38 taxonomic groups (see below) were created.

Around this time we decided that we needed to start thinking about 'rules', as a lot of new pan-species listers were joining us. The 'guiding principles' as we now call them are described on page 37.

Taxonomy of the old website

Now that we had a bespoke website, we were able to list not only the total number of species per person in the rankings, but also the subtotals – for instance, we could now see who the top beetle, butterfly or mammal listers were. Mark Telfer did a sterling job of creating standardised taxonomic categories that were both practical and comprehensive, and we made a few modifications to his original list. It seemed reasonable to separate butterflies from moths, despite the fact that they all sit in the same order. We also felt that although dragonflies and damselflies, and crickets, grasshoppers and allies are 'small orders', they needed to be pulled out of the catch-all group of insects at the end into their own groups, but we had to draw the line somewhere. The largest order of insects that didn't get its own group was the caddisflies.

The original 37 categories that we eventually decided upon (as a community) are listed here (with an additional category, the cyanobacteria, added to the beginning of the list in 2023).

1. Cyanobacteria (blue-green algae)
2. Algae (photosynthetic eukaryotes other than plants)
3. Eumycetozoa (slime moulds)
4. Protists other than algae and slime moulds (a polyphyletic group)
5. Lichens (lichenised fungi)
6. Fungi other than lichens (a polyphyletic group)
7. Bryophytes (mosses, hornworts and liverworts)
8. Vascular plants (clubmosses, horsetails, ferns, naked-seeded plants and flowering plants)

9. Porifera (sponges)
10. Ctenophorans (comb jellies)
11. Cnidarians (jellyfish, sea anemones, corals and hydras, etc.)
12. Molluscs (slugs, snails, limpets, bivalves, squids, chitons and sea slugs, etc.)
13. Bryozoans (sea mats, hornwracks and lace corals, etc.)
14. Annelids (earthworms, marine worms and leeches, etc.)
15. Platyhelminths (flatworms, tapeworms and flukes, etc.)
16. Pycnogonids (sea spiders)
17. Arachnids (spiders, harvestmen, scorpions, pseudoscorpions, ticks and mites, etc.)
18. Myriapods (millipedes, centipedes, pauropods and symphylans)
19. Crustaceans (woodlice, amphipods, crabs, lobsters, crayfish, barnacles, shrimps and copepods, etc.)
20. Entognatha (springtails, proturans and two-tailed bristletails)
21. Insects: Archaeognatha and Zygentoma (three-tailed bristletails and silverfish)
22. Insects: Odonata (dragonflies and damselflies)
23. Insects: Orthopteroids (grasshoppers, bush-crickets, crickets, groundhoppers, stick-insects, cockroaches, earwigs and mantids)
24. Insects: Hemipteroids (true bugs, leafhoppers, aphids, whiteflies, scale insects, psyllids, psocids, thrips and lice, etc.)
25. Insects: Hymenoptera (bees, ants, wasps and sawflies, etc.)
26. Insects: Coleoptera (beetles)
27. Insects: Diptera (true flies)
28. Insects: Lepidoptera: butterflies
29. Insects: Lepidoptera: moths
30. Insects: Remaining small orders (mayflies, stoneflies, caddisflies, lacewings, scorpionflies, snakeflies, alderflies, stylops, web-spinners, fleas)
31. Echinoderms (sea urchins, sea stars and sea cucumbers, etc.)
32. Tunicates (sea squirts and salps)
33. Fish (a paraphyletic group)
34. Reptiles (a paraphyletic group)
35. Amphibians
36. Birds
37. Mammals
38. Other animals (nematodes, tardigrades, rotifers, other worms, hemichordates, etc.)

The UK Species Inventory

The UK Species Inventory (UKSI), which is a database of all wildlife species in the UK, is a vitally important resource for biological recording of those species. Recording tools such as iRecord and Recorder 6 run off the UKSI. The database is kept as up to date as possible (in terms of species that are new to the UK, changes in taxonomy and changes in nomenclature) in order to maximise its usefulness and relevance, so it is constantly evolving. Chris Raper at the Natural History Museum (NHM), whose job it is to keep this list up to date, kindly provided me with a short history of the UKSI:

> The UK Species Inventory has been evolving over many years, and the current version reflects the thoughts and intellectual input from a number of people, including Stuart Ball (Joint Nature Conservation Committee, JNCC), Dave Mills (JNCC), George Boobyer (JNCC), Charles Copp (Environmental Information Management, EIM), Charles Hussey (NHM), John Tweddle (NHM), Mike Weideli and Chris Raper (NHM). The original generic data model was constructed by Charles Copp in 1998, and was physically implemented in the National Biodiversity Network's Recorder 2000 software (a precursor to Recorder 6) by Steve Wilkinson and Andrew Cottam (JNCC). The project is now managed and maintained by Chris Raper at the Natural History Museum in the Angela Marmont Centre for UK Nature. The UKSI is updated with the kind help and support of a wide range of amateur and professional taxonomists.

PSL and BUBO listing

In October 2023 the PSL Facebook group discussed the possibility of putting the UKSI on the old website; this subject has come up many times over the last decade but due to the cost and skills needed, it was never a viable option. Then Andy Musgrove announced that he and Mike Prince were thinking about putting the entire UKSI on to their long-established BUBO Listing website.

BUBO Listing is a completely free website that has been around since 2006. It was designed, built and maintained by a few friends who wanted to create a place where birders can input their life lists in a standardised way by ticking birds off the most up-to-date and taxonomically accurate list available, with complete transparency. It has been very popular, and at the time of writing has around 5,400 users.

Although this sounded like a brilliant idea, I realised there was a risk that it could make the old website obsolete. In order for it to be possible to update the species totals on the old PSL site and keep that running in parallel with BUBO, the taxonomic groups in BUBO would need to be exactly the same. Moreover, the transition would need to be not only smooth but swift, as I was in the middle of writing this book and desperately wanted it to remain relevant. I explained all of this to Andy Musgrove and we set up a working group, though Mike Prince and Andy did all the important work.

We started by adding a few of the smaller orders to BUBO. The first batch consisted of the butterflies, orthopteroids and Odonata. This worked really well. Not only does BUBO Listing allow you to tick off your list from the UKSI, but because it's completely transparent everyone else can see your species lists (something that

A collage I put together in October 2023 to promote the first batch of insects added to BUBO. Left to right: Purple Emperor *Apatura iris*, Southern Migrant Hawker *Aeshna affinis*, Large Cone-head *Ruspolia nitidula*.

had not been possible before). It also has several very useful functions, such as 'targets' and 'blockers'. The target function shows the species that you have not seen, but that most other listers have seen. The blocker function shows the species that the least number of people have seen, starting with a list of species that only you have seen. I shall consider targets and blockers in more detail on pages 33 and 35 respectively.

Between October and December 2023 we added more taxonomic groups. It was a very smooth transition, and provided the perfect opportunity to go through my list with a fine-toothed comb. I added quite a few species in some taxa – vascular plants and moths I did very well from, I'd seen far more species than I had listed on the old website, whereas with fungi I did very badly, my poor record keeping here meant that I lost quite a few species.

A new situation emerged shortly after the second or third update. BUBO Listing had been a birding site for the last 19 years. Many new recruits were coming in to PSL via this route, so it seemed completely unacceptable that they also had to sign up to the existing PSL site, where they would have to enter all of their subtotals manually. It was hard enough to convince those on the old site who had been doing this for ten years to keep doing so. It was clear that this situation could not continue, and that putting the UKSI on BUBO was going to make the old website obsolete, as predicted.

Yet this wasn't necessarily a bad thing. Out of the Ash *Fraxinus excelsior* of the old website, a Phoenix *Eulithis prunata* was born. We started to consider whether we could rebuild the old website, with *at least* all of its existing useful functionality, inside BUBO. It became clear that this might be an option, and I jumped at the chance. There were a couple of major deal-breakers. First, the new website would need to have its own unique branding and web address, so that PSL did not lose its own identity as a result of being embedded within another website. Secondly, some of the key functionality of the homepage would need to be moved across, so that for the existing listers it didn't feel like a completely different approach. Before I knew it, I had bought the following website:

www.panspecieslisting.com

We launched the new site in mid-January 2024, and Mark Lawlor has designed a stylish and clever logo for it. The site is working extremely well – at the time of writing, nearly two years after the launch, around 409 people are on the main rankings. To put this in perspective, it took ten years to get up to 300 people on the rankings on the old website, though when I last checked the old site, only around 100 people had updated in the last two years. It is clear why the new site is so popular – the UK Species Inventory is right there at your fingertips, managed very smoothly by the guys at BUBO.

Chapter 5

The PSL website and how to take part

The new website

It's worth emphasising that the website is not a means of submitting records. Instead it offers the user something that is fun to do in addition to entering records. We are not competing with iRecord or other similar recording platforms.

In this chapter I shall describe some of the key parts of the site. The homepage shows the rankings, the latest few listing milestones, news and a statistics box – along with Mark Lawlor's smart new logo in the top left-hand corner.

In order to appear on the main rankings, it's important to set your list up as being in the location 'Britain, Ireland, Isle of Man & Channel Islands', even if you've never been outside the UK, or never intend to go outside it, as this is the area that we've decided PSL can operate within.

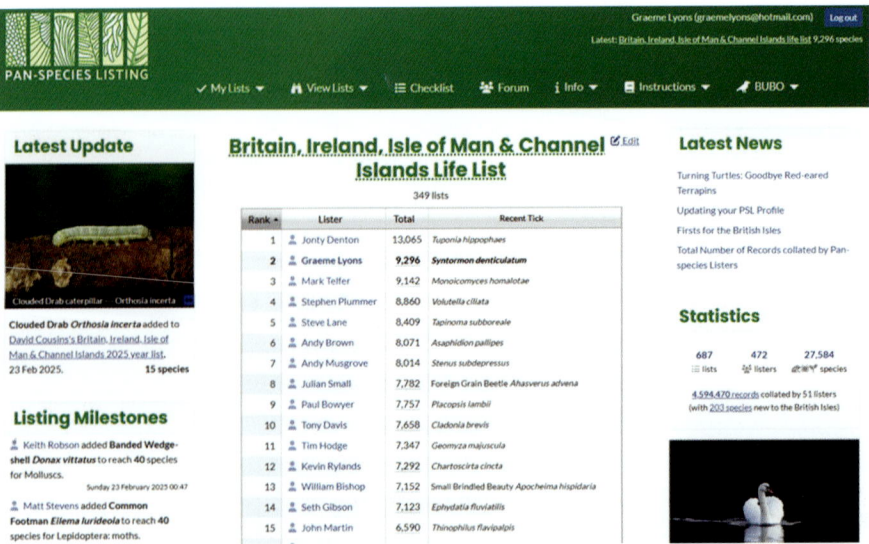

A screengrab of the homepage on 22 December 2024.

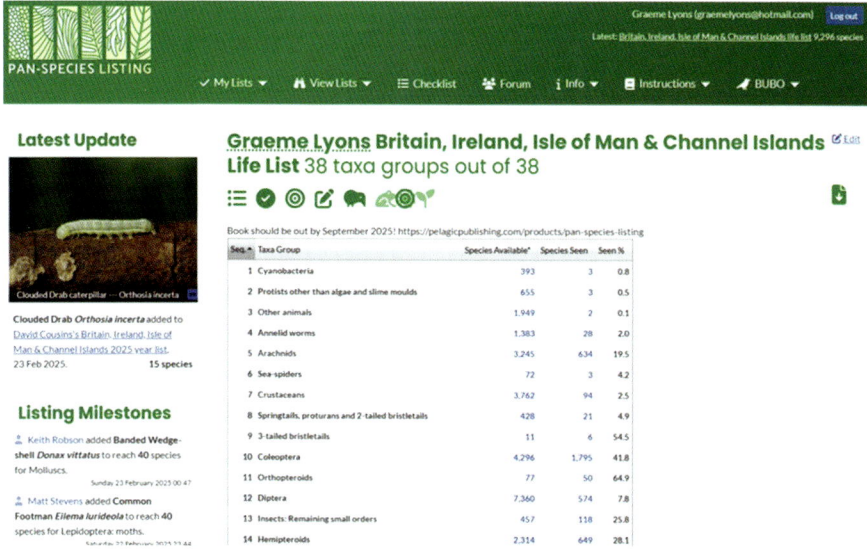

The top part of my list from 22 December 2024.

My list

If I now click on my name, it takes me to my whole list. This is too unwieldly to show in its entirety, so here I have just included a section of my summary table. If you click on the person icon next to your name, it will take you to your profile. We encourage everyone to complete this, as it's really helpful for those who are new to listing to see what the more experienced listers are doing in terms of their careers, etc.

Statistics box

A great new feature is the statistics box, which shows what the collective species list is for the current listers. At the time of writing there are 552 listers, and the statistics box has just reached 28,746 species – compare this to the number of species recorded on iRecord (33,860) and on iNaturalist (24,342) for the UK. There are 75,474 species up for grabs on the website (in addition to all of the new species that turn up annually), so we have collectively recorded just 38.1% of the species on the UKSI! Over the last two years, people have been ticking off an average of 12 new species a day on the website, which means that around 8,216 species have been ticked off for the first time in this period.

You can also enter in your profile how many biological records you have made (see page 259 for a quick introduction to biological recording). This part of the website is still in the early stages of development, but we already have 5,133,400 records from just 70 listers! There are two reasons for presenting this information:

1. It demonstrates the scale of the collective recording effort of pan-species listers, which already rivals that of a local record centre. For example, Sussex Biodiversity Record Centre currently holds over 12.6 million records submitted by over 22,564 recorders. We have generated over 40% of that total with

only 0.3% of the number of recorders! Of course, not all of those records can be attributed directly to PSL, but many of them can, and in some cases (such as mine) almost all of them can.

2. It encourages record keeping among pan-species listers, especially with so many new people joining over the last year who perhaps do not come from a biological recording background. This seemed like a great opportunity to bring biological recording into the heart of what we are doing, and allow everyone to contribute to the steadily growing total – whether they have half a million records or just a handful. It is hoped that everyone will update their number of records at least annually.

Ideally, these records will then be further shared with iRecord, local record centres and/or recording schemes. My working assumption is that anyone who has made the effort to digitise a lifetime's natural history experiences will want those records shared and used as much as possible. However, there are sometimes contractual and logistical limitations to sharing records, in which case you can still provide a total of everything you have digitised, and it is hoped that such records will find their way into wider use eventually.

It is fairly straightforward for me to obtain the figure from my Recorder 6 database, but you might have records in, for example, Excel, iRecord, iNaturalist, BirdTrack and/or one or more local record centres. In this case, rough figures are totally fine. If you have years' worth of data in notebooks, this is a gentle prompt to get them digitised and shared. If you pan-species list but don't record, hopefully this can be the start of an exciting new chapter in your life as you finally take the leap into biological recording.

Although it's vitally important that biological records are submitted, shared and used more widely, this is not the only reason for recording. There are also huge benefits to the individual who is making the records, bringing structure and rigour to a naturalist's life whether, professional or amateur. The purpose of generating my own species maps daily in my #speciesaday feature on Bluesky is as much for me to gain insight into my own records (such as revealing distributions of species I was unaware of, or highlighting gaps in my recording efforts) as it is to inspire others. However, biological recording is most enjoyable and rewarding – and, more importantly, *nature will benefit most from it* – if you share your records so that others can

Statistics

827 lists **552** listers **28,746** species

5,133,400 records collated by 70 listers
(with 203 species new to the British Isles)

Statistics box on 22 October 2025.

use them. I personally take great comfort in knowing that my records will continue to be of value long after I am gone.

The PSL website and the UKSI

One major and unexpected benefit of the new website is that, with so many experienced and passionate taxonomists scrutinising the UKSI, we are now in a position to feed changes and updates to the team at BUBO. They in turn forward this information to Chris Raper in carefully considered batches. As a result, everyone benefits – not just the world of PSL but also the whole world of biological recording. There are 379 pending additions to the UKSI from the first two years of the website alone.

Species new to the British Isles

I have been fortunate enough to find five species new to the British Isles, four of which I found in the last two years. Collectively, pan-species listers have recorded at least 203 species new to the British Isles.

I have yet to find a species new to science, but I was amazed to discover that so many people had, notably Finley Hutchinson, who had already found at least three by the age of 20! In total around 11 species new to science appear to have been recorded by pan-species listers.

The bug *Geocoris megacephalus*, that I found new to the British Isles on Jersey.

You can enter in your profile how many species you've seen that are new to the British Isles, and write some text about what they are. There are clearly many different levels at which people contribute to such firsts, but they all count – whether you found the species and identified it alone, or as part of a group, or identified it for someone else. It's expected that multiple people may report having found the same species, and this is fine. If there are any reasons why a species that you've found that is new to Britain might be shared by other listers, please highlight this on your list of species. We will probably have to do some editing of the final total to reflect this, but it won't affect your total.

My first four new species were fairly clear cut, as they appeared in my records alone, but a bug I recently recorded as new to the British Isles (on Jersey) is an example of a situation where things might get complicated. Someone else can claim that they have recorded this species when it eventually turns up in the UK, which is fine, and we would edit the total to reflect this. However, when a new species turns up and spreads rapidly, multiple people might find it in the same year, but it is likely that one person found it first, in which case they alone should claim that record (unless one of the others made some significant contribution to its identification).

'Listing Milestones'

This is the new name for what was called 'Ranking News' on the old website. The website is set so that it announces on the homepage:

1. when you reach a multiple of 10 for any taxonomic group
2. when you reach a multiple of 100 for any taxonomic group (this milestone is displayed in a pink box)
3. when you reach a multiple of 1,000 (this milestone is displayed in a much-coveted grey box!)

Bill Urwin added **Knotty Shining Claw** *Lamprochernes nodosus* to reach 150 species for Arachnids.	2,499	6 Nov 2024 10:02
Dave Pearson's record of *Grateloupia subpectinata* is the first for PSL!	3,263	6 Nov 2024 07:50
Derek Mays added **Lesser Weever** *Echiichthys vipera* to reach 30 species for Fish.	2,109	5 Nov 2024 23:02
Derek Mays added **Speckled Wood** *Pararge aegeria* to reach 30 species for Lepidoptera: butterflies.	2,024	5 Nov 2024 19:16
Derek Mays has reached 2,000 species with the addition of *Pthirus pubis*!	2,000	5 Nov 2024 18:18
Derek Mays added **Red-necked Phalarope** *Phalaropus lobatus* to reach 260 species for Birds.	1,979	5 Nov 2024 17:34
Sally Luker added **Pine Carpet** *Pennithera firmata* to reach 590 species for Lepidoptera: moths.	5,286	5 Nov 2024 16:44
Martin Gray's record of **Dune Navel** *Omphalina galericolor* is the first for PSL!	4,169	5 Nov 2024 15:40
Lloyd Davies has reached 1,000 species with the addition of **Slender-horned Leatherbug** *Ceraleptus lividus*!	1,000	5 Nov 2024 15:33
Lloyd Davies added **Broad-clawed Porcelain Crab** *Porcellana platycheles* to reach 30 species for Crustaceans.	968	5 Nov 2024 15:01
Paul Clack added *Favolaschia claudopus* to reach 130 species for Fungi other than Lichens, including fungoid organisms.	3,422	5 Nov 2024 10:59
Derek Mays's record of **Chinese-lantern** *Abutilon pictum* is the first for PSL!	1,666	5 Nov 2024 09:23
Brian Hedley added *Coremacera marginata* to reach 90 species for Diptera.	3,426	5 Nov 2024 07:32
Graeme Lyons's record of **Midas Tree-weaver** *Midia midas* is the first for PSL!	9,195	4 Nov 2024 20:34
Derek Mays added **Four-spotted Chaser** *Libellula quadrimaculata* to reach 20 species for Odonata.	1,391	4 Nov 2024 20:12

A selection of listing milestones from 4 and 5 November 2024.

4. when you reach 10,000, which we have witnessed only once on the new site, something suitably celebratory is planned (you'll need to put in about 20 years of work to find out what this is)
5. the first time that a species has been listed by anyone on the website (this milestone is displayed in a bright yellow box).

Targets

The targets feature is denoted by the bull's-eye icon. On your list, click on this icon and it will show you the species that you have not seen, but that most other listers have seen. The targets are arguably the main benefit to people who dislike the competitive nature of the rankings, as they teach you which species you should be looking out for. My top ten targets are shown in Table 1. You may be surprised to see that the dragonfly Common Hawker *Aeshna juncea* is my top target! This is because I'm a southerner now and this species is actually very scarce in the south-east. Each and every one of us has a number one target, and it's totally transparent – you can see everyone else's targets as well as your own.

As so many birders came across from BUBO at the launch of the website, the proportion of birds (even very rare vagrants if they were widely twitched by many birders) high up in anyone's targets is going to be greater than it would have been if we had just moved over the old group of pan-species listers. It is expected that this will change over time, but it does mean that, without some editing, the targets are not at present very 'pan'. Table 2 shows my top ten targets with the birds removed.

If you select a taxa group (or sub taxa group) from the drop-down box on your main targets page, it will give you your targets for that tax group or sub group. A few months ago something happened which demonstrates that PSL is much more collaborative than it is competitive. I had decided to spend a day doing some recording with my friend Tony Davis at Shortheath Common in Hampshire. Before we set off, Tony (top moth lister and micro-moth expert) went on to my list to see what common leafminers I needed; he noticed that I was missing a guild of species

Table 1. My personal top ten targets (including birds) at the end of October 2025.

Rank	Species	Seen by other listers	
1	Common Hawker *Aeshna juncea*	190	47%
2	Feral Goat *Capra hircus*	173	43%
3	Barred Warbler *Curruca nisoria*	155	39%
4	Spanish Bluebell *Hyacinthoides hispanica*	136	34%
5	Common Rosefinch *Carpodacus erythrinus*	133	33%
6	Snow Goose *Anser caerulescens*	130	32%
7	Spotted Sandpiper *Actitis macularius*	129	32%
8	Pine Marten *Martes martes*	126	31%
9	Caspian Tern *Hydroprogne caspia*	117	29%
10	Stilt Sandpiper *Calidris himantopus*	117	29%

Table 2. My personal top ten targets with the birds removed.

Rank	Species	Seen by other listers	
1	Common Hawker *Aeshna juncea*	190	47%
2	Feral Goat *Capra hircus*	173	43%
3	Spanish Bluebell *Hyacinthoides hispanica**	136	34%
4	Pine Marten *Martes martes*	126	31%
5	*Bryotropha affinis*	109	27%
6	Common Bistort *Bistorta officinalis*	101	25%
7	Grass-of-Parnassus *Parnassia palustris*	100	25%
8	*Pammene fasciana*	99	24%
9	Knotted Wrack *Ascophyllum nodosum*	98	24%
10	Atlantic Salmon *Salmo salar*	97	24%

* Many people are incorrectly listing Hybrid Bluebell *Hyacinthoides* x *massartiana* as Spanish Bluebell.

Table 3. The number one targets of the top 20 pan-species listers.

Rank	Lister	Species	Seen by other listers	
1	Jonty Denton	Minke Whale *Balaenoptera acutorostrata*	153	38%
2	Graeme Lyons	Common Hawker *Aeshna juncea*	190	47%
3	Stephen Plummer	Arctic Skua *Stercorarius parasiticus*	210	52%
4	Mark Telfer	Snowdrop *Galanthus nivalis*	209	52%
5	Steve Lane	Bottle-nosed Dolphin *Tursiops truncatus*	222	55%
6	Andy Musgrove	Palmate Newt *Lissotriton helveticus*	209	52%
7	Andy Brown	Human *Homo sapiens*	261	65%
8	Julian Small	Grayling *Hipparchia semele*	252	63%
9	Paul Bowyer	Bottle-nosed Dolphin *Tursiops truncatus*	222	55%
10	Tony Davis	Common Pipistrelle *Pipistrellus pipistrellus*	264	66%
11	Tim Hodge	Palmate Newt *Lissotriton helveticus*	209	52%
12	William Bishop	Mole *Talpa europaea*	232	58%
13	Kevin Rylands	*Pterostichus madidus*	144	36%
14	Seth Gibson	Mottled Rustic *Caradrina morpheus*	174	43%
15	John Martin	Rudd *Scardinius erythrophthalmus*	140	35%
16	Calum Urquhart	White Admiral *Limenitis camilla*	216	54%
17	Sam Thomas	Chinese Water Deer *Hydropotes inermis*	206	51%
18	John Poland	*Agapeta hamana*	184	46%
19	Sarah Patton	Golden Eagle *Aquila chrysaetos*	195	49%
20	Simon Van Toller	Lesser Black-backed Gull *Larus fuscus*	267	67%

on Blackthorn *Prunus spinosa*, and so we spent half an hour looking for them. I was even able to find one myself, and recorded and ticked *Stigmella plagicolella*. I have since recorded it elsewhere this autumn, all thanks to the new website and Tony's creative thinking.

I can hear the top 20 pan-species listers laughing at my lack of a Common Hawker, so Table 3 is a list of each of their own number one targets (not all of them serious – some are clearly down to people not adding all their species on yet).

Most people would expect to have seen everyone else's top target here – I certainly have. Several people have clearly not entered their entire list yet, some have clearly opted out of counting Human (an approach I strongly disagree with, for the reasons given on page 238) and others might have omitted some species in error but I think most of these are genuinely things people have not yet encountered. I was surprised that 13 of the 20 species listed here are vertebrates, with Bottle-nosed Dolphin and Palmate Newt the only species featuring more than once.

This exercise serves three purposes:

1. It shows how enjoyable it is to use this feature. You will be endlessly surprised as you look through other people's lists and find out what they haven't seen.

2. It's a good way to demonstrate how useful this feature is. For instance, it shows you things you might have missed off your list or that you thought you had already ticked. Using the targets feature regularly helps you to keep your list accurate, as well as giving you inspiration about what to look for next.

3. It shows that no matter how many species someone has seen, they will always have a top target and that target will probably always be a surprise to most other listers. No one will ever see every common species, let alone the rare and scarce ones. I still can't believe that the sixth most frequently listed invertebrate that I have not seen is Monarch *Danaus plexippus*!

Blockers

The blockers feature is denoted by the kiwi icon. On your list, click on this icon and it will show you the species that the least number of people have seen, starting with a list of species that only you have seen. I have seen 126 species that no one else on the website has listed, of which 36 species are spiders. After this, there are around 148 species that only one other person has listed, and so on. This is another handy feature for exploring your list. Maybe one day we will have rankings for blockers, showing who is listing the highest number of these 'unique' species (at the time of writing it is Julian Small, at 742 species, followed by Jonty Denton, at 616 species). Looking at the more difficult end of the Hymenoptera certainly pays dividends in this area.

Another way to use this feature is to reverse the filter, showing which species have been seen by the most people. Quite unexpectedly, the species that more people have seen than any other is Grey Squirrel *Sciurus carolinensis*, with 78.9% of listers noting this, followed closely by Red Admiral *Vanessa atalanta* and Peacock *Aglais io*. Confusingly, the first bird (Jackdaw *Coloeus monedula*) does not come in until 32nd place. This is almost certainly an artefact of a small but significant

One of my blockers, the shieldbug *Carpocoris purpureipennis*.

number of listers only adding in a few small groups of taxa and then giving up, or using the site to list, say, just their Lepidoptera. Of course, even though I would expect everyone to have seen Jackdaw, there is only one species that everyone has definitely seen, Human (yet only around two-thirds of pan-species listers have observed this species).

Pan-species listing in October 2025

At the time of writing the new site is not yet two years old, and a few of the big listers who have been active since the early days have not come across to it yet, including the lister who holds the highest species total for protists (see Table 4).

What counts?

'Your list, your rules' has become something of a mantra in PSL circles. However, we do all keep our lists according to broadly similar principles. Some people are a little harder on themselves than others, counting only species that they have recorded since they started keeping their list (i.e. discounting all of their historical observations), or only including species that they have identified themselves. Most of what I add falls into the latter category by necessity, but I also add species that I am shown on the rare occasions when I get to spend time with other naturalists. One area that most (but certainly not all) of us agree on is the 'fridge tick' – ticking a moth that has been caught in a light trap and stored for an indefinite period in a fridge for viewing at your convenience. This practice is generally frowned upon as being the entomological equivalent of ticking a wild-caught animal that you see in a zoo.

My friend Penny Green recently presented me with a Dark Crimson Underwing *Catocala sponsa* in a pot, freshly emerged from a fridge. Although I learned what this species looked like (smaller than I was expecting, and very beautiful), I drew the line at adding it to my list, because I had done nothing to contribute to it. My rule when we ran a moth trap at Woods Mill was that, as I ran that trap myself most days, if there was the occasional day when I didn't set the trap myself, I would still tick the species present if I was there to open it. However, if it was someone else's trap, or if someone brought a moth in from another site, I would not have done anything to contribute directly to such records, so wouldn't list them.

In contrast, I am very relaxed about galls and mines. Some people insist on opening them to see the tenants, even if identifying the gall or mine is not only unnecessary for the identification of the species, but also destructive to the individuals. To me, this seems pointless. It is the record that counts. However, I'm not consistent across all taxa with this – for example, I wouldn't count a dead bird if I found one. For some people, consistency across all taxa is more important than anything else, whereas for me that is not the case – for example, I would tick a dead spider that I had trapped in a pitfall trap, but I wouldn't tick a dead vertebrate.

Some people have very hard and fast rules about what they count, and stick rigidly to them. I was once chided by a friend for putting an as yet unidentified dead-wood beetle into a glass tube (if I had not done so, the beetle would probably have dropped off the fallen tree it was found on, never to be seen again), as it would then effectively be in captivity and therefore could not be ticked (even though he was only on the other side of the tree!). This was important to my friend, and it's important that we respect each other's views. Many of these differences in approach are more philosophical than scientific, which can make them hard to understand from another's perspective, especially when there are so many potential points of conflict, and our emotions are involved. These things matter to us.

The guiding principles of pan-species listing

Rather than opting for hard and fast rules, we have a set of 'guiding principles', and the glue that binds us together is the 'your list, your rules' mantra. I have reproduced here a section of text from the new PSL website (the text was originally written by Mark Telfer in 2014 for the old website, and I edited it in early 2024 for the new site).

> As a community, we've opted for a relaxed approach to the 'rules', with every pan-species lister approaching their list in a slightly different way. The mantra 'your list, your rules' has emerged over the years to remind us that, although we are all doing roughly the same things, there are subtly different ways that we do them, and all of them are valid (within reason!). Hence the term 'principles' makes more sense than 'rules'. Some people count things that others wouldn't. But nobody minds – we realise that we're not competing against each other, and that PSL is a personal challenge to get a grip on the immense biodiversity of these islands.

Table 4. Summary of the highest number of species recorded in each group, and the listers who achieved them, as of 22 October 2025.

Category	Top lister	Available species	Collective effort	Percentage	Most species listed	My list
Cyanobacteria	Jonty Denton	393	23	5.9	17	3
Algae	Jonty Denton	5320	533	10.0	287	56
Slime moulds	Stephen Plummer	511	107	20.9	78	12
Protists[1]	Jonty Denton	656	58	8.8	22	3
Lichens[2]	Paul Bowyer	2274	1038	45.6	549	198
Fungi	Stephen Plummer	15388	4055	26.4	2036	639
Bryophytes	Paul Bowyer	1121	820	73.1	607	265
Vascular plants	John Poland	7371	4097	55.6	2915	1523
Sponges	Nathan Jackson	412	45	10.9	21	19
Comb jellies	Kian Hayles-Cotton, Geoff Morgan	5	3	60.0	3	1
Cnidarians	Yolanda Evans	541	91	16.8	45	30
Molluscs	Graeme Lyons	2079	496	23.9	254	254
Bryozoans	Nathan Jackson	326	50	15.3	24	13
Annelid worms	Nathan Jackson	1384	152	11.0	53	33
Platyhelminth worms	Seth Gibson	831	48	5.8	20	13
Sea spiders	Nathan Jackson, Graeme Lyons	72	9	12.5	5	5
Arachnids	Graeme Lyons	3245	929	28.6	649	649
Myriapods	Steve Gregory	190	143	75.3	122	56
Crustaceans	Graeme Lyons	3760	337	9.0	114	114
Springtails etc.	James McCulloch	427	168	39.3	103	23
Three-tailed bristletails	Mark Telfer	11	10	90.9	8	7

Taxon	Lister				
Odonata	Philip Rhodes	60	54	90.0	50	41
Orthopteroids	John Poland	79	72	91.1	61	51
Hemipteroids	Jonty Denton	2320	1527	65.8	956	438
Hymenoptera	Julian Small	8105	2358	29.1	1135	679
Coleoptera	Peter Hodge	4301	3735	86.8	3091	1825
Diptera[3]	Jonty Denton	7372	3967	53.8	1727	606
Butterflies	Dave Horton	84	76	90.5	67	60
Moths	Tony Davis	2624	2302	87.7	1783	1315
Remaining small orders	Jonty Denton	457	329	72.0	196	124
Echinoderms	Kian Hayles-Cotton	160	31	19.4	18	11
Tunicates	Nathan Jackson	134	51	38.1	30	27
Fish	Graeme Lyons	718	180	25.1	101	101
Reptiles	Chris Griffin, James Harding-Morris, Robert Jacques, John Poland, Philip Rhodes, Will Soar and Simon West	16	13	81.3	9	7
Amphibians	John Poland	20	17	85.0	15	11
Birds	Matthew Deans	649	624	96.1	565	364
Mammals	Simon West	109	94	86.2	71	59
Other animals	Jonty Denton	1949	64	3.3	25	2
Total species		75474	28706	38.0	N/A	9637

[1] Although the top lister for protists is Jonty Denton at 58 species, Brian Eversham has listed more at 118 species but he has not yet had chance to add his species to the new website.

[2] The late Simon Davey's lichen list contained 1,128 species. Sadly, Simon passed away in 2018, before his list could be moved across to the new website.

[3] Dave Gibbs's last total for Diptera was 3,184 species as of 2018. Dave has decided not to come across to the new website; he is still involved with listing but from a more global perspective.

Each species counts as one and should be seen (or trapped) in the wild, by the recorder.

Geography: The biogeographical unit of 'the British Isles', i.e. Britain, Ireland and the Isle of Man, including the seas around the isles (defined for the UK as the UK Economic Exclusion Zone of 200 nautical miles (370 km) or midpoint between the UK and any neighbouring country). The Channel Islands also count, even if they are biogeographically part of France.

This is one of the non-negotiable guiding principles. That is, you can decide to record in a smaller geographical area if you wish (such as your home county), but you won't appear in the main PSL rankings [on the website], which will always refer to lists of the full British Isles. Equally, you may never choose to go to Ireland, but if you want to appear in the main rankings, please select

Top left, map of Britain. Bottom left, Britain and its territorial waters. Top right, the British Isles (Britain, Ireland, the Isle of Man and the Channel Islands). Bottom right, the full recording area mapped for the first time: the British Isles and the territorial waters of the UK, Republic of Ireland, Guernsey, Jersey and the Isle of Man. (Bob Foreman, Sussex Biodiversity Record Centre)

the correct location when creating your pan-species list. Obviously, for the same reason, you can't add any additional countries to your recording area, but who knows, world pan-species listing might take off one day! You can, of course, have multiple PSL lists for different regions.

The PSL recording area covers a total land area of 314,653 km^2 (0.06% of the surface area of our planet). The territorial seas cover a total area of approximately 1,827,483 km^2. Adding the land and sea areas together (see bottom-right image) gives a combined area of 2,142,136 km^2. This means that only 14.7% of the recording area is land, and the total PSL recording area is 0.42% of the surface area of the entire planet, and even includes some abyssal plain in Irish Waters!

Taxonomy: All species in the animal, plant, fungus and protist Kingdoms, i.e. everything except bacteria and viruses. Bacteria are excluded partly because they have not been traditionally covered by naturalists and partly because the very concept of a species is difficult to apply to Bacteria. However, we have recently decided to include Cyanobacteria.

Taxonomic level: Only species count as ticks. No subspecies or variants. And aggregates don't count either. If you want a dandelion or a common rustic species on your list, you need to identify it to species.

Sensory: Ideally, you should see each species, but in some rare cases you may wish to count species that are heard only, especially if the sounds are diagnostic, the species is difficult or impossible to see and/or doing so would cause it harm. It doesn't count if you see something only via television or digital camera. Some pan-species listers count gall-causers and leafminers only when they've seen a living occupant, but some are happy just to see the diagnostic characters of the gall or leafmine without laying eyes on the species itself. No one should feel pressured to break open a gall and kill the insect inside just for a tick, but if you feel you have to, so be it. Just remember, it could just as likely be a parasitoid anyway.

Alive or dead: Ideally it should be seen alive, but there are exceptions. Many entomologists count species that they have only seen dead, such as those they have caught in lethal trap samples (e.g. pitfall traps, Malaise traps). Of course, trapping should only be carried out where the results are likely to justify the casualties, e.g. on a site that is scheduled for development, as part of research and monitoring or when targeting the recording of species that are impossible to find by any other means. Counting other people's herbarium specimens, for example, absolutely does not count.

Free or captive: Although many moth-ers will count moths in pots kept in fridges, most pan-species listers look down on this as a way to boost your moth list. We encourage you to find your own rare moths. However, pan-species listers can tick things that are being temporarily held captive if they wish. It's worth pointing out, though, that no one is going to 'fridge tick' their way to becoming top moth lister! Species in long-term captivity (e.g. zoo or farm animals) or culture (e.g. crops, garden plants) don't count.

Developmental stage: All developmental stages count. For example, eggs, larvae, nymphs and pupae count just the same as adult insects, if they can be confidently identified; however, few pan-species listers would count a bird from its egg. Likewise, plants that are not in flower can be counted (even though not everyone chooses to do so). Even seeds can be counted as long as they are alive, which raises the possibility of adding sea beans, nickar nuts and other marine drift seeds if you can find them and get them to germinate.

Aliens: This is the area, more than any other, where attempts to make a simple set of rules that can be applied consistently across all taxonomic groups are doomed to failure! There is a spectrum from native species that have lived in Britain from before the time that humans started to make their mark on the planet, up to alien species that have just been intercepted on arrival at one of our ports. Although we value the true natives above all others, we find the whole spectrum fascinating. So, the 'line in the sand' is drawn to include the majority of aliens as long as they have established, or seem capable of establishing, without deliberate human assistance. You can count any garden plant that has dispersed and established beyond the garden fence. You can count any invertebrate that has established, even if only in highly human-modified environments (e.g. in warehouses or heated greenhouses). But you can't count any of the invertebrates that you can occasionally find in your groceries as primary imports, but which are unlikely to be able to survive here unaided.

Hybrids: Ordinarily, interspecific hybrids are evolutionary dead-ends with little or no fertility, so not countable. But amongst plants at least, there are numerous species which have a hybrid origin, usually formed by polyploid hybrid speciation. Pan-species listers won't want to count a hybrid unless it can reproduce and persist in the absence of one or both parent species. The wording 'one or both' is deliberately chosen: Edible Frogs and various *Sorbus* species are countable but have a hybrid origin and need reproductive contact with one of the parent species to be able to reproduce themselves. In practice, when adding hybrids for you to tick on the website, we have decided that stable hybrids, those separately numbered in Clive Stace's *New Flora of the British Isles*, are a good starting point for what hybrids are tickable. For example, the familiar garden escape Montbretia is listed as '4. *Crocosmia* × *crocosmiiflora* Montbretia' and is available to tick on the website, while the hybrid listed as '2 × 4. *C. masoniorum* × *C.* × *crocosmiiflora*' is not, as it neither has its own number nor is listed only by its parents' names and, as such, is not considered to be a stable hybrid.

We recognise that Stace did not necessarily cover all the hybrids in the same way and that some hybrids, not numbered, might be better considered for inclusion here as stable hybrids. We suggest that, if anyone has a candidate hybrid not already listed on the website that might fall into this category, they approach us and we will consider it for inclusion.

Pending species not yet on the website: If you are lucky enough to find a first for Britain or, even better, something new to science, well done! Once confirmed, let us know and we will add it to the website so that it can be ticked. Most people are now tallying up their 'pending species' or listing them all in the comments section on their list, until they are available to tick. In some cases, this will even involve species that are as yet unnamed. Any such species will be relayed to the UK Species Inventory (UKSI), as this is the cornerstone for taxonomy and nomenclature for biological recording in the UK.

Contribution: Perhaps the most important principle of all is this: how involved were you with the encounter? Did you find it? Did you record it? Did you set the trap? Did you do the research? Did you somehow contribute to our collective knowledge of that particular species in the UK? Now there is no way we would expect you to do this all of the time (unless you are a purist, self-found and self-identified lister), but if most or all of your list is made up of encounters that you have not contributed to in some way, you've got the wrong idea about PSL. It's not about ticking off as many things as possible for the sake of it to rise up the rankings. It's about becoming a better naturalist, putting something back into the natural history of the British Isles and earning your place among your fellow pan-species listers.

Twitching is not frowned upon in PSL, it's encouraged even, but it should be part of a balanced diet of natural history. So make sure you get plenty of 'self-found fibre' in there too.

Be your own judge: Most pan-species listers have their own rigid standards about what they can and can't count but, at its heart, PSL is about recording as much wildlife as you can in the British Isles over a lifetime – so we shouldn't make this too hard for ourselves. If you want to take the purist approach on all of the above principles, you're going to make life very difficult for yourself – nor would it be that much fun. If you go the other way, you can amass a huge list in a short time, learn absolutely nothing and contribute even less, while pumping vast amounts of CO_2 into the atmosphere in the process. In reality, we all sit somewhere on this spectrum and none of this is policeable anyway. You have to be your own judge. We only need a set of rules if we're competing against each other to get the biggest list. But PSL is more about the personal challenge to get a grip on the immense biodiversity of these islands.

The concept of 'contribution'

The penultimate guiding principle of 'contribution' is perhaps the hardest to pin down, and we only added it to the list for the new website. In some ways it's the most important principle of all, as at its core this is about guiding you towards becoming a better naturalist and diametrically away from just ticking off other people's finds. I used to say 'if it's good enough for a record, it's good enough for a tick', but there are more and more scenarios where this doesn't stack up for me. Sending a load of pitfall traps that you've set to be identified by someone else, and then ticking the list yourself, really goes against this core principle. What are you contributing to the record here? Yes, it's your record, but you shouldn't really count it on your pan-species list. I have many spider records in my database from pitfall traps I set at

Broadwater Warren in 2007, all identified by someone else (and including a spider I have not ticked because I didn't do enough to warrant counting it on my list – not only did I not see it alive, but I didn't identify it either).

Imagine another scenario. You are one of two entomologists working together on a survey, and your colleague calls you over to show you a rare wasp that they think they've found. They have only found the one, and they're obviously going to take it home and secure its identification at the microscope. I'm not going to have any problems ticking this eventually, but I wouldn't tick everything they had recorded on the survey at the end of the year. I might not have seen half of these species, as I was on the other side of the field for half of that survey. Similarly, I am happy to be shown things by an expert on the handful of days I get out in the autumn looking for fungi and/or lichens, but even here I put in the time searching for them, and learning to key them out or field identify them with the experts. I will often take photos and write about these days online, too, so my contribution here is far from zero.

Building and maintaining your list

There is perhaps a reason why PSL did not take off until around 2010 – several things had to come together at this point. By this stage we'd all been computer literate for about 15 years, and you needed to know your way around Excel and/or databases, even with the new site. You also needed to have heard about PSL, and that all came through social media. Without Facebook, in particular, PSL would have found it very difficult to get going. Although the software in our brains has clearly been able to do the identifications, and store all that information for a very long time, it was the digital technology that needed to get to a key point before PSL could really take off, as managing large lists across multiple taxa, especially the more obscure groups, requires all of these things to come together. There are now also better keys to and texts on many groups than ever before. PSL is a movement of our time.

Preparatory work

If all of your data are in one place and neatly curated, you will have very little difficulty adding them to the website. If you have lists all over the place, use this as an opportunity to digitise your backlog of records and to put everything into a standardised format. Whatever way you go about this, aim to start with the smaller groups first – butterflies, dragonflies or birds – as you probably know these smaller groups well. From there I would work towards doing your largest groups last, perhaps with the exception of vascular plants. The latter are probably for most beginners their largest, best-known group, and are a little different in that the majority of them have familiar common names, unlike the other very large orders of Coleoptera, Hymenoptera, Diptera and Fungi. Don't wait too long to tackle these large groups, though, as it can take a while if you have over 1,000 species to add.

Perhaps over the years you have submitted many records to your local record centre but have not kept copies of them. If this is the case, they should be happy to send them back to you on request. I am lucky enough to have a fantastic centre in the Sussex Biodiversity Record Centre and I have a great relationship with them, but this might not be the situation in every county.

Building your list on the PSL website

Once you have signed up and logged in (you can use your BUBO login to access the PSL site, too), all you need to do is create a new list, select the right location (Britain, Ireland, Isle of Man and the Channel Islands) and then select UKSI from the drop-down menu. You can then search by taxonomic group and start entering your species.

There are two ways to do this.

1. *Add species*. Start typing the species in the box and it will come up. This was how I ticked off the vast majority of my species.
2. *Batch edit*. This opens up the entire UKSI, so you'll have to focus down into a group to make it work. I found this a more difficult way to enter my list, but it works for some people.

Listers often think that they have found 'missing species' during this phase. There are two possible scenarios that you need to consider before you assume that this has happened:

1. You've already entered the species (as it won't show on the drop-down menu if it has already been ticked off).
2. The species has had a name change.

A useful tip for finding the latest version of a scientific name is to type it into the NBN atlas. Old synonyms will be picked up, and the latest version for the species will be displayed.

Starting from scratch or pulling all your old records together

A few experienced naturalists have gone down the route of only counting species they have recorded from the time when they decided to start PSL. This is a good way of avoiding the hassle of going through all your old records, especially if they are not in an easily accessible format, or possibly you don't even have any records. However, I don't really recommend this approach. I think PSL works best when your list captures all of your natural history encounters – both in the past and in the present.

Back in 2010, I pulled all of my records together from different areas and totalled them up in Excel. Back then, my plant list existed only as ticked-off species in a book, and I didn't manage my own database, so this was the only way I could calculate my vascular plant total. Having the kind of brain that remembers tiny details does help, though. I pulled my list together in August 2010, and it took me an entire day – possibly one of the most worthwhile and important days of my life.

There is a place to enter your first location and date for each species, and although ideally these should be filled in, don't worry if you can't do it all at once. Back in the winter of 2023–24 I had to prioritise getting my list entered quickly, so that I could troubleshoot each taxonomic group's idiosyncrasies in turn and, more importantly, save myself enough time to write this book. Since then, I have entered the details of all my new species (this was in addition to submitting the biological records – it's important to reiterate that we are not a recording platform). When I have a spare moment, I will go through my list and add in all my first locations and dates.

Using a personal offline database

The two most popular offline databases are Recorder 6 (which I use and recommend) and MapMate (which is now defunct and no longer available for purchase, although a 'legacy support portal' is being made available for existing users), and some people also use purpose-built databases. The big advantage of using a database that runs off the UKSI, such as Recorder 6, is that the nomenclature is all sorted out for you in the species dictionaries, and from a PSL point of view there are far more species groups in Recorder 6 than there are in MapMate, such as all of the marine groups. However, the downside of using an offline database is that it will take time for the species dictionary updates to reach you. There is speculation that this side of Recorder 6 will eventually be managed online, where updates will then be much quicker, like a hybrid approach that combines Recorder 6 and iRecord but this is seemingly on the back burner at present.

I never quite managed to get everything from the first 32 years of my life into Recorder 6, so for many years I managed my list as a kind of hybrid approach between my database, Excel and the old website, which was really messy. Now my list is entirely on the new website but backed up with my records in my database, which is a far more satisfying and accurate arrangement. If you don't have a database you should be able to manage perfectly well using Excel.

Managing a personal database or records in Excel

This is probably the easiest and most popular method for people who are new to PSL. I wish I had been advised at the start of my biological recording/PSL journey *to think in terms of records*. Lists of species might be useful to you, but they won't be very useful to the world of natural history, and they might even prevent you from noting down when you see species after your first encounter with them. Ideally, if I was going to store my species list in Excel, I would create one huge list of records, and then use pivot tables to generate my actual species list (or any meaningful subset of the overall data set). This might seem counterintuitive, involving more work than is necessary, but it will actually save you time in the long term.

Even during my invertebrate surveys, which involve a similar level of effort across many different compartments, during my field season I have fallen into the trap of storing the records in a matrix in Excel (i.e. recording compartments in columns, and species in rows). This was mainly due to my impatience at wanting to see how the species list was developing during the survey. In fact I would have saved a great deal of time by entering the data in record form first, and then generating the matrix afterwards from the records (only when all identifications had been made and all data entered), using a pivot table. The same is true of storing your own personal data. If you don't have access to a database, this is the second-best thing, and it's all ready to be imported into a database or uploaded into iRecord if you choose to do this in the future.

Table 5. An example of the 11 columns that I use to enter my data in Excel, ready for importing into Recorder 6.

Recorders	Determiner	Date	Location	Grid reference	Species	Abundance	Record type	Sample type	Survey name	Notes
Graeme Lyons; Oliver Froom	Graeme Lyons	08/01/2017	Holywell	TV60189678	Acanthodoris pilosa	1 Present	Under stone	Field observation	Graeme Lyons' casual records	Grid reference is exact from GPS
Graeme Lyons	Graeme Lyons	25/03/2018	Seaford Head sea	TV51039723	Aeolidia filomenae	1 Present	Under stone	Field observation	Graeme Lyons' casual records	Grid reference is approximate from GIS
Karen McDermott; Graeme Lyons	Graeme Lyons	22/03/2023	Freshwater Bay	SZ345855	Edmundsella pedata	1 Present	Under stone	Field observation	Graeme Lyons' casual records	Grid reference is approximate from GIS
Graeme Lyons	Julie Hatcher	30/09/2023	Kimmeridge Bay	SY90227898	Pelagella castanea	1 Immature	Unknown	Field observation	Graeme Lyons' casual records	Grid reference is approximate from GIS
Finley Hutchinson; Louis Parkerson; Graeme Lyons	Finley Hutchinson	16/10/2023	Silver Steps Rockpools	SW821318	Aeolidiella alderi	Present taxon	Under stone	Field observation	Graeme Lyons' casual records	Grid reference is approximate from GIS
Evan Jones; Graeme Lyons	Graeme Lyons	21/03/2019	The Pound	TV60299681	Polycera quadrilineata	1 Pair	Netted	Field observation	Graeme Lyons' casual records	Grid reference is exact from GIS. Swept from Wireweed
Evan Jones; Oliver Froom; Tony Davis; Graeme Lyons	Evan Jones	28/03/2017	The Pound	TV603968	Archidoris pseudoargus	1 Pair	Under stone	Field observation	Graeme Lyons' casual records	Grid reference is approximate from GIS
Evan Jones; Graeme Lyons	Graeme Lyons	21/03/2019	The Pound	TV603968	Facelina bostoniensis	1 Present	Under stone	Field observation	Graeme Lyons' casual records	Grid reference is approximate from GIS

Table 6. The first record from Table 5 pulled out and analysed (my first ever encounter with a nudibranch).

Field	Example	Other examples/notes
Recorder(s)	Graeme Lyons; Oliver Froom	Who found it (use semi-colon and a space to separate recorders)
Determiner	Graeme Lyons	Who identified it
Date	08/01/2017	20/06/2013–21/06/2013 (date range)
Species	*Acanthodoris pilosa*	
Abundance	1 Present	Examples include: 'Present Taxon' (when I have no further detail), 'Present Adult', '4 AdultMale'*, 'c.50 Immatures', etc.
Location	Holywell	An additional column, 'Location name', is useful if you have many sample points (e.g. quadrat, pitfall or plot numbers)
Grid reference	TV60189678	Always use British National Grid for Great Britain, Irish Grid in Ireland, and UTM in the Channel Islands
Sample type	Field observation	MV light, suction sampler, etc.
Record type	Under stone	Swept, beaten, unknown, etc.
Notes	Grid reference is exact from GPS	Use this for anything additional. I use it to show clearly whether my record is using a site centroid of a compartment, or a more accurate grid reference
Survey	Graeme Lyons' casual records	This ties the records to a particular survey (not a visit). Surveys are usually always made up of a series of visits

* This is a peculiarity of Recorder 6, which needs to read the quantifier (here '4') separately from the qualifier (here 'AdultMale'). It does this by inserting a space between them (but this means that no other spaces are allowed in this field – hence 'AdultMale').

An example of the 11 columns that I use to store records in Excel, which can then be easily imported into a recording database (in this case, specifically set up to work well with Recorder 6) is shown in Table 5. You can easily add additional columns on family and order which will help you to order your list (though they are not so important in a recording sense). A single record from this table has been pulled out and scrutinised in more detail in Table 6.

Of course, this is really just a way to get records into either an online or offline database, but some people use it as a way to store their entire list and/or records, instead of an offline database, in addition to sharing them with iRecord, etc.

Don't use multiple worksheets unnecessarily – try to keep everything in a single spreadsheet. Excel displays just over 1 million rows, so you are unlikely to fill it! I can export all of my records (there are over 350,000 of them) out of Recorder 6 into a single Excel spreadsheet very easily. It is also very easy to import data into my database, if the format is correct.

Worksheets in Excel should only be used if there is no logical alternative. Obviously they can be useful at times (e.g. when storing information of very different kinds), but not when collating records. If you want to further classify your data, just add in additional columns as and when they are needed. When I receive survey data from people who have never manipulated or handled data before, they will often use separate worksheets for each sample, transect or visit. This means that I then have to (quite unnecessarily) sort all the data into one place first before I am able to do anything meaningful with them. I suspect that people use worksheets because they are there, and they feel familiar – like physical sheets of paper. However, it's better to avoid them altogether.

Updating your list and keeping records

It's important to remember that the website is not a recording platform, and records should always be submitted via the routes mentioned earlier in this chapter. Adding a date and location of a sighting to the website does not mean that a record has been submitted.

Maintaining your pan-species list is something you do *as well as* recording, not instead of it. I would hate to lose even one biological recorder in this way. Even ticking things off on a garden list in PSL should be done as well as submitting those records, not instead of it.

PSL is something that you can dip in and out of throughout your life, but that really does require you to keep rigorous records of what you find. It's never been easier to do this, thanks to the new website, iRecord and other digital technology.

Some people update their list in batches, others do it just once a year, but most of us now update our lists as and when we see a new species. This is probably the best approach for PSL as a movement. It also keeps you keen, and it keeps the homepage of the website fresh and engaging. I have noticed that the people who update less regularly are often those who bounce in and out of PSL. Therefore, if at all possible, you should update as you go along. It's not good sportsmanship to withhold months' or even years' worth of additions, so that the actual rankings are only accurate for a few days each year. With the creation of the new website, it's now so much easier to update as you go.

Whether you keep your records in Excel, iRecord or a database, you need to be rigorous. I would encourage anyone planning a career in conservation, ecology or biodiversity, and definitely anyone who is intending to become an entomologist, to manage their own database of records. This is vital for processing and curating large amounts of data, and it is a great transferable skill too.

Filling in the gaps and deleting species

I recently discovered an aphid that I had not added to my list but that I almost certainly found in 2014 (as it was not in Recorder 6 when I found it, so I couldn't record it). This kind of situation will occur regularly. The 'target' feature is really useful for scrutinising your list and seeing what you have forgotten to add. You will be amazed how much stuff you can miss.

Don't be afraid to take species off your list if you learn new information about a specific record or a species; it's okay to make mistakes and/or change your mind. I have taken a couple of species off recently. One I had identified incorrectly. The other I had assumed to be native, in view of where I found it, but I have since come to believe that it was most probably planted. *Quality is better than quantity*, both in maintaining your pan-species list and in making accurate biological records.

Back up your data

Wherever and however you store your list, make sure that you back it up! I back up my database every time I add new data to it. Transferring my database to a new laptop after losing the old one is always really nerve-racking, as it inevitably follows a period during which I have had no access to it. What if this time I can't restore it? Reinstalling the software and the database can be quite laborious. Of course a further way to ensure that you don't lose data is to make sure that you have shared it with the recording schemes, iRecord and/or local record centres.

PSL is for life, not just for Christmas

Although we do always get a flush of new recruits in the New Year, PSL is a hobby, a calling, a movement or a bit of fun (depending on your viewpoint) that can be with you for the rest of your life. Literally everywhere you go in the British Isles there is a chance that you might stumble upon a lifer. I've even found new beetles in local pubs or hand caught in my local park, and I once found a species of spider I had never seen before in my living room. So the best way to get a huge list over many years is to start as early as you can and capture all of these moments before they are lost for ever.

Some of the youngest listers are already amassing huge lists. Finley Hutchinson is currently ranked 29th, and by the age of 21 had found over 5,314 species. For comparison, when I pulled my list together at the age of 32, around nine years into my career in conservation, it only contained 2,748 species! Starting early pays off.

Chapter 6

Accessing the different taxonomic groups

In this chapter I shall give a brief account of how to 'break into' each of the 38 taxonomic groups, and will list useful advice, equipment, texts, websites, Facebook groups, societies and more. Although not comprehensive, the text is full of practical tips that I have used over the years to compile a list of over 9,637 species that I have seen in the British Isles.

Many of the taxonomic groups are further divided into subgroups on the new website (see Chapter 5), but these are beyond the scope of this book.

I do not profess to be an expert on all taxa – far from it. In fact any pan-species lister will, by definition, become a generalist. Obviously we each have our own particular areas of interest, and in my case there is an inevitable bias towards invertebrates. Furthermore, I have not spent as much time on some of the better-known groups, such as vascular plants, birds and mammals, as these are generally much easier to access with limited effort and cost. In the spirit of inclusivity that defines the movement, where necessary I have reached out to other pan-species listers (or further afield into the wider natural history community) to add some flesh to these accounts. Hopefully this will be of use to all naturalists, whether or not they are pan-species listers.

Number of species available on the UK Species Inventory (UKSI)

Here the total number of species currently on the website for the entire group is given. Further down in the subgroups, the same relevant figure will be given, or will be replaced with 'number of species on the UKSI'. There is a subtle but distinct difference, as the number of species on the website might have been edited; some species may have been removed that do not qualify for inclusion on your pan-species list, and other species may have been added. In some subgroups it is not clear how many species are present, in which case I have provided an estimate.

Broad habitat

The name of the group is followed by one or more coloured dots – for example:
● ● ●

These dots indicate which of the following broad habitats the groups cover, with the first dot denoting the habitat with the highest number of species in that group or subgroup, the second dot indicating the second most species-rich habitat, and so on:

● = terrestrial   ● = freshwater   ● = marine.

Here marine habitats include only species that live in the sea, with this broad habitat comprising all marine life found in the intertidal, pelagic and abyssal zones. Freshwater habitats include only species that are obligated to spend part or all of their life in the water. Terrestrial habitats include all other species.

If all of the species in a particular group (and therefore in its associated subgroups) occupy the same broad habitat, the coloured dot will only appear in the group heading and not in the subgroup headings.

Scale

For each taxonomic group and subgroup, the maximum number of species available on the website or the UKSI is given (which in most cases is the maximum number of species in the PSL region, but this is not always so – aphids, scale insects and protists are poorly represented on the UKSI), and graded from very small to very large.

Number of species	1–10	11–100	101–1,000	1,001–5,000	> 5,000
Size	Very small	Small	Medium	Large	Very large
Rank	1	2	3	4	5

Ease

To help new listers to decide where to focus their efforts, each of the taxonomic groups and subgroups has been assessed in terms of its ease according to a five-point scale ranging from very easy to very difficult. This is a subjective scale – effort and difficulty are not quite the same thing, but are intrinsically linked. Nothing seems difficult once you have put in the effort to master it, but until you have acquired the necessary knowledge, the level of effort required can seem overwhelming. Difficulty is often associated with the use of out-of-date or very confusing keys, or literature that is difficult to access (or, in some of the very awkward groups, non-existent).

Effort	Very easy	Easy	Moderate	Difficult	Very difficult
Rank	1	2	3	4	5

Although this is a subjective scale, I have applied some logic to it.

Very easy: readily identifiable by picture matching and typically identifiable from a photo – therefore usually field identifiable.

Easy: typically involves some keying out in the field, but usually or often field identifiable with experience, and may or may not be identifiable from a photo.

Moderate: involves keying out, but keys work well and are readily available and up to date.

Difficult: involves keying out, genitalia examination and/or complex work at high magnification. All species are described well and the literature is up to date and readily available for the entire group, but could be spread over different texts, some of which are out of print, unpublished and/or shared privately between naturalists.

Very difficult: all of the same issues as described for Difficult, but keys are often non-existent for significant portions of the group, or if they do exist are often not widely published.

Unknown: in a few very rare cases there appears to be no literature available. If you know of any, please get in touch.

Accessibility

Scale × ease = accessibility

If you have ever had to write a risk assessment, you will see that this follows a very similar principle. Accessibility is graded on a scale of 1 to 25.

Accessibility	1–5: highly accessible	6–12: intermediate accessibility	13–25: difficult to access

Collective effort and top lister

For each taxonomic group, the total number of species recorded by pan-species listers (at the time of writing this book) is provided, together with that number as a percentage of the total number of species available. The individual who has recorded the most species (known as the top lister) is also given, together with the total number of species they have recorded in that group.

Essential texts

I have generally avoided including beginners' guides here, unless they are particularly useful, as they are often not comprehensive enough. I have also, with a few exceptions, avoided recommending general natural history books.

In general, only atlases that contain particularly useful identification aids are included. Increasingly sophisticated atlases are being produced across a wide range of taxa. They will all deepen your understanding of the ecology and distribution of species, but many of them are beyond the scope of this book.

Perhaps the largest financial commitment of the pan-species lister will be their library. I have tried to list only the essential books in the following sections, but even so it would cost over £5,000 to buy new copies of all of the books mentioned here, so consider looking for second-hand copies.

Places to buy books include:

- NHBS (formerly the Natural History Book Service).
- Pemberley Books (particularly useful for older natural history titles and those that are out of print).
- Summerfield Books (particularly for more obscure natural history titles, older books and those that are out of print).
- AbeBooks (for reasonably priced second-hand books on all subjects).

- second-hand bookshops and book stalls at events, stands and exhibitions
- eBay (it's surprising what you can find there).

Shop around, and don't give up if you can't find an out-of-print book straight away. It might be disheartening to see so many out-of-print books listed here, but simply knowing that they exist is the first step towards tracking them down. Moreover, many of them – for example, the old Royal Entomological Society (RES) keys – can be found online, and some are shared between individuals digitally.

Vital equipment

The bare minimum of kit typically needed to start you off in each group is listed. Useful but non-essential equipment that might help you take things to the next level is written in italics.

Further breakdown of taxonomic groups

Rather than divide each of the 38 groups entirely along taxonomic lines, I have split them according to the available literature, resources and recording groups. For instance, I have provided details of a wide selection of fly and beetle families because the keys and resources needed to identify them can be found in different books and/or on different online resources. Following the same logic, the arachnids have been split into spiders, harvestmen, pseudoscorpions, scorpions, ticks, gall mites, halacrid mites, water mites, soil mites and other mites, for example.

For some taxa, these distinctions are less relevant to aiding identification – for instance, the fungi are only split as far as the Basidiomycetes and Ascomycetes. Furthermore, it has only been possible to provide a flavour of each group, with beginner-level through to intermediate-level families typically presented. The main aim of this chapter is to help you to identify species yourselves, rather than to provide exhaustive coverage of all taxonomy.

For the larger orders, such as the beetles and flies, to enable you to tackle the easiest and smallest families and subgroups first, families and subgroups have first been listed in order of accessibility score, and then by number of species.

Sources of keys, papers and other material to aid identification

The Field Studies Council's ID Resource Finder: www.fscbiodiversity.uk/idsignpost
    Please be aware that there may be some omissions and broken links, as the catalogue is crowd-sourced, but this is nevertheless an extremely useful resource.

Mike's Insect Keys: https://sites.google.com/view/mikes-insect-keys/mikes-insect-keys
    These are very useful aids to insect identification. Mike Hackston has done an immense amount of work to make keys accessible to everyone, especially 'the next generation of entomologists'. He covers the Coleoptera, Dermaptera, Diptera, Ephemeroptera, Hemiptera, Hymenoptera and Lepidoptera, as well as the Thysanoptera and various other small orders.

APHOTOMARINE: www.aphotomarine.com
    David Fenwick's comprehensive photo resource covers the marine life to be found in the intertidal areas and coastal waters of Britain and Ireland.

https://www.flickr.com/photos/56388191@N08/collections/
> Ian Smith's Flickr account (Morddyn) is really good for a number of marine groups, especially marine gastropods, nudibranchs and crustaceans.

NatureSpot: www.naturespot.org
> Although it focuses primarily on identification and recording of wildlife in Leicestershire and Rutland, this website provides a huge amount of useful information about British natural history in general.

NBN Atlas: www.nbnatlas.org
> The NBN Atlas is the largest source of biodiversity data for the UK, and is well worth consulting. You can always try applying a filter, too (e.g. using the search term 'Conchological Society' when viewing a mollusc's distribution).

The group accounts in this chapter are suitable for everyone – from the complete beginner to the more experienced naturalist who is trying to get a foothold in a new area. The aim is to show you where to start, so that you can then take the next steps yourself. I would have found this incredibly useful ten years ago, so my hope is that it will be of use to you now.

6.1 Cyanobacteria (blue-green algae): 393 available species

Scale: Large Effort: Very difficult Accessibility: 20

Collective effort = 23 (5.9%) species **Top lister** = Jonty Denton (17 species)

Vital equipment: compound microscope, hand lens, water-tight plastic containers, plastic bags, pipettes, waterproof notebooks, *grapnel, pond net, immersion oils*.

Essential text

David M. John, Brian A. Whitton and Alan J. Brook (2021) *The Freshwater Algal Flora of the British Isles: An Identification Guide to Freshwater and Terrestrial Algae*, 2nd edn. Cambridge University Press.
Although the bulk of this book covers the freshwater algae, it also includes species that are in fact classed as cyanobacteria, of which there are some 364 species.

This is a 'new' group that has been added to pan-species listing since the new website was created. Although we have always drawn the line at bacteria and viruses, there was a growing consensus that we should include the unrelated cyanobacteria. In fact, many of us were already listing them incorrectly within the algae.

The taxonomy is therefore rather confusing. Even the well-known *Nostoc commune* (covered by *The Freshwater Algal Flora of the British Isles*) is in fact listed as a cyanobacterium, and the genus *Rivularia* is at least mainly maritime, and resembles a small, primitive seaweed. The following section on freshwater algae is therefore very relevant to this group, too.

The delicious-looking *Nostoc commune*, coming soon to some wet gravel near you.

Rivularia bullata on intertidal rocks on Jersey.

6.2 Algae (photosynthetic eukaryotes other than plants): 5,320 available species

Scale **Very large**   Effort **Very difficult**   Accessibility **25**

Collective effort = 533 (10.0%) species   **Top lister** = Jonty Denton (287 species)

Vital equipment: compound microscope, hand lens, water-tight plastic containers, plastic bags, pipettes, waterproof notebooks, *grapnel, pond net, immersion oils*.

Here I have divided the algae into freshwater algae, diatoms, charophytes and seaweeds. The first two subgroups are almost entirely microscopic, whereas the last two are mostly macroscopic, but microscope work is needed at the more involved end.

Freshwater algae (excluding the diatoms): 2,480 available species

Scale **Large**   Effort **Very difficult**   Accessibility **20**

Essential texts

Allan Pentecost (1984) *Introduction to Freshwater Algae*. Richmond Publishing.
   Despite the fact that this book is out of print, it is not difficult to find second-hand copies. It provides a great introduction to freshwater algae and contains some fantastic line drawings.

David M. John, Brian A. Whitton and Alan J. Brook (2021) *The Freshwater Algal Flora of the British Isles: An Identification Guide to Freshwater and Terrestrial Algae*. 2nd edn. Cambridge University Press.
   Consider getting this lovely book if you want to delve deeper into the world of freshwater algae. It is very comprehensive and covers *c*.2,600 species (including the 364 cyanobacteria mentioned earlier).

I am grateful to my friend Howard Matcham for his help in this area, of which I had no experience until he got me hooked. An important thing to know about freshwater algae is that although you will find them in most larger waterbodies, they can also be found in any place where water gathers. Perhaps the best way to sample freshwater algae is to squeeze wet moss into a water-tight container. However, you are just as likely to find them in ponds, lakes, puddles or wheel ruts. The samples can be stored for a few days if kept cool, before you look at them with a compound microscope.

Another group of algae are the sub-aerial species. These form green stains on bark, walls and windows.

A power of up to 450× should be sufficient, so the oil immersion techniques required for higher magnifications of around 1,000× are unlikely to be necessary. Howard told me that when he was already an extremely experienced bryologist and mycologist he suddenly became quite taken with freshwater algae after a chance

Micrograph of *Micrasterias rotata*. (Chris Carter)

encounter with a species new to England. He then went on to focus almost exclusively on algae for a decade, becoming the County Recorder for this extremely under-recorded group. However, he did warn me that they are extremely difficult and very time-consuming, so it seems unlikely that a pan-species lister would ever be able to just 'dip into' this group.

Website

AlgaeBase: www.algaebase.org
> This website starts with the heading 'Listing the Word's Algae'. Full of microscopic photographs of freshwater algae, it is also an invaluable resource for studying them here in the UK.

Diatoms (Bacillariophyceae): 2,320 species on the UKSI

Scale | Large   Effort | Very difficult   Accessibility | 20

Diatoms are found almost everywhere, in freshwater, sea water and soil, and they present a huge identification challenge – even many algae specialists avoid them. These silicate-encased algae (see page 63 for an image of them photo-bombing a protist) are also extremely fiddly to identify, as they often need to be cleaned in order to see them properly. I have no experience of this group, but have included several texts and a website, though I've been told it's unlikely that any pan-species lister will ever really cover this group in detail. Now to me that sounds like a challenge!

Essential texts

Martyn G. Kelly (2000) *Identification of Common Benthic Diatoms in Rivers*. Field Studies Council.

David G. Seamer (2019) *Beginners Guide to Freshwater Diatoms*. David Seamer.
If you want to take things to the next level, this book contains incredibly detailed photo micrographs of many species, and has nearly 1,000 pages.

Marco Cantonati, Martyn G. Kelly and Horst Lange-Bertalot (2017) *Freshwater Benthic Diatoms of Central Europe: Over 800 Common Species Used in Ecological Assessment*. Koeltz Botanical Books.

Website

Of Microscopes and Monsters: www.microscopesandmonsters.wordpress.com
This blog includes a lot of information about diatoms and is written by Martyn Kelly, who is an expert on Europe's diatoms.

Stoneworts (Charophyceae): 37 species on the UKSI

Scale: Small Effort: Difficult Accessibility: 8

Being macroscopic and resembling vascular plants in certain respects, members of this small group of around 40 unusual freshwater algae are most often encountered when one is surveying aquatic macrophytes. Typically, I find them when using a grapnel to pull out other submerged aquatic plants. Although 946 species sit in the phylum Charophyta, the vast majority of these are microscopic freshwater algae and are covered by John et al. (2021).

Essential texts

David M. John, Brian A. Whitton and Alan J. Brook (2021) *The Freshwater Algal Flora of the British Isles: An Identification Guide to Freshwater and Terrestrial Algae*, 2nd edn. Cambridge University Press.
Stoneworts are included in this book, as well as a few additional species recently added to the British list which are not covered in the following text.

Jenny A. Moore and A. Tebbs (1986) *Charophytes of Great Britain and Ireland*. Botanical Society of Britain and Ireland.
This is a good text with some really helpful line drawings.

N.F. Stewart and J.M. Church (1992) *Red Data Books of Britain and Ireland: Stoneworts*. Joint Nature Conservation Committee.
This book includes a key to all of the British species, as well as a simple key to the dozen commonest species (which works 95% of the time). Available as a free download from the JNCC website.

Seaweeds: > 650 available species

Scale: Medium   Effort: Difficult   Accessibility: 12

Vital equipment: compound microscope, hand lens, plastic bags/containers.

Essential texts

Francis Bunker, Juliet A. Brodie, Christine A. Maggs and Anne R. Bunker (2017) *Seaweeds of Britain and Ireland*, 2nd edn. Wild Nature Press.
There are over 650 species of seaweed around the UK, roughly divided into *c.*350 red seaweeds, *c.*200 brown seaweeds and *c.*100 green seaweeds. This book covers around 235 of these species. As such, it is not comprehensive but it is a great place to start and does not feel like a generic beginner's guide. It includes a comprehensive review of the modern literature for each of the three main seaweed groups

If you want to take seaweeds to the next level (including detailed, high-powered microscope drawings), these two texts are the way forward:

J.A. Brodie, C.A. Maggs and D.M. John (2008) *Green Seaweeds of Britain and Ireland*. British Phycological Society.
This book covers around 100 species.

R.L. Fletcher (2024) *Brown Seaweeds (Phaeophyceae) of Britain and Ireland*. Pelagic Publishing.
Covers nearly 200 species.

Website

AlgaeVision: www.nhm.ac.uk/our-science/data/algaevision.html
This site is particularly helpful for issues regarding classification of algae.

Society

British Phycological Society (BPS): www.brphycsoc.org
Membership of this society is essential for anyone who wants to take algae seriously. Its benefits include annual meetings, training and a number of journals published by the BPS.

Facebook group

Seaweeds and Algae of Britain and Ireland.
The main focus of this group is seaweed identification.

6.3 Eumycetozoa (slime moulds): 511 available species

Scale: Medium   Effort: Very difficult   Accessibility: 15

Collective effort = 107 (20.9%) species

Top lister = Stephen Plummer (78 species)

Vital equipment: compound microscope, containers for collection and cultures, *head torch*.

Essential text

Bruce Ing (2020) *The Myxomycetes of Britain and Ireland: An Identification Handbook*, 2nd edn. Richmond Publishing.
  Both hardback editions of this vital book went out of print quickly. A paperback edition of this title has since been republished by the University of Chester.

Although slime moulds are not fungi, they are most often encountered when surveying for fungi, and this means that many mycologists also cover this fascinating group. By far the best way to find them is by turning over logs and searching fallen trees, especially in dark and damp environments. The same species can look wildly

Metatrichia floriformis, a slime mould that is fairly commonly found under logs in winter.

different as they age, in just a short period of time. They are often present through the winter, and so, like money spiders, add some much needed excitement to winter field work!

Stephen Plummer regularly takes sections of damp bark back to the lab, where he cultures them by placing them in a humid petri dish and seeing what grows.

Website

https://www.barrywebbimages.co.uk/Images/Macro/Slime-Moulds-Myxomycetes
- The truly spectacular images of British slime moulds on Barry Webb's site are not only otherworldly and beautiful, but can also be really useful for securing an identification.

Facebook groups

British Slime Moulds (Myxogastria)
- This small group has useful identification tools.

Slime Mold Identification & Appreciation
- This global group, with tens of thousands of members, also has useful identification tools.

6.4 Protists other than algae and slime moulds (a polyphyletic group): 656 available species (under-represented on the UKSI) ● ●

Scale Medium   Effort Very difficult   Accessibility 15

Collective effort = 58 (8.8%) species

Top lister = Brian Eversham (118 species, not on the new website yet)

Vital equipment: compound microscope, water-tight containers, pipettes, slides, etc.

The best way to find protists is to squeeze a sample of wet moss (especially *Sphagnum*) into a water-tight container, in much the same way as you would record freshwater algae. Place a droplet of this water on a cover slide with a cover slip over the top, and view it. At the time of writing, this was the only one of the 38 groups for which I had no species listed, so I collected some *Sphagnum* in the New Forest and managed to identify two species from it. Brian Eversham recommends using a cavity well slide and placing a small amount of *Sphagnum* in the depression before covering it, as some species will be attached to or hiding within the moss itself. Apart from the testate amoebae, there are very few protists that can be identified to species level.

The following text was provided by Brian Eversham:

A more sophisticated approach to microscopy is probably needed here, with a built-in or attached camera that can capture images to aid identification. A basic digital camera that slots into one eyepiece or the third turret of the microscope (if it has one) should come with easy-to-use software that makes measurement easy,

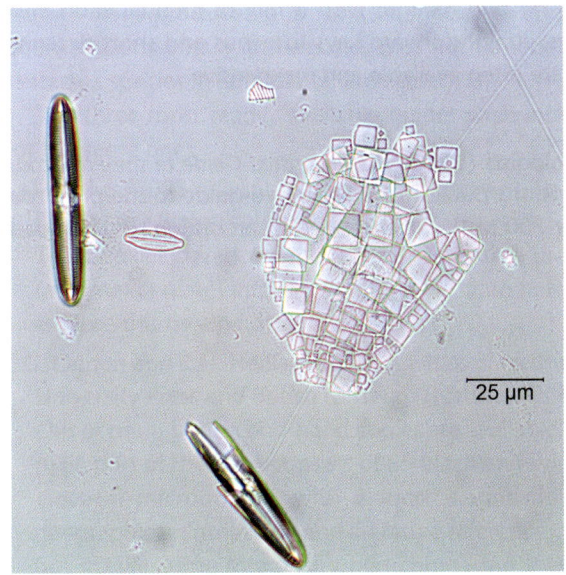

Quadrulella symmetrica, which is a testate amoeba (right), and a couple of diatoms (left). (Chris Carter)

resolved images viewed down a light microscope. Although very useful, this book needs to be used alongside additional publications that cover the other species.

M. Todorov and N. Bankov (2019) *An Atlas of* Sphagnum-*Dwelling Testate Amoebae in Bulgaria*. Pensoft and Bulgarian Academy of Sciences.
After a short introduction and a checklist, each species has a page of text and a page of illustrations, which helpfully include light-microscope images as well as electron micrographs. A total of 142 species are covered, 93 of which are currently on the British list. Don't be put off by the title, as many of the species covered occur in other places as well as *Sphagnum*. A download is also available (see Websites section).

There is a five-volume set of Ray Society Monographs on the rhizopod and heliozoan amoebae (testate amoebae are a shell-making subset of the rhizopods). Despite being over a century old, these volumes are still very useful, and are well illustrated with colour drawings that show what you can expect to see with a light microscope. Long out of print, but second-hand copies can occasionally be found.

J. Cash and J. Hopkinson (1905) *The British Freshwater Rhizopoda and Heliozoa. Vol. I: Rhizopoda Part 1*. Ray Society.
This volume includes 16 plates, and mainly covers non-testate species.

J. Cash and J. Hopkinson (1909) *The British Freshwater Rhizopoda and Heliozoa. Vol. II: Rhizopoda Part 2*. Ray Society.
This volume includes 32 plates, and covers the first half of the testate species.

J. Cash, G.H. Wailes and J. Hopkinson (1915) *The British Freshwater Rhizopoda and Heliozoa. Vol. III: Rhizopoda Part 3*. Ray Society.
This volume includes 24 plates, and covers the rest of the testate species.

J. Cash, G.H. Wailes and J. Hopkinson (1919) *The British Freshwater Rhizopoda and Heliozoa. Vol. IV: Supplement to the Rhizopoda and Bibliography*. Ray Society.
This volume includes six plates, and covers several more testate species.

J. Cash, G.H. Wailes and J. Hopkinson (1921) *The British Freshwater Rhizopoda and Heliozoa. Vol. V: Heliozoa*. Ray Society.
No testate species are included.

Collectively, this set includes 184 species and varieties of testate amoeba, and 32 other species with a less solid test, which are not included in other guides.

Websites
There is a rapidly growing online resource on amoeboid species, and downloadable teaching resources are widely available.

Microworld – world of amoeboid organisms: www.arcella.nl/lobose-testate-amoebae
Microworld is a good starting point. Before you dive in, it's really useful to see how the species look down the microscope, as opposed to their appearance in the illustrations.

The International Society for Testate Amoeba Research (ISTAR) – Identification Keys and Illustrated Monographs: http://istar.wikidot.com/id-keys
This site has put many out-of-print papers and monographs online, and shares unpublished draft keys.

www.ab.pensoft.net/articles.php?id=38685&journal_name=ab

Ciliate protozoa: number of species unknown

Scale Unknown   Effort Very difficult   Accessibility ?

Essential texts

The following two volumes key, illustrate and describe the genera, and provide references to other works to get to species level.

C.R. Curds (1982) *British and Other Freshwater Ciliated Protozoa. Part I. Ciliophora: Kinetofragminophora.* Synopses of the British Fauna (New Series), No. 22. Cambridge University Press.

C.R. Curds, M.A. Gates and D. McRoberts (1982) *British and Other Freshwater Ciliated Protozoa.* Synopses of the British Fauna (New Series), No. 22. Cambridge University Press.

W. Foissner (1999) Protist diversity: estimates of the near-impossible. *Protist* 150: 363–68.

Global estimates of ciliate diversity range from 3,000 to 5,000 described species, and maybe ten times as many have yet to be described. I have not been able to find a modern estimate of the number of British species, and the work of Wilhelm Foissner (see below) explains why. This Austrian protist taxonomist reckons that he finds, on average, one undescribed species in every sample he examines, and that about half of the 2,000 species he has seen were undescribed.

Website

Wilhelm Foissner – Publications: www.wfoissner.at/publications.htm

There is a large and growing online resource on the identification and ecology of ciliates. A good starting point is the website of Wilhelm Foissner. Many of his long and detailed monographs are also free to download (though the text is largely in German).

6.5 Lichens (lichenised fungi): 2,274 available species

Scale: Large | Effort: Very difficult | Accessibility: 20

Collective effort = 1,038 (45.6%) species

Top lister = the late Simon Davey (1,228 species on the old website); Paul Bowyer (549 species on the new website).

Vital equipment: 10× magnification hand lens (preferably with a light), UV torch, lichenologist's chemicals, dissecting and compound microscopes, knife, hammer and chisel for taking samples.

Essential texts

Frank S. Dobson (2018) *Lichens: An illustrated Guide to the British and Irish Species*, 7th edn. British Lichen Society.
An essential text for anyone with an interest in lichens. A huge amount of information is crammed into this book. The seventh edition covers around 1,000 species, and includes keys, chemical reactions, photographs, microscopic illustrations and distribution maps.

Paul Whelan (2024) *Lichens of Ireland & Great Britain: A Visual Guide to Their Identification*. Holm Oak Press.
This new two-volume book nicely complements Dobson's classic text. It covers around 700 species, includes distribution maps and is beautifully laid out with a huge amount of information. I particularly like the small inset photos that show close-ups of salient features.

C.W. Smith, A. Aptroot, B.J. Coppins et al. (2009) *Lichens of Great Britain and Ireland*, 2nd edn. British Lichen Society.
This is currently the 'bible' for lichenologists, and you'll need to get hold of a copy if you want to do lichens seriously. The keys and descriptions for each lichen family are gradually being superseded by revisions available to download from the British Lichen Society website.

Lichens can be found on numerous substrates, including trees, rocks, soil and human-made substrates in a variety of habitats, but open woodland and the uplands are among the best places to find them. The south, west and north of the UK are generally more species rich than the Midlands and East Anglia.

Lichenologists' chemicals

These include the following:

- potassium hydroxide (KOH)
- sodium hypochlorite (bleach) (NaOCl)

Small containers of these chemicals are used to dispense a tiny amount on to the lichen, and then the reaction, if any, is observed. If a reaction occurs, it will happen quite quickly and result in a change in colour of the lichen; different chemicals will show different reactions on different species, and can therefore help aid identification.

Golden-eye Lichen *Teloschistes chrysophthalmus* is one of the few lichens that I know (with the more yellow and ubiquitous *Xanthoria parietina* in the background).

Recent advances in lichenology have shown that chemical reactions when used with a UV torch can greatly aid identification. However, UV light with a wavelength of 365 nanometres is essential for this.

Website
Lichens Marins, Maritimes, Oceaniques et Ressemblants: www.lichensmaritimes.org
   Based in Brittany, this resource is particularly useful if you are in a coastal area. It has many good photos, including images of side-by-side chemical reactions of difficult-to-separate species.

Society
British Lichen Society: www.britishlichensociety.org.uk
   The benefits of membership include field meetings, talks, courses, workshops and the opportunity to meet other lichenologists. The BLS website has species accounts of all British species with distribution maps, ID tips including chemical reactions and more that are available to all.

Facebook group
British and Irish Lichens
   This is a surprisingly small group with little traffic.

6.6 Fungi (a polyphyletic group): 15,388 available species

Scale: Very Large   Effort: Very difficult   Accessibility: 25

Collective effort = 4,055 (26.4%) species

Top lister = Stephen Plummer (2,036 species)

Vital equipment: collecting vessel for specimens (e.g. trug, plastic tool box, screw sorter), knife, compound microscope, *mycologist's chemicals such as iron sulphate crystals and guaiac for testing reactions in the field, immersion oils* and a good sense of smell!

This is the largest group of all. Not only is it the only group with more than 10,000 species, but it actually exceeds the 15,000 species mark. In fact, calling it a group doesn't do it justice – this is an entire kingdom! For this reason, breaking into fungi can be difficult. I strongly recommend that, at least to start with, you focus on one or two families or habitats rather than spreading yourself too thinly.

The huge number of species is expected to rise as advances in mycology, especially those involving DNA technology, result in many more species being discovered than were previously thought to be present. As a result, fungal taxonomy is currently in a state of extreme flux. In Sussex, there are County Recorders for both the basidiomycetes and the ascomycetes, and I shall consider these two classes of fungi here.

The distinction between basidiomycetes and ascomycetes is microscopic and is concerned with spore production. In fact attempts to separate the two classes at the macroscopic level are not particularly helpful, so I shall refrain from doing this. All of the mushrooms and toadstools with gills or spores are basidiomycetes in the phylum Basidiomycota, although basidiomycetes include many other kinds of fungi, too, such as rusts. Ascomycetes, which belong to the phylum Ascomycota, include many microfungi. The disc and cup fungi are all ascomycetes, as are morels, mildews and *Cordyceps*, among many others.

Basidiomycetes: 4,891 species on the UKSI

Scale: Large   Effort: Very difficult   Accessibility: 20

Woodlands are typically the best place to see the largest number of species of fungi, but all terrestrial habitats support them. Ancient woodlands with a range of tree species of different ages are generally best for overall diversity. Old grasslands that are cut regularly and have the clippings (arisings) removed, such as churchyards and cricket pitches, can be very good for grassland fungi, with the colourful waxcaps being particularly well represented here.

Essential texts

Paul Sterry and Barry Hughes (2009) *Collins Complete Guide to British Mushrooms and Toadstools*. Collins.

One of my favourite species of fungi, the spectacular Magpie Inkcap *Coprinopsis picacea*.

This a great beginner's guide. Although it is far from complete, there are good photos and identification tips, and it is particularly relevant if you are based in south-east England. Features such as the 100 commonest species and some sets of fungi displayed by habitat are useful for field identification, but to take it to the next level you'll soon need something more detailed.

Thomas Læssøe and Jens H. Petersen (2019) *Fungi of Temperate Europe*. Princeton University Press.
This two-volume set has beautiful colour photos and a great deal of useful text, but it is expensive, and both volumes are fairly large coffee-table-style books. There is novel use of 'wheels' of similar species to aid identification before delving into those species more deeply.

Geoffrey Kibby (2017–2023) *Mushrooms and Toadstools of Britain and Europe*. Geoffrey Kibby.
This set of four volumes, all written, illustrated and published by Geoffrey Kibby, covers just under 2,500 species of fungi, and for me these are the best illustrations of fungi I have ever seen. Spores and other microscopic features are also illustrated. The four volumes are an essential investment for anyone who wants to take fungi seriously.

Henning Knudsen and Jan Vesterholt (eds) (2012) *Funga Nordica [English]*. Nordsvamp (out of print).
Even with the above texts, you'll need to key species out and that will involve getting hold of keys such as those in *Funga Nordica* (a two-volume book). Many mycologists will invest in monographs of their interest groups, and a significant number of these monographs are also published in *Field Mycology*.

A selection of 11 species of waxcap, most of which can be identified in the field with experience. Left-hand column, from top: Crimson Waxcap *Hygrocybe punicea*, Golden Waxcap *H. chlorophana*, Honey Waxcap *H. reidii*, Slimy Waxcap *Gliophorus irrigatus*; centre column: Pink Waxcap *Porpolomopsis calyptriformis*, Fibrous Waxcap *Hygrocybe intermedia*, Splendid Waxcap *H. splendidissima*; right-hand column: Goblet Waxcap *H. cantharellus*, Heath Waxcap *Gliophorus laetus*, Scarlet Waxcap *Hygrocybe coccinea*, Yellow-foot Waxcap *Cuphophyllus flavipes*.

There's a lot to bear in mind when identifying fungi. You will not just be concerned with the physical appearance of the species or its microscopy – mycology is also a sensory experience, with smell, taste and texture all playing a role in the identification of many species. You'll also need to know your plants, as many mycorrhizal species are associated with the roots of particular plant species. Be prepared for a long journey before you become a really competent mycologist, as there is just so much to learn – even the associated microscopy is in a different league to that practised by entomologists.

It is sometimes useful to see how fungi react with certain chemicals in the field, such as ferrous sulphate ($FeSO_4$), or to cut the flesh open (or bruise it) to see if it reacts, and how. Spore prints made at home can also help with identification.

Rusts and smuts (Pucciniomycotina): 526 species on the UKSI

Scale Medium   Effort Difficult   Accessibility 12

Rusts are an ideal group for pan-species listers, as they require a sound botanical knowledge of the host plants that they grow on before you can even get started on their identification. Along with galls and mines, it is a good idea to have rusts in the back of your mind when you find certain plants. There are two ways to come across rusts – you may stumble on them by chance when you are looking at plants, or you may have an idea of where you might find them, and then go and look for them.

Essential texts

A.J. Termorshuizen and C.A. Swertz (2011) *Dutch Rust Fungi*. A.J. Termorshuizen.
   The text of this very useful book is in Dutch, followed by an English translation. An online version can also be found at www.bodemplant.nl/en/roesten.

Martin B. Ellis and J. Pamela Ellis (1997) *Microfungi on Land Plants: An Identification Handbook*. Richmond Publishing.
   Although out of print, this book is essential for identifying rust and smuts.

Ray G. Woods, R. Nigel Stringer, Debbie A. Evans and Arthur O. Chater (2015) *Rust Fungus Red Data List and Census Catalogue for Wales*. Unpublished.
   This useful text is free to download from ResearchGate.

A selection of monographs

There are far more monographs on our fungi than can be listed here, but a selection of some of the more accessible groups or genera are provided.

Geoffrey Kibby (2017) *The Genus Russula in Great Britain: with Synoptic Keys to Species*. Geoffrey Kibby.

Geoffrey Kibby (2017) *British Boletes, with keys to species, 8th edn*. Geoffrey Kibby.

Geoffrey Kibby (2017) *British Milkcaps* Lactarius *and* Lactifluus. Geoffrey Kibby.

Bluebell Rust *Uromyces muscari* on Hybrid Bluebell *Hyacinthoides* × *massartiana*.

Arched Earthstar *Geastrum fornicatum* and Vaulted Earthstar *Geastrum britannicum* growing in a churchyard in East Sussex. Not all basidiomycetes are the traditional mushroom shape.

Leif Ryvarden, Irenela Melo & Tuomo Niemelä (2022) *Synopsis Fungorum volume 37: Poroid Fungi of Europe, 3rd edn.* FUNGIFLORA.

David Boertmann (2010) *Fungi of Northern Europe, Volume 1: The Genus Hygrocybe.* Svampetryk. Out of print.

Arne Aronson & Thomas Læssøe (2016) *Fungi of Northern Europe vol. 5: The Genus Mycena s.l.* Svampetryk.

Ascomycetes: 11,326 species on the UKSI

| Scale | Very large | Effort | Very difficult | Accessibility | 25 |

With over 11,500 species the ascomycetes are a daunting group, and it is not surprising that so few people record them compared with the basidiomycetes. They are mostly found in the same kinds of places, with a large tranche being host specific, growing directly out of a specific plant, or part or stage of a plant – for example, *Lanzia echinophila* on last year's Sweet Chestnut *Castanea sativa* husks.

Essential texts

Ascomycetes are partially covered in Volume 2 of Læssøe and Petersen (2019).

Martin B. Ellis and J. Pamela Ellis (1997) *Microfungi on Land Plants: An Identification Handbook.* Richmond Publishing.

Martin B. Ellis, J. Pamela Ellis and David L. Hawksworth (1998) *Microfungi on Miscellaneous Substrates: An Identification Handbook.* Richmond Publishing.
These two volumes are out of print (and now quite out of date), and second-hand copies are quite expensive but they are still well respected by mycologists.

Powdery mildews: 166 species available ●

Scale Medium   Effort Difficult   Accessibility 12

There are around 166 species in Britain, and again you will need a good botanical knowledge of vascular plants.

Arthur O. Chater (2019) *The Powdery Mildews (Erysiphales) of Wales: An Identification Guide and Census Catalogue.* Privately published.
This book covers around 122 species.

Websites

Plant Parasites of Europe (leafminers, galls and fungi): www.bladmineerders.nl
This is very useful for rusts and mildews.

Bucks Fungus Group: www.bucksfungusgroup.org.uk
An extremely useful website with a species image library and a wealth of useful resources for beginner and expert alike.

The Quekett Microscopical Club: www.quekett.org
This website provides extremely useful advice on some of the complex microscopy techniques that are used in mycology, courses on which are not easy to access.

Society

British Mycological Society (BMS): www.britmycolsoc.org.uk
The society's journal, *Field Mycology*, contains a great deal of useful information and regular updates. Unpublished keys are often shared directly among mycologists, so it is well worth joining the BMS and your local recording group to

Hairy Nuts Disco *Lanzia echinophila*. Seriously, that's what it's called!

participate in this. The journal is now publicly available online on the BMS website, and back issues (before 2022) can be found on the ScienceDirect website: www.sciencedirect.com/journal/field-mycology

Facebook groups

British Mycological Society
This is a very large group, with over 70,000 members, and it is also the best place for detailed and technical questions and support. This group could not be more different to the many Facebook foraging groups that you need to avoid if you are serious about reliable identification.

Ascomycetes of the World

Rusts, Smuts and other Phytoparasitic Fungi.

Fungus recording groups
One of the best ways to get experience of fungi in the field is to go on excursions with experienced field mycologists. There are many fungi recording groups around the UK. My local one is the West Weald Fungus Recording Group, of which I am a member. I regularly join their excursions and have had some fantastic experiences with them, as well as bringing an element of pan-species listing to the group.

Laboulbeniales: 129 species on the UKSI

Scale: Medium   Effort: Difficult   Accessibility: 15

Small fungi that infect chitinous invertebrates, many of which are host specific. The fungi is passed on by contact between generations, so species that are adult most of the year and hibernate over the winter are likely hosts (such as carabids, ladybirds and millipedes, for example). There is no up-to-date checklist for the British Isles, therefore it's seemingly quite easy to find new species. This group is crying out for an entomologist to sort out the checklist! The following is downloadable from Research Gate.

Andre De Kesel, Cyrille Gerstmans and Danny Haelwaters (2020) *Catalogue of the Laboulbeniomycetes of Belgium.* STERBEECKIA (2020) 36: 3–143.

Aquatic Hyphomycetes: at least 100 species available

Scale: Medium   Effort: Difficult   Accessibility: 15

You can find these microscopic fungi in waterbodies wherever foam forms (such as on a wooded stream or after a flood). Take a sample of the foam and look at it under a compound microscope. They can also be found on submerged leaves as the autumn progresses.

C. T. Ingold (1975) *Guide to Aquatic Hyphomycetes – SP30.* Freshwater Biological Association.
Although old, this book is still in print and reasonably priced.

6.7 Bryophytes (mosses, hornworts and liverworts): 1,121 available species

Scale Large   Effort Difficult   Accessibility 16

Collective effort = 820 (73.1%) species   **Top lister** = Paul Bowyer (607 species)

Vital equipment: hand lens, paper envelopes (for specimens), compound microscope.

Essential text

Ian Atherton, Samuel D.S. Bosanquet and Mark Lawley (2010) *Mosses and Liverworts of Britain and Ireland: A Field Guide*. British Bryological Society.

An essential field guide with most of the more common species you are likely to encounter before you take bryology to the next level. It contains excellent photographs, distribution maps and descriptions of similar species. Lists of species by habitat at the back of the book are also really useful. This is a very accessible, up-to-date guide.

Like grasses, sedges and rushes, bryophytes have to be identified without recourse to petals and large leaves. A hand lens is vital for all but the largest species, and a microscope is essential for many, especially the smallest species. You will need to

Hylocomium splendens (a pleurocarpous moss).

learn a whole new nomenclature associated with the plants, and you'll require a compound microscope to get into them in detail.

It's very easy to store bryophytes in a reference collection in paper envelopes, and even after many years they can be rehydrated and viewed under the microscope.

Mosses: 814 species on the UKSI

Scale	Medium	Effort	Difficult	Accessibility	12
Acrocarps	579 – Medium	Effort	Difficult	Accessibility	12
Pleurocarps	196 – Medium	Effort	Moderate	Accessibility	9
Sphagnales	39 – Small	Effort	Very difficult	Accessibility	10

The first thing to learn about mosses is that they include pleurocarps, which usually resemble tiny ferns, typically found in large prostrate patches, and acrocarps, which are often smaller but can form big tufts, and tend to look like little upright trees. Then there are the Sphagnales, which include *Sphagnum* species. Generally speaking, the pleurocarps are much easier to access as a beginner, and there are also fewer of them, so I would recommend that you start with these.

Essential texts

J.E. Smith (2004) *The Moss Flora of Britain and Ireland, 2nd edn.* Cambridge University Press.

This is a large and very comprehensive book designed to be used at the microscope, not as a field guide. It covers all UK species and includes high-quality line drawings of mosses and their cells, as well as keys to species. Note, though, that there has been some major nomenclatural revision in recent years.

The great sweeping shoots of *Dicranum majus* (an acrocarpous moss).

Pleurozia purpurea, a leafy liverwort that I encountered in the west of Ireland.

Liverworts: 249 species on the UKSI

Scale Medium   Effort Difficult   Accessibility 12

Liverworts are subdivided into the thallose liverworts (Metzgeriales), which are a relatively small group but easier to identify due to being larger and fewer in number, and the leafy liverworts (Jungermanniales), which are considerably more numerous and can be a challenge to identify.

Essential texts

Jean A. Paton MBE (2014) *The Liverwort Flora of the British Isles*. E.J. Brill.

Jean A. Paton MBE (2022) *A Supplement to the Liverwort Flora of the British Isles*. Self-published.
   Like Smith (2004), the above are not a field guide. They are companion volumes, and you will need both in order to cover all bryophytes, but they are not cheap.

Society

British Bryological Society: www.britishbryologicalsociety.org.uk
   The benefits of membership include many county-based groups that run regular field meetings, as well as two national meetings each year, technical indoor meetings, various projects, and copies of the *Journal of Bryology* and the membership bulletin *Field Bryology*.

The large thallose liverwort *Conocephalum conicum*. (David Hawkins)

Facebook group

Bryophytes of Britain and Ireland
 This is a small (private) Facebook group, so you will need to receive an invitation from an existing member to join.

VASCULAR PLANTS   81

6.8 Vascular plants (clubmosses, horsetails, ferns, naked-seeded plants and flowering plants): 7,371 available species

Scale **Very large**   Effort **Easy**   Accessibility **10**

Collective effort = 4,097 (55.6%) species   **Top lister** = John Poland (2,915 species)

Vital equipment: 10× (or better still 20×) magnification hand lens; a dissecting microscope is sometimes helpful.

All ecologists, entomologists and pan-species listers will need to learn their plants – without botanical knowledge, many invertebrates, galls, fungi, mines, mildews, rusts and other groups will be very difficult to find and identify. I learnt my plants long before I learnt my invertebrates, but it makes a huge difference to entomologists if they know their plants, too. It's very hard to find a good list of rare invertebrates that feed on specific plants if you don't know how to identify those plants. Botanical knowledge is also important for conservation. For example, advice on the grazing of a wet heath dominated by Purple Moor-grass *Molinia caerulea* will be very different from that for a drier heath dominated by Wavy Hair-grass *Avenella flexuosa*, so you need to know the difference between these grasses in the first place, before you can ever give advice on how to manage the habitats they grow in.

In most open places, but especially grasslands, the dominant species are typically grasses, sedges and rushes, which often drive the ecology of the landscape. It is vital to learn them, yet they are often learnt last, after the more showy and accessible 'flowering' plants.

Pasqueflower *Pulsatilla vulgaris*.

Alpine Bartsia *Bartsia alpina* on the banks of the River Tees, seen on a long trip during a specific time window.

Essential texts

There are many different texts on vascular plants.

David Streeter and Ian Garrard (1998) *Wild Flowers of the British Isles*. Midsummer Books (out of print).
This is my favourite plant book, and if picture-matching is your route into botany, this is the one to get. You can easily find second-hand copies. It's a large, A4-size book with excellent illustrations, accurate colours and concise, useful text. The size of the book means that the illustrations are large enough to really capture the 'jizz' of the species. However, it does not cover ferns, grasses, sedges, rushes or trees.

David Streeter, Christina Hart-Davies, Audrey Hardcastle, Felicity Cole and Lizzie Harper (2016) *Collins Wildflower Guide*, 2nd edn. Collins.
This book covers all of the groups and has good keys, though coverage of aliens is limited and the quality of the illustrations is rather variable.

Clive A. Stace (2021) *New Flora of the British Isles*, 4th edn. C&M Floristics.
When you want to get really technical, you need a good solid key like this one. Stace is excellent but heavy, so I have both the full version and the much smaller abridged version. The full version includes many more non-native species than the smaller version. However, even Stace doesn't include all of the naturalised species in the British Isles.

At less than 1p, this is Britain's smallest and cheapest grass, Early Sand-grass *Mibora minima*.

Day 1 of my course was always the grasses, with sedges and rushes on day 2.

Francis Rose (1989) *Colour Identification Guide to the Grasses, Sedges, Rushes and Ferns of the British Isles and North-Western Europe*. Midsummer Books.
Although it's now out of print, I used this book for many years to teach grasses, sedges and rushes. I really like the keys, and the illustrations of graminoids are the best I have ever seen. I was first shown this book back in the winter of 2001–2002, but found it bewildering to begin with. Unlike plants with colourful, showy flowers, grasses all appear to be confusingly similar at first glance. There is a lot of new terminology (e.g. glume, ligule, spike, spikelet, panicle, raceme, floret) that you need to learn just in order to be able to describe a grass. Then I got my lucky break with the ecology team at the RSPB, and found myself on a course with Malcolm Ausden, the late James Cadbury and the late John Day. A year later I joined them in teaching RSPB staff about grasses, sedges and rushes, I went on to teach the same subject for 15 years at the RSPB and then Sussex Wildlife Trust. The key to learning this group, and indeed to getting the most out of any wildlife course, is to get out in the field with your key identifying species on your own *straight after the course*, so that you can consolidate what you have learned.

James Merryweather (2020) *Ferns of Britain and Ireland*. Princeton University Press.
This is an excellent photographic guide to ferns.

I started botanising at the age of around 13 or 14, when Ewart Gardner gave me a copy of the *Wildflowers of Britain and Northern Europe* by Richard Fitter, Alastair Fitter and Marjorie Blamey. It was the first detailed botanical book I had ever seen, it felt like I was unlocking the secrets to the universe. Things I had been standing on my whole life suddenly had names. I shall never forget trips up to the limestone of the Peak District and being shown species like Wild Marjoram *Origanum vulgare* and Meadow Crane's-bill *Geranium pratense* for the first time. The smells of these plants really linger in my mind. And even after 30 years of living down in Sussex, the chalk grassland of the South Downs still feels exotic to me. I'll never forget the joy of getting to grips with common species on my garden lawn, such as Cat's-ear *Hypochaeris radicata*, Procumbent Pearlwort *Sagina procumbens* and Thyme-leaved Speedwell *Veronica serpyllifolia*. Those lazy summer evenings with my head buried in a plant book were just magical – an experience that could well be taken away from people who now use ID apps. Further afield up San's Brook there were single patches of Devil's-bit Scabious *Succisa pratensis*, Wood Sage *Teucrium scorodonia* and Hedge Woundwort *Stachys sylvatica*, with yet more intoxicating smells to captivate the senses. Meanwhile a huge patch of Pignut *Conopodium majus* would feed the caterpillars of the day-flying Chimney Sweeper *Odezia atrata*, a lovely sooty black moth.

During my first summer at university, I wandered up on to the chalk grassland of the South Downs and stumbled upon my first ever Round-headed Rampion *Phyteuma orbiculare*. Fast forward to summer 2024, when I was surveying the same area for the local council, and I found an entirely new population of Nottingham Catchfly *Silene nutans*, a rare plant in Sussex.

Chalk grassland at Castle Hill, Sussex in 2024, with a riot of colour produced by Nottingham Catchfly, Round-headed Rampion, Eyebright *Euphrasia* agg., Rough Hawkbit *Leontodon hispidus*, Wild Thyme *Thymus drucei*, Heath-grass *Danthonia decumbens*, Oxeye Daisy *Leucanthemum vulgare*, Yellow-rattle *Rhinanthus minor*, Common Knapweed *Centaurea nigra s.l.*, Squinancywort *Asperula cynanchica* and Dwarf Thistle *Cirsium acaule*, among others.

The rare native Orange Bird's-foot *Ornithopus pinnatus* was only in leaf when I was on the Isles of Scilly, and some effort was needed to find it, whereas great big showy plants like Jersey Lily *Amaryllis belladonna* were a little easier to spot. However, they both have the same value on your pan-species list (I know which one I prefer, though).

I remember carrying Streeter and Garrard (1998), Rose (1989) and the large version of an early edition of Stace in my rucksack to the top of Glen Feshie in 2004. I shall never forget finding a single plant of Rock Speedwell *Veronica fruticans* with one flower on it. As my old boss Malcolm Ausden and I looked at it, the single flower fell off in front of us, all the more memorable for the fact that neither of us had a camera then. If we had been there one minute later, we would not have seen it. Malcolm wouldn't leave the mountain until we had found Alpine Cinquefoil *Potentilla crantzii*. I have not seen either species since that day.

Once you get to about 1,000 species, it becomes more difficult to stumble upon new plants and you really need to go looking for them. This is especially true of natives and long-established aliens. There are also rare species that you have to make a specific journey to see at a specific time of year (such as Alpine Bartsia *Bartsia alpina*). Then of course there are the non-natives. Ticking these off might seem like an easy ride, but recording them is actually vital. Non-native species are the second-largest cause of extinctions globally, and we can do nothing to change this if we don't know which plants are here in the first place.

A trip to the Isles of Scilly is a must for any pan-species lister. I have only been once, in 2020 with Tony Davis and Seth Gibson, who are both extremely knowledgeable about the Isles, especially the rich and varied naturalised flora that abounds there. I came back with around 100 new plant species, but it was the rare natives – in fact only a small proportion of the species that I added from that trip – which made the deepest impression on me.

Vegetative botany

Vegetative plants count just as much as those in flower. The first edition of this very helpful book was co-authored by PSL's top vascular plant lister, John Poland, when he was still in his twenties:

John Poland and Eric J. Clement (2020) *The Vegetative Key to the British Flora*, 2nd edn. John Poland.

Another way to learn your vegetative botany is to spend a long time looking at the vegetative parts of a plant when it is in flower, so that you can learn to recognise them in the future when the flowers are missing. Quadrat bashing is a very good way of doing this – I know my Sussex chalk-grassland plants in leaf better than any other plants due to this.

Microspecies and difficult families

The Botanical Society of Britain & Ireland (BSBI) has produced a huge range of monographs on microspecies (such as dandelions and hawkweeds) and difficult families or genera (such as umbellifers, broomrapes, pondweeds, sedges, lady's-mantles, water-starworts and fumitories). The former are especially important for 'unlocking' many plants that are otherwise difficult to get to species level.

E.F. Dees and A. Newton (1988) *Brambles of the British Isles*. CD-book edition. Pisces Conservation.
   This covers 307 species of *Rubus* but may well be out of date.

A. Hannah (2025) *Brambles of Scotland – BSBI Handbook No. 2025*. BSBI.
   While focusing on Scotland, this new book will actually be useful for aspiring batologists all over the UK.

Website

Plant Atlas 2020: www.plantatlas2020.org
   Published by the BSBI, this is an extremely useful modern distribution map with valuable information on species' ecology and trends.

Societies

Botanical Society of Britain & Ireland: www.bsbi.org
   The benefits of membership include three issues of the magazine *BSBI News* each year, access to a network of plant experts for help with identification, and access to the BSBI's Distribution Database.

The Wild Flower Society: www.thewildflowersociety.org.uk
   This organisation holds a wide range of field meetings across the UK, as well as online lectures.

Regional botanical societies
   These often have field meetings in your local area.

Facebook groups

There are Facebook groups for many of the complex genera of vascular plants. For example:

Hawkweeds (*Hieracium*) of Britain and Ireland

Euphrasia (Eyebrights) of Britain and Ireland

Brambles (*Rubus*) of Britain and Ireland.

Wildflowers of Britain and Ireland.
   A large group with nearly 25,000 members at the time of writing.

6.9 Porifera (sponges): 412 species available

Scale Medium Effort Very difficult Accessibility 15

Collective effort = 45 (10.9%) species   **Top lister** = Nathan Jackson (21 species)

Vital equipment: waterproof camera, hand lens, *snorkelling/scuba diving gear, compound and dissecting microscopes, dissolving agent for microscopy.*

Marine sponges: 406 species available

Scale Medium Effort Very difficult Accessibility 15

Essential texts

Sarah Bowe, Claire Goodwin, David Kipling and Bernard Picton (2018) *Sea Squirts and Sponges of Britain and Ireland*. Wild Nature Press.
   Although it only covers around 60 species, this is the most comprehensive photographic guide available. Its many useful features include underwater images, distribution maps and the species' preferred substrates and depths, as well as identification tips.

A selection of sponges or a Jackson Pollock? The blue species is *Terpios gelatinosa*, the blood red one is *Clathria atrasanguinea*, the beige one is Breadcrumb Sponge *Halichondria panicea* and the orange one is Shredded Carrot Sponge *Amphilectus fucorum*. Identified by Dr Gerald Legg on the old Victorian pipe at Worthing (known as Worthing Pipe) in 2016, before it was removed.

Tethya citrina, a large, spherical sponge the size of a tennis ball, in a rock pool in Sussex.

P.J. Hayward and J.S. Ryland (1990) *The Marine Fauna of the British Isles and North West Europe*. Oxford Science Publications.
This is available both as a comprehensive two-volume set and as an abridged handbook (itself a large book). The two-volume version covers many more species, and it is well worth getting this book if you spend any time in marine habitats, especially the intertidal zone. Both versions include technical keys with macroscopic and microscopic line drawings, but cover less than 20% of the sponge species on the UKSI.

R. Graham Ackers, David Moss, Bernard E. Picton, Shirley M.K. Stone and Christine C. Morrow (2007) *Sponges of the British Isles ("Sponge V")*. Marine Conservation Society.
This is also available to download for free at:
www.habitas.org.uk/marinelife/sponge_guide/sponge5.pdf
The PDF contains very technical and detailed accounts of over 100 species, making it the most comprehensive text on sponges available.

Marine sponges can be found in intertidal rock pools, under rocks, on submerged rocks, and on seaweeds, spider crab shells and human-made substrates. Different species grow at different depths, so if you are really serious about sponges you will need to take up scuba diving and/or snorkelling. As there are many gaps in our knowledge of them, there is a good chance of finding species new to science.

Freshwater sponges: 6 species available

Scale: Very small | Effort: Moderate | Accessibility: 3

Essential text

J. Briggs (2023) Freshwater sponges: our native species and their inhabitants. *British Wildlife* 35(2): 105–14.
This article describes six freshwater sponge species, as well as their habitats and distribution.

Society

Porcupine Marine Natural History Society: www.pmnhs.co.uk
Benefits of membership include two issues of the Bulletin each year and attendance at field meetings. This society would be relevant to anyone with an interest in any of the marine groups covered in this chapter (or groups that contain a significant marine element).

Facebook groups

Porcupine Marine Natural History Society
The Facebook page of the above society, which includes many highly knowledgeable and helpful members.

Rockpooling & Shrimping (British Isles)

NE Atlantic Porifera

Download

Claire Goodwin and Bernard Picton (2011) *Sponge Biodiversity of the United Kingdom*. National Museums Northern Ireland.
This report, which contains a large number of underwater photographs identified to species level, can be downloaded from ResearchGate.

6.10 Ctenophorans (comb jellies): 5 species available

Scale Very small   Effort Easy   Accessibility 2

Collective effort = 3 (60.0%) species

Top listers = Kian Hayles-Cotton & Geoff Morgan (3 species each)

Vital equipment: pond net, tray, waterproof camera.

Essential text

Hayward and Ryland (1990) cover only three of the five species listed on the UKSI.

As there are only five species on the UKSI, there's enough space here to describe them all.

Sea Gooseberry *Pleurobrachia pileus*. This is the species that most people have seen, and the only one on my list. It's almost entirely spherical, and the common name is an apt description of it.

Beroe cucumis. A slightly more elongate and opaque animal, and the second most frequently listed species, this is larger than the Sea Gooseberry, and its presence is given away by the oscillating cilia along its length.

Beroe gracilis. Currently there are no records on the NBN Atlas for this species. It is similar to *Beroe cucumis*, but even more elongate.

Sea Gooseberry *Pleurobrachia pileus*. (James Emerson)

Beroe cucumis. (Seth Gibson)

Bolinopsis infundibulum. The shape of this large species resembles an inverted wing-nut. Only one person has noted it so far on the new PSL website.

Sea Walnut *Mnemiopsis leidyi.* This species is neither recorded on the NBN Atlas nor listed on the new PSL website. It is predicted to colonise our waters in the near future. It is most like the above species.

These gelatinous, ovoid, translucent, jellyfish-like organisms tend to appear in significant numbers on the coast in some years, but appear to be entirely absent in other years. Sea Gooseberries are almost invisible in the water, and are only likely to be spotted when washed up on the beach, where they soon dry out. I have netted Sea Gooseberries during falls of Blue Jellyfish *Cyanea lamarckii* on the Sussex coast, and watched them apparently disappear when placed in a container of water. In contrast, *Beroe* species are much more opaque, so can be spotted motoring around in the water, and apparently they are also bioluminescent.

Ctenophorans can be found all around the UK coastline, but there appear to be fewer records on the south coast than in most other places.

6.11 Cnidarians (jellyfish, sea anemones, corals and hydras, etc.): 541 species available

| Scale | Medium | Effort | Difficult | Accessibility | 12 |

Collective effort = 91 (16.8%) species   **Top lister** = Yolanda Evans (45 species)

Vital equipment: waterproof camera, snorkelling/scuba diving gear, dissecting microscope.

Essential text

Hayward and Ryland (1990) cover many of these groups, but coverage within groups is very patchy.

Jellyfish (Scyphozoa): 16 species on the UKSI

| Scale | Very small | Effort | Very easy | Accessibility | 1 |

There are very few jellyfish species and there is little overlap between them, so identification should not pose too much difficulty. There are six common species around the British Isles, namely Moon Jellyfish *Aurelia aurita*, Barrel Jellyfish *Rhizostoma octopus*, Compass Jellyfish *Chrysaora hysoscella*, Blue Jellyfish *Cyanea lamarckii*, Lion's Mane Jellyfish *Cyanea capillata* and the Mauve Stinger *Pelagia noctiluca*. I have seen all but the last two. There are a further 10 species on the UKSI, all of which are rare pelagic species, and if there are any records on the NBN they are in the middle of the Atlantic Ocean, so no pan-species lister has recorded them.

Portuguese Man o' War *Physalia physalis* is in fact a hydrozoan in the order Siphonophorae.

CNIDARIANS 93

In the water, jellyfish are incredibly beautiful animals, like this Lion's Mane Jellyfish in a fjord in Norway (it's not on my list, I just needed a photo).

Out of the water they are less attractive – this Compass Jellyfish looks like an alien's attempt at a dessert.

Stalked jellyfish (Staurozoa): 10 species on the UKSI

Scale: Very small   Effort: Moderate   Accessibility: 3

Essential text
Hayward and Ryland (1990) cover seven species of stalked jellyfish with a detailed key.

You can find stalked jellyfish attached to vegetation in rock pools, and although they occur around the entire coast of the UK, they are significantly more abundant in the west. Another way to find them is to sweep net weed, then transfer the material to a tray and sort though it with close-focus binoculars. These creatures are smaller than you might expect!

Website
Stalked Jellyfish/Stauromedusae of the UK: www.stauromedusae.co.uk
This very useful site for helping to identify stalked jellyfish was created by David Fenwick.

Calvadosia campanulata is one of only two stalked jellyfish I have seen so far, this was on the Isles of Scilly in 2020. This seems to be the most frequently recorded species.

Sea anemones and corals (Anthozoa): 217 species on the UKSI

Scale: Medium   Effort: Difficult   Accessibility: 12

Essential text

Chris Wood (2013) *Sea Anemones and Corals of Britain and Ireland*. Wild Nature Press. This excellent photographic guide includes beautiful underwater photographs, useful identification tips, information on species' preferred substrates and depths, and distribution maps.

Sea anemones: 75 species on the UKSI

Scale: Small   Effort: Moderate   Accessibility: 9

Over 55 species are covered by Wood (2013), making this an invaluable guide to anemones. Hayward and Ryland (1990) cover just over 30 species.

Anemones are easiest to find in rock pools, both out in the open and under stones. There are also two rare species that occur in brackish lagoons, and some species are only found in deeper water. Many species are identifiable from a good photograph, so an underwater camera is essential.

The incredible Parasitic Anemone *Calliactis parasitica* in a rock pool on Jersey (top left), Snakelocks Anemone *Anemonia viridis* by night under UV light (top right), Plumose Anemone *Metridium dianthus*, which is usually seen in deeper water (bottom left) and the large Dahlia Anemone *Urticina felina* (bottom right).

Soft corals and sea fans (Alcyonacea): 60 species available ●

Scale Small   Effort Very difficult   Accessibility 10

Wood (2013) covers just five species in this group (three soft corals and two sea fans). Hayward and Ryland (1990) cover just two soft corals and two sea fans.

Websites
These sites are likely to provide more comprehensive help with identification.
Habitas: www.habitas.org.uk/marinelife/index.asp?item=D10300&group=CNIDARIA
APHOTOMARINE: www.aphotomarine.com

Dead Men's Fingers *Alcyonium digitatum*, a soft coral out of the water, photographed by me (left), and the same species looking wildly different in the water, photographed by Gerald Legg (right).

Sea pens (Pennatulacea): 8 species available on the UKSI ●

Scale Very small   Effort Moderate   Accessibility 3

Both Wood (2013) and Hayward and Ryland (1990) cover just three of these species, all of which are only likely to be encountered at depth. The remaining five species have no records on the NBN and must be quite scarce. Of the three species covered in the texts, the Slender Sea Pen *Virgularia mirabilis* is the most widespread, followed by the Phosphorescent Sea Pen *Pennatula phosphorea*. The Tall Sea Pen *Funiculina quadrangularis* appears to be restricted to north-west Scotland.

Hard corals (Scleractinia): 58 species on the UKSI ●

Scale Medium   Effort Very difficult   Accessibility 15

Including the only species in the Corallimorpharia, which is Jewel Anemone *Corynactis viridis*, Wood (2013) covers just eight species in the photographic guide, and Hayward and Ryland (1990) cover only two.

CNIDARIANS 97

Jewel Anemone *Corynactis viridis* and Devonshire Cup Coral *Caryophyllia smithii* (a hard coral) at the Outer Mulberry, Selsey. (Gerald Legg)

Website
www.aphotomarine.com/corals_stoney_hard_scleractinia.html
This includes photographs of six species, but there appears to be no other literature on identification of the majority of the species on the UKSI.

Hydroids (Hydrozoa): 300 species available

| Scale | Medium | Effort | Very difficult | Accessibility | 15 |

Hydroids make up more than half of the cnidarians in the region, and their identification poses the biggest challenge. They are a very difficult group, with a variety of possible growth stages (and intermediates) complicating matters further. The vast majority of species are marine, and only a small number are found in freshwater.

Essential texts

Joanne Porter (2012) *Bryozoans and Hydroids of Britain and Ireland*. Wild Nature Press. Although it only covers around 50 species, this guide is a great place to start, based on underwater photography. However, it will be difficult to cover hydroids in detail without the following two keys:

Peter Schuchert (2012) *SBF Volume 59: North-West European Athecate Hydroids and Their Medusae*. Field Studies Council.

Paul F.S. Cornelius (1995) *SBF Volume 50: North-West European Thecate Hydroids and Their Medusae* (two volumes). Field Studies Council.

The synopses mentioned here all contain keys to tackling this difficult group of marine invertebrates, and are not for the faint-hearted. Volume 59 covers around 113 species, while Volume 50 (a two-volume set) covers the remaining species.
The following texts may also be helpful.

The worm-like *Candelabrum cocksii* – an athecate hydroid, which is quite atypical. Rainbow Sea Slug *Babakina anadoni* eats these, so if you find this hydroid you might encounter the nudibranch too!

Hayward and Ryland (1990) cover just over 100 species of hydrozoan.

Peter J. Hayward (1987) *Animals on Seaweed*. Naturalists' Handbooks 9. Richmond Publishing (out of print).

Freshwater hydrozoans: 6 species available ●

Scale Very small   Effort Moderate   Accessibility 3

Essential text

R. Fitter and R. Manuel (1986) *Collins Field Guide to Freshwater Life*. Collins (out of print).

There are four species of freshwater hydrozoan in the genus *Hydra* currently on the UKSI (although a fifth species may be present). These are *Hydra circumcincta*, Brown Hydra *Hydra oligactis*, Green Hydra *Hydra viridissima* and Common Hydra *Hydra vulgaris*. The Green Hydra is the species most frequently recorded by pan-species listers, though I have never seen one. There is also the colonial species *Cordylophora caspia*, which seems to be most common in the east of East Anglia, and can tolerate brackish conditions. The sixth species is the tiny non-native freshwater jellyfish *Craspedacusta sowerbii*, which appears to be rather rare – no one has listed it yet and there are very few records on the NBN.

Website

The World Hydrozoa Database: www.marinespecies.org/hydrozoa

6.12 Molluscs (slugs, snails, limpets, bivalves, squids, chitons and sea slugs, etc.): 2,079 species available

| Scale | Large | Effort | Difficult | Accessibility | 16 |

Collective effort = 496 (23.9%) species   **Top lister** = Graeme Lyons (254 species)

Vital equipment: metal-rimmed pond net, sieve, tray, hand lens, dissecting microscope, suction sampler, *alcohol*.

Molluscs are such a mixed bag (there are eight classes and 50 orders on the UKSI) that you do need quite a few texts to cover all of them, and the vast majority of species are marine.

Terrestrial snails: *c.*100 species available

| Scale | Medium | Effort | Moderate | Accessibility | 9 |

Essential texts

Fred Naggs, Richard C. Preece, Roy Anderson, Amritha Peiris, Harold Taylor and Tom S. White (2014) *An Illustrated Guide to the Land Snails of the British Isles.* London Natural History Museum.
This is a really useful foldout photo guide with very good comparable images, and it is excellent value for money.

The wonderfully creamy, caramel swirl that is the Carthusian Snail *Monacha cartusiana*, a rare species of chalk grassland and dunes in the south-east.

Dwarf Snail *Punctum pygmaeum* is easy to miss – here it's on a 1 mm grid!

Robert Cameron and Gordon Riley (2003) *Land Snails in the British Isles*. Field Studies Council.
This book is a little out of date, but still great value for money, especially when combined with the above photo guide. I suggest that you tuck that inside this book so that you don't lose it!

M.P. Kerney and R.A.D. Cameron (1994) *A Field Guide to the Land Snails of Britain and North-west Europe*. Collins.
This book also covers slugs, and although it is out of print, second-hand copies are easy to find.

All terrestrial habitats support land snails in some way, but the best sites for native snails tend to be chalk grassland and base-rich woodland, with the lowest diversity found on acidic sites. Non-native species are typically most often found close to habitation. It is easy to keep a reference collection of terrestrial snail shells.

Many snails are sensitive to change. For example, the Heath Snail *Helicella itala* is easily grazed out if sheep are introduced to a site, and snails like the Carthusian Snail *Monacha cartusiana* can also be lost through the abandonment of grazing.

Many species are extremely small and need to be identified at the microscope. The suite of tiny snails on calcareous grassland are a particularly pleasurable to identify, the Dwarf Snail *Punctum pygmaeum* being the smallest of these.

Terrestrial slugs: 53 species on the UKSI

Scale: Small | Effort: Moderate | Accessibility: 6

Essential text

Ben Rowson, James Turner, Roy Anderson and Bill Symondson (2014) *Slugs of Britain and Ireland: Identification, Understanding and Control.* Field Studies Council.
This joy of a book has excellent photos that look more like illustrations because of the way they are presented. The text and identification tips are spot on. This guide has completely changed my attitude to slugs.

The easiest way to find slugs is under logs and stones. Some really good places to find them are around houses and buildings, in dumped waste, and in other synanthropic habitats. Going out at night with a torch can also yield results. A few species are associated with ancient woodland, and everyone should try to visit the south-west of Ireland to see the Kerry Slug *Geomalacus maculosus*. The large *Arion* species seem to be the most problematic – they and some other groups are only identifiable by dissection. I personally draw the line at dissecting slugs!

The spectacular Kerry Slug *Geomalacus maculosus*.

Giant Pea Mussel
Pisidium pulchellum.

Freshwater molluscs: *c.*86 species available ●

Scale Small   Effort Difficult   Accessibility 8

Freshwater bivalves: 32 species available ●

Scale Small   Effort Difficult   Accessibility 8

A different approach to most aquatic invertebrate sampling methods is needed here, as you will want to actively sample the substrate rather than just the vegetation. This is very messy and will require a sieve to clean the sample. Hand dredging is also used to monitor molluscs in the sediment of larger waterbodies.

Essential text

Ian Killeen, David Aldridge and Graham Oliver (2004) *Freshwater Bivalves of Britain and Ireland*. Field Studies Council.
   This spiral-bound, A4-sized book covers 32 species. It has excellent photos and very useful keys, species accounts and distribution maps. The most problematic bivalves are the pea mussels (which account for around half of all species), as the majority of these are extremely small and difficult to identify.

Freshwater snails: *c.*54 species available ●

Scale Small   Effort Moderate   Accessibility 6

Essential text

Ben Rowson, Harry Powell, Martin Willing, Michael Dobson and Hannah Shaw (2021) *Freshwater Snails of Britain and Ireland*. Field Studies Council.
   This book is crammed with information and covers all of the freshwater snails. The species accounts are among the most comprehensive I have ever seen in a field guide.

Radix auricularia (left) and the much commoner *Ampullaceana balthica* (right).

The best way to find aquatic snails is by sampling the vegetation with a sturdy pond net. In contrast to bivalve sampling you should try to avoid the substrate.

Marine molluscs: *c*.1,840 species available ●

Scale `Large`   Effort `Difficult`   Accessibility `16`

Things get much more complicated when you are dealing with saltwater-dwelling molluscs, as there are so many marine mollusc species across several classes, and a bewildering number of orders and families.

Blue-rayed Limpet *Patella pellucida* in the 'kelp zone'.

Flat Topshell *Steromphala umbilicalis* and Grey Topshell *S. cineraria* – two of the commonest marine gastropods to be found in rock pools.

Marine gastropods: *c*.1,175 species on the UKSI

Scale: Large   Effort: Difficult   Accessibility: 16

These are the largest group of marine molluscs. They can be found almost anywhere around the coasts, particularly on rocky sites with plenty of weed, although many species also occur on sandy shores.

Essential texts

Hayward and Ryland (1990) cover over 230 species of marine gastropod (over 70 of which are sea slugs, which are not covered in the three texts listed below), but coverage is patchy.

Geoffrey D. Wigham and Alastair Graham (2017) *Marine Gastropods 1: Patellogastropoda and Vetigastropoda*. Field Studies Council.

Geoffrey D. Wigham and Alastair Graham (2017) *Marine Gastropods 2: Littorinimorpha and Other Unassigned Caenogastropoda*. Field Studies Council.

Geoffrey D. Wigham and Alastair Graham (2017) *Marine Gastropods 3: Neogastropoda*. Field Studies Council.

These three synopses cover the majority of shelled marine gastropods in the British Isles in detail, with very good keys, detailed species accounts and exquisite line drawings that really capture the detail and texture of the shells. I thoroughly recommend them.

Sea slugs (Nudibranchia): 168 species on the UKSI

Scale: Medium   Effort: Difficult   Accessibility: 12

At the time of writing, I have made fewer than 100 records of 31 species of nudibranch, but the excitement of each encounter would rival that of the most high-octane bird twitch. Although I have always done a lot of rock-pooling, I didn't see my first sea slug until January 2017, at the Pound in Eastbourne. I think I

The incredible *Polycera quadrilineata*.

Solar-powered Sea Slug *Elysia viridis*.

must have assumed that they were tropical creatures, or would only rarely be met with on these shores. Shortly after that, with the help of Evan Jones, who is very knowledgeable about where to find sea slugs and how to identify them, I swept my first *Polycera quadrilineata* from Wireweed *Sargassum muticum* in the shallow lagoon at the Pound. Transferring the contents to a white tray and then searching the tray with close-focus binoculars also helps. In fact, several of the nudibranchs I have seen I have only spotted this way.

However, by far the most productive method of finding them is to turn over rocks in suitable habitat. Although this is a numbers game (the more rocks you turn, the greater the chance you'll find something), it's also worth being thorough and taking your time, especially on rocks covered in hydroids or other appropriate food. Make sure you return any rocks gently to their original position. Surprisingly, I have also encountered a few sea slugs quite high up the beach, even in places where one would only expect to see really common species such as Shanny *Lipophrys pholis*, but you will also find them at the extreme low-tide mark.

It's important to remember that many of the sea slugs in rock pools are extremely small. Many of those I have seen must have been immature, as the upper limits of the size ranges cited in the texts are all a great deal larger, but you do also regularly see adults – such as the *Discodoris rosi* overleaf. Nudibranchs lay distinctive eggs; these can sometimes be identified to species, or indicate that the animals are nearby. So when you are searching under a rock, you are looking for something really small, often less than 5 mm in length. They are very delicate, so getting them off the rock into a container for viewing needs to be done carefully with a suitable implement. As soon as they are out of the water, nudibranchs bunch up and look completely different from in photographs (so viewing the underside of a rock when it is submerged will be more productive if possible), yet they always seem to remain shiny and slimy looking, and the bright colours often stand out as being quite distinct from the surrounding substrate and from other plants and animals.

Discodoris rosi, found new to the Channel Islands on Jersey by Nicolas Jouault. I was lucky enough to be there and take this shot.

Nudibranchs often feed only on one species or type of marine life, with hydroids and bryozoans being the most popular food. Look for sheltered areas with ideally clear and flowing water rich in this kind of marine life and you've got a good chance of finding nudibranchs. A bit like a species-rich grassland holding lots of invertebrates!

There are a number of other sea slugs and sea slug-like creatures that are not actually nudibranchs, such as the Sea Hare *Aplysia punctata*, the Yellow-plumed Sea Slug *Berthella plumula* and the Solar-powered Sea Slug *Elysia viridis*.

Essential text

Bernard Picton and Christine Morrow (2023) *Nudibranchs of Britain, Ireland and Northwest Europe*, 2nd edn. Princeton University Press.
This a glorious new book, with outstanding photos of nearly 200 species of exotic looking sea slugs. I feel like a kid in a sweet shop every time I open it, the Rainbow Sea Slug *Babakina anadoni* is now my most coveted species.

Facebook group

NE Atlantic Nudibranchs
An exceptionally useful group for all your sea slug queries with knowledgeable experts and exquisite photos.

Marine bivalves: *c.*556 species on the UKSI

| Scale | Medium | Effort | Difficult | Accessibility | 12 |

Typically, the diversity of marine bivalves is higher on more sandy substrates, but they occur in a range of habitats, including rock pools.

Essential texts

Hayward and Ryland (1990) cover around 135 species of marine bivalve.

Norman Tebble (1966) *British Bivalve Seashells*. Pisces Publications.
This book is out of print but available on CD, though it is now quite dated.

Website

Marine Bivalve Shells of the British Isles: www.naturalhistory.museumwales.ac.uk/britishbivalves/id.php

This free online resource is the best tool for identifying marine bivalves in the British Isles. Species are sorted by depth, from the intertidal zone right down to the abyssal plain. The site includes keys, excellent species accounts, clear photos and useful distribution maps.

Other marine molluscs

Essential text

A.M. Jones and J.M. Baxter (1987) *Molluscs: Caudofoveata, Solenogastres, Polyplacophora and Scaphopoda*. Field Studies Council (out of print).

This text covers a range of primitive molluscs, including chitons and tusk shells.

Cephalopods (squid, cuttlefish and octopi): 62 species on the UKSI

Scale: Small   Effort: Moderate   Accessibility: 6

Essential text

Hayward and Ryland (1990) cover 15 species (two cuttlefish, five bobtail squid, six squid and two octopus species), and this should take care of the vast majority of your cephalopod encounters.

Most species are relatively large and are generally only encountered frequently further out to sea. I dream of finding an octopus one day!

Society

The Conchological Society of Great Britain and Ireland: www.conchsoc.org

A great society for those interested in identifying, recording and conserving molluscs. It publishes a journal and a magazine, and runs a range of recording schemes and field and indoor events, as well as training courses.

Common Squid *Loligo vulgaris* line caught off Brighton beach, and an adult Little Cuttlefish *Sepiola atlantica* caught by net off Rye Bay, Sussex during a survey for the Inshore Fisheries and Conservation Authority (IFCA).

6.13 Bryozoans (sea mats, hornwracks and lace corals, etc.): 326 species available

Scale: Medium　　Effort: Very difficult　　Accessibility: 15

Collective effort = 50 (15.3%) species

Top listers = Nathan Jackson (24 species)

Vital equipment: waterproof camera, dissecting microscope, waterproof containers.

You can find bryozoans most easily in rock pools, where they cover a number of substrates but are most obvious on seaweeds and rocks. They are often predated by sea slugs, so a site rich in bryozoans might have nudibranchs, too. A few species can be identified in the field or from a photo, but most will require a specimen to be keyed out under a dissecting microscope.

Essential texts

Joanne Porter (2012) *Guide to Bryozoans and Hydroids of Britain and Ireland*. Wild Nature Press.
    This excellent photographic guide covers around 80 species. Hayward and Ryland (1990) also provide a really helpful guide, especially with regard to the microscopic features, and they cover over 100 species.

Flustrellidra hispida found on seaweed in Plymouth, Devon, and placed in a plastic container to observe it opening up.

For a more in-depth view that covers all marine species, four volumes of the *Synopses of the British Fauna* are very useful, although the last two are out of print:

Peter Joseph Hayward and John S. Ryland (1998) *Cheilostomatous Bryozoa, Part 1: Aeteoidea-Cribrilinoidea*. Synopses of the British Fauna (New Series), No. 10. Field Studies Council.

Peter Joseph Hayward and John S. Ryland (1999) *Cheilostomatous Bryozoa, Part 2: Hippothooidea – Celleporoidea*. Synopses of the British Fauna (New Series), No. 14. Field Studies Council.

Peter Joseph Hayward, R.S.K. Barnes and Doris M. Kermack (1985) *Ctenostome Bryozoans*. Synopses of the British Fauna (New Series), No. 33. Field Studies Council.

Peter Joseph Hayward, John S. Ryland, R.S.K. Barnes and Doris M. Kermack (1985) *Cyclostome Bryozoans*. Synopses of the British Fauna (New Series), No. 34. Field Studies Council.

Freshwater bryozoans: 19 species available

Look for them on submerged detritus, such as old traffic cones!

Timothy S. Wood and Beth Okamura (2005) *A New Key to the Freshwater Bryozoans of Britain, Ireland, and Continental Europe*. Freshwater Biological Association. This exhaustive work covers all 19 freshwater species.

6.14 Annelids (earthworms, marine worms and leeches, etc.): 1,384 species available 🔵 🟤 🟢

| Scale | Large | Effort | Difficult | Accessibility | 16 |

Collective effort = 152 (11.0%) species

Top lister = Nathan Jackson (53 species)

This is a disparate group of animals that occupy some widely different ecological niches. Here they have been grouped mainly according to these niches.

Earthworms (Lumbricidae, Acanthodrilidae and Sparganophilidae): 31 species available 🟤

| Scale | Small | Effort | Moderate | Accessibility | 6 |

A surprisingly small family of well-known terrestrial oligochaetes, these are well covered in the literature. They are easy to keep as a wet reference collection. Note that only adult earthworms (those with saddles) can be identified to species level.

Vital equipment: spade/trowel, digital camera, dissecting microscope, alcohol.

Earthworms are easily found when digging and also when turning over rocks and logs. Some are found high up in decaying timber, and some live in compost.

Essential texts

R.W. Sims and B.M. Gerard (1999) *Earthworms*. Synopses of the British Fauna (New Series), No. 39. Linnean Society of London.

Emma Sherlock (2018) *Key to the Earthworms of the UK and Ireland*, 2nd edn. Field Studies Council.

Allolobophora chlorotica. (Lloyd Davis)

Society

The Earthworm Society of Britain: www.earthwormsoc.org.uk
 This extremely useful resource provides much information on the British species. It includes several different approaches to keying out earthworms, such as multi-access keys.

Leeches (Hirudinea): 45 species on the UKSI

| Scale | Small | Effort | Difficult | Accessibility | 8 |

Leeches are in a different subclass and are well covered in the texts. They include both freshwater and marine species.

Freshwater leeches: 17 species available

| Scale | Small | Effort | Easy | Accessibility | 4 |

These are easily found when pond netting for aquatic invertebrates in general.

Vital equipment: metal-rimmed pond net, tray, digital camera.

The large freshwater leeches are fairly easy to identify. Most leeches can be identified from a good digital macro shot. Make sure that you get a good focus on the head end, as the number and arrangement of the eyes are key features for identifying many species.

Essential text

J.M. Elliot and Michael Dobson (2015) *Freshwater Leeches of Britain and Ireland: Keys to the Hirudinea and a Review of their Ecology*. Freshwater Biological Association. The updated key covers all 17 freshwater species. Unlike most other invertebrates, leeches do not store well in alcohol, so it is best to identify them in the field or put a much higher percentage of water in the alcohol. I take a laminated colour copy of the single plate from this guide into the field when I'm doing aquatic invertebrate surveys, as most of the species can be identified from this.

Hemiclepsis marginata, a fish parasite.

Marine leeches: c.27 species available

Scale: Small | Effort: Difficult | Accessibility: 8

Around 22 species are covered by Hayward and Ryland (1990). All of the marine species are fish parasites, so you are most likely to find them if you handle a lot of sea fish.

Freshwater oligochaetes: number of species, scale, effort and accessibility unknown

Essential text

R. Fitter and R. Manuel (1986) *Collins Field Guide to Freshwater Life*. Collins (out of print).
   This guide gives a useful overview of nine families of freshwater oligochaetes.

Marine worms: c.1,300 species on the UKSI

Scale: Large | Effort: Difficult | Accessibility: 16

Vital equipment: dissecting microscope, containers, trowel, forceps.

Essential text

Hayward and Ryland (1990) provide good coverage of nearly 300 species of marine annelids (around 250 polychaetes, 25 oligochaetes and 22 marine leeches).

Tube worms are found attached to the substrate, under rocks or in rock pools. Armoured scale worms such as *Harmothoe* species are commonly found on the underside of rocks, too, and are well covered in the SBF volume 54.

APHOTOMARINE features nearly 200 polychaetes and 11 oligochaetes.

The beautifully elegant Twin Fan Worm *Bispira volutacornis*, seen in Jersey (left) is a stark contrast to the wonderfully ugly underside of the Sea Mouse *Aphrodita aculeata*, seen in Rye Bay, Sussex (right).

A downright bizarre strawberry worm *Eupolymnia* sp.

Polychaetes: 1,106 species on the UKSI

Scale Large   Effort Difficult   Accessibility 16

Only a couple of species of polychaete stray into the upper reaches of saltmarshes, so for the sake of simplicity all polychaetes are treated as marine species here.

The following four volumes of the SBF cover the marine polychaetes, although only two are in print.

J. David George, Gesa Hartmann-Schröder, R.S.K. Barnes and Doris M. Kermack (1985) *Polychaetes: British Amphinomida, Spintherida and Eunicida*. Synopses of the British Fauna (New Series), No. 32. Field Studies Council.

Frederik Pleijel and R.P. Dale (1991) *Polychaetes: British Phyllodocoideans, Typhloscolecoideans and Tomopteroideans*. Synopses of the British Fauna (New Series), No. 45. Backhuys.

S.J. Chambers and A.I. Muir (1997) *Polychaetes: British Chrysopetaloidea, Pisionoidea and Aphroditoidea*. Synopses of the British Fauna (New Series), No. 54. Field Studies Council.

W. Wetheide (2008) *Polychaetes: Interstitial Families*. Synopses of the British Fauna (New Series), No. 44. Field Studies Council.

Peanut worms (Sipuncula): 20 species on the UKSI

Scale Small   Effort Difficult   Accessibility 8

This is a small family of unsegmented marine annelids.

Essential texts

P.E. Gibbs (2001) *Sipunculans*. Synopses of the British Fauna (New Series), No. 12. Field Studies Council.

Hayward and Ryland (1990) cover around 12 species.

Download

R. Barnich (2011) *Identification of Scale Worms in British and Irish Waters*. NMBAQC 2010 Taxonomic Workshop, Dove Marine Laboratory. www.nmbaqcs.org/media/rr4eji50/identification-of-scale-worms-in-british-waters.pdf

Marine oligochaetes: number of species, scale, effort and accessibility unknown

Essential text

J.S. Ash and J.E. Miskell (1982) *British and Other Marine Estuarine Oligochaetes*. Synopses of the British Fauna (New Series), No. 21. Cambridge University Press (out of print).

Website

APHOTOMARINE is a helpful website for all of the marine groups.

6.15 Platyhelminths (flatworms, tapeworms and flukes, etc.): 831 species available

| Scale | Medium | Effort | Very difficult | Accessibility | 15 |

Collective effort = 48 (5.8%) species   **Top lister** = Seth Gibson (20 species)

Vital equipment: waterproof camera, hand lens, pond net, tray.

Five classes are covered here: Turbellaria (flatworms), with 15 species (some of which are freshwater); Trematoda (flukes), with 395 species (mixture of habitats); Cestoda (tapeworms), with 299 species (mixture of habitats but many marine species); Monogenea, with 5 species (all marine); Rhabditophora, with 119 species (mixed habitats).

Terrestrial flatworms: 23 species available

| Scale | Small | Effort | Easy | Accessibility | 4 |

Essential text

Hugh D. Jones (2005) British land flatworms. *British Wildlife* 16: 189–94.

Although a few more species have been found in the UK since this was published, this identification guide is still very useful, covering 14 species of terrestrial flatworms. It includes identification advice, taxonomy details and colour photos of all 14 species. Since 2005 the total number of species has risen to 23, with *Caenoplana variegata* being among the recent introductions.

The large and impressive Southampton Flatworm *Caenoplana variegata* in London.

Dendrocoelum lacteum, a freshwater planarian.

Freshwater flatworms: *c.*11 species available ●

Scale Small   Effort Easy   Accessibility 4

By far the commonest freshwater flatworm that I see when pond dipping is the *Polycelis nigra* or *tenuis* species aggregate.

Essential texts

T.B. Reynoldson and J.O. Young (2000) *A Key to the Freshwater Triclads of Britain and Ireland with Notes on Their Ecology*. Freshwater Biological Association.
There are just 11 species in this small publication, all of which are illustrated towards the front of the key.

J.O. Young (2001) *Keys to the Freshwater Microturbellarians of Britain and Ireland with Notes on Their Ecology*. Freshwater Biological Association.

Candy-striped Flatworm *Prostheceraeus vittatus* in the class Rhabditophora, on Jersey.

Cycloporus sp. (possibly *papillosus*) from Felpham, also in the Rhabditophora.

Marine Turbellaria and Planarians: number of species, scale, effort and accessibility unknown

Essential texts

I.R. Ball and T.B. Reynoldson (1981) *British Planarians*. Synopses of the British Fauna (New Series) No. 19. Cambridge University Press.

S. Prudhoe (1982) *British Polyclad Turbellarians*. Cambridge University Press.
Both of these books are out of print, and Hayward and Ryland (1990) do not tackle any species directly.

Website

APHOTOMARINE covers around 30 species of marine flatworm.

The few species I have encountered have typically been found under rocks on the south and west coasts. An up-to-date key is needed for this group

Cestoda (tapeworms) and Trematoda (flukes): 299 and 395 species, respectively; 696 species on the UKSI

Scale Medium   Effort Unknown   Accessibility ?

The vast majority of these species are marine, and there appears to be very little literature available on identifying individuals to species level. Terrestrial parasites of animals are another important group, but again the literature on their identification seems to be sparse.

6.16 Pycnogonids (sea spiders): 72 species available

Scale: Small Effort: Difficult Accessibility: 8

Collective effort = 9 (12.5%) species

Top listers = Nathan Jackson and Graeme Lyons (5 species each)

Vital equipment: metal-rimmed pond net, plastic containers, dissecting microscope.

Essential texts

Hayward and Ryland (1990) cover just 19 species (and the handbook version only covers 10 species).

Roger N. Bamber (2010) *Sea-Spiders (Pycnogonida) of the North-East Atlantic*. Synopses of the British Fauna (New Series), No. 5. Field Studies Council.
This book covers all 84 species found in the region, with detailed keys and species accounts.

You can find sea spiders under rocks, out at sea in nets and by picking through samples of seaweed and hydrozoans, etc. As they are rather small animals it's likely that none of them are field-identifiable.

A Gangly Lancer *Nymphon gracile*. (Sally Luker)

6.17 Arachnids (spiders, harvestmen, scorpions, pseudoscorpions, ticks and mites, etc.): 3,245 species available

Scale: Large   Effort: Very difficult   Accessibility: 20

Collective effort = 929 (28.6%) species   **Top lister** = Graeme Lyons (649 species)

Vital equipment: dissecting microscope, hand lens, suction sampler, sweep net, beating tray, sieve, *electric toothbrush, small paintbrush, UV light (scorpions only), compound microscope (mites and pseudoscorpions only).*

The vast majority (perhaps around 75%) of the available species are mites, making much of this group extremely difficult to access, though many of the smaller subgroups are very accessible. The best place to start is with the spiders, especially the large ones, and gall mites are also fairly easy if you know your plants.

Spiders (Araneae): 735 species on the UKSI

Scale: Medium   Effort: Moderate   Accessibility: 9

At the time of writing, spiders are my main focus, and there are around 735 species on the UKSI, though this number is rising all the time. We have around 440 'larger' spiders, while the other 295 species all belong to one family – the money spiders (the 'micro-moths' of the spider world, many of them being less than 2 mm long as adults). These are not as well known, but they are fascinating to look for and identify under the microscope, and many of them are winter active.

The larger spiders are much less of a challenge to identify, with small manageable families like the orb weavers and jumping spiders being very popular with beginners. You can find spiders in any habitat, but some of the best places to look are heaths, bogs and woods – anywhere with lots of structure, moss and/or litter.

A very rare money spider, *Walckenaeria mitrata*, showing its novelty secondary sexual characteristic of the 'bed-knob'-shaped structure on its cephalothorax.

Micrommata virescens is one impressive spider.

Essential texts

Lawrence Bee, Geoff Oxford and Helen Smith (2020) *Britain's Spiders: A Field Guide*, 2nd edn. Princeton University Press.

The first edition of this book was ground breaking for our larger spiders – the text is like a field guide version of the Spider Recording Scheme (see below), but with many cutting-edge identification tips also included. It's a must-have even for serious arachnologists, covering nearly 400 species, although it does not include all the money spiders (it has a selection of 11 of the largest ones). It also contains valuable information about conservation status, which tends to be omitted even from most modern field guides.

Michael J. Roberts (1993) *The Spiders of Great Britain and Ireland*. Apollo Books.

This is usually referred to as 'Big Roberts'. The two-volume set is still in print, and you can do all (bar the very recent additions) of the UK spiders from this book – the exquisite illustrations of spider palps and epigynes are mind-blowingly detailed but you'll have to learn to interpret a 2D illustration as a 3D structure. The three-volume set has all the same info in it, it's just presented much more pleasingly in three hardback books (though there are some long-established species that are not included in this set) and is less likely to fall apart.

M.J. Roberts (1995) *Spiders of Britain and Northern Europe*. HarperCollins.

The small field guide version only covers the larger spiders and a selection of larger money spiders, and is affectionately known as 'Little Roberts'. The missing money spiders will soon become frustrating if you want to do all spiders though.

However, it does include a number of species you'll find in mainland Europe that are not in Big Roberts, plus many additional species found on the Channel Islands. It's worth getting both if you are serious about spiders.

Website

Spider Recording Scheme (SRS): www.britishspiders.org.uk

This is perhaps the best online atlas for any taxa. All of the data that drive the distribution maps, phenology and conservation status of our spiders come from this excellent recording scheme. If you join the British Arachnological Society (see below) this will give you access to all of the data behind the SRS at the county and site level. At the time of writing the scheme is being updated, and readers will be able to access it by going to the 'recording' tab on the British Arachnological Society website. It is now also possible to add iRecord and/or iNaturalist data to the SRS dataset when looking at species distributions, especially useful for recent additions to our fauna or species that are spreading rapidly.

Once you start to look at spiders more closely you will realise that at first they are not very field-identifiable, and that you need to take specimens and examine their genitalia under a dissecting microscope. In many spider species it is only possible to ID the adults, but unlike a lot of other invertebrates, both males and females are identifiable and the genitalia are external, which means that you rarely have to do messy dissections. This, and the fact that all of the species are covered in one book, means that spiders are actually one of the easiest medium to large groups of invertebrates to tackle. It's surprising that they are not more popular (especially with pan-species listers – my annual spidering efforts alone would put me in third place on the all-time rankings for spiders), as it is so easy to get new site and county records, even in the south-east.

From an ecological point of view, all spiders have one thing in common: they are all predators. Unlike phytophagous species, they don't reveal much about vegetation

The adorable *Pellenes tripunctatus*, just look at those eyelashes! I got into a lot of trouble on my street WhatsApp group for posting this picture.

The Domino Sun-jumper *Heliophanus kochii*.

composition, but they do tell you a great deal about a site's structure. As a result, they are really useful for monitoring changes in rewilding projects, for example, where changes in the structure of vegetation are often as important as changes in its composition.

Most of the joy I get from spidering involves the use of a suction sampler and sieve to collect tiny money spiders. Many of these are winter-active adults, which means that (in contrast to most insect groups) you can carry on spidering all year round. Sieving *Sphagnum* on a freezing cold day in a bog in winter can be extremely productive and provides a valuable escape from the desk during the long winter days!

Keeping a collection is easy but a little disappointing, as the specimens need to be kept wet, in alcohol.

I have been fortunate enough to find one spider new to the British Isles (*Heliophanus kochii* on Brighton beach in 2023) and one new to the UK (*Enoplognatha mandibularis* in Cornwall in 2013 and 2023). I named *Heliophanus kochii* the Domino Sun-jumper due to its domino-like pattern. Natural colonists and accidental introductions are turning up at an increasing frequency (see the section on hothouses, botanical gardens and garden centres on page 251 for more information on finding spiders in such locations).

Websites

Araneae Spiders of Europe: www.araneae.nmbe.ch
This site is ideal if you need a European perspective on spiders. It has a variety of genitalia diagrams, and is especially useful when you might have found something new to the British Isles. It also has many of the genitalia drawings from Roberts (1995), though the quality of these images is better in the book, and it is much easier to compare species there, too.

World Spider Catalogue: www.wsc.nmbe.ch
This is a really useful site.

www.britishspiders.org.uk/spiders-not-in-roberts
This section of the website includes 'Roberts-like' accounts of all the species that have been added to the British list since its publication.

Society

British Arachnological Society (BAS): www.britishspiders.org.uk

Membership benefits include access to all of the data in the SRS, which is an immensely valuable resource, as well as annual meetings, field events, newsletters and journals.

Facebook group

UK Spiders

This large and lively group includes many of the UK's most active arachnologists, as well as some very helpful European experts. It is heavily moderated due to the number of members. There are some similarly named groups, but 'UK Spiders' is by far the best one.

Opiliones (harvestmen): 34 species on the UKSI

Scale Small   Effort Moderate   Accessibility 6

Harvestmen are covered by the same platform as that described on page 121 for spiders, under a different recording scheme. The species accounts follow a similar format, except for the fact that as yet no conservation status data have been assigned to this group. They are found in the same way as spiders.

The harvestman *Megabunus diadema*. Check out those eye spines!

Essential texts

P.D. Hillyard (2005) *Harvestmen*. Field Studies Council.

Paul Richards (2022) *Harvestmen of the British Isles*. Field Studies Council.
   Both of these publications have clear photos that back up more technical keys, so you should be able to get most harvestmen to species level fairly easily. Avoid wasting time on immature specimens, as the vast majority of species will reach adulthood later in the year.

Download

The following earlier version of the SBF is available as a PDF:

P.D. Hillyard and J.H.P. Sankey (1989) *British Harvestmen: Arachnida – Opiliones*. Synopses of the British Fauna (New Series), No. 4. Linnean Society.
   https://mndi.museunacional.ufrj.br/aracnologia/pdfliteratura/Hillyard%20&%20Sankey%201989%20British%20Opiliones.pdf

Scorpions (Scorpiones): 2 species on the UKSI ●

The only established species of scorpion in the UK is the Yellow-tailed Scorpion *Euscorpius flavicaudis*. A UV light and a night-time trip to the docks at Sheerness are all that is needed to see it. Not only is this species quite spectacular and well worth the pilgrimage, but you are almost guaranteed to see the impressive spider *Segestria florentina*, too. Apparently this scorpion rarely uses its sting, and when it does it's less painful than a bee sting. The most dangerous part of this trip then, is hanging around Sheerness docks after dark!

Yellow-tailed Scorpion glowing under UV light at Sheerness.

Pseudoscorpions (Pseudoscorpiones): 28 species on the UKSI ●

Scale Small   Effort Moderate   Accessibility 6

The best way to find pseudoscorpions is by sieving litter, moss or compost, or by using a suction sampler. Another cluster of species can be found in dead wood, typically under bark or by pulling wood apart. You never see many of them, and they are overwhelmingly dominated by a couple of very common species, *Neobisium carcinioides* and *Chthonius ischnocheles*, which between them make up over 70% of my records.

Essential texts

Gerald Legg, Francis Farr-Cox and Richard Jones (2016) *Illustrated Key to the British False Scorpions (Pseudoscorpions)*. Field Studies Council.
This handy little fold-out guide covers the 27 UK species quite well, and is very useful for beginners.

Gerald Legg and Richard E. Jones (1998) *Pseudoscorpions*. Synopses of the British Fauna (New Series), No. 40. Field Studies Council (out of print).
This is a very good book, but second-hand copies are hard to find.

Website

Chelifer.Com: www.chelifer.com
Gerald Legg's excellent website has species accounts for all of our pseudoscorpions, including distribution maps.

Marram Grass Chelifer *Dactylochelifer latreillii* is a large and local pseudoscorpion that can be quite abundant where it occurs, in dense tussocks of Marram *Ammophila arenaria* in sand dunes.

Mites: 2,454 species on the UKSI

Scale: Large | Effort: Very difficult | Accessibility: 20

Vital equipment: compound microscope.

This is the most challenging group of arachnids. There are four orders and one 'cohort', which makes identification a near impossible challenge in many cases. Most mites in the UK are soil invertebrates, though there also a number of freshwater and marine species. Another group of mites cause galls, and most pan-species listers are likely to encounter mites this way. Ticks also sit here, and are usually a little easier to identify.

Halacaridae (a family of marine mites): 83 species on the UKSI

Scale: Small | Effort: Very difficult | Accessibility: 10

Vital equipment: fine sieve (for sieving samples of sea water), compound microscope.

Essential text

J. Green and M. Macquitty (1987) *Halacarid Mites*. Synopses of the British Fauna (New Series), No. 36. Linnean Society.

These microscopic marine mites can only be identified using a compound microscope, and are almost invisible to the naked eye. Attempting their identification is not for the faint-hearted.

Ash Key Gall *Aceria fraxinivora*.

Gall mites (Eriophyidae and Tarsonemidae): 282 species on the UKSI

| Scale | Medium | Effort | Easy | Accessibility | 6 |

Mites are probably the most speciose gall-causing invertebrates. The Eriophyidae account for the majority of the species, with 256 species on the UKSI, whereas there are just 27 species belonging to the Tarsonemidae (for more detailed information about galls, see page 244).

Essential texts

Margaret Redfern, Peter Shirley and Michael Bloxham (2023) *British Plant Galls: Identification of Galls on Plants and Fungi*, 3rd edn. Field Studies Council.

M. Chinery (2011) *Britain's Plant Galls: A Photographic Guide*. WILDGuides Ltd.

Soil mites and other terrestrial mites: *c.*1,750 species on the UKSI

| Scale | Large | Effort | Very difficult | Accessibility | 20 |

Be aware that there is a plethora of other types of mite that are not soil mites, including bird, feather, skin, face and eyelash mites, among others. Some mites occur on vegetation but do not cause galls, such as the well-known Gorse Mite *Tetranychus lintearius*.

That said, the majority of the almost 2,500 species of mite are soil mites. They are readily found by sieving litter, suction sampling and looking under stones. Every suction sample that I collect contains a few mites, none of which I can identify!

The literature on mites is very patchy. The best text, by Matthew Shepherd, who is a national specialist on these species, is very much a work in progress.

Matthew Shepherd (2024) *A Key to the Soil Mites of Britain and Ireland. Version 11*. Unpublished key.
This is really only a key to genera, but there are some genera that contain only one species, so you will be able to identify some soil mites to species level. The key is available directly from Matthew at Natural England.

Website

Soil Biodiversity UK: www.soilbiodiversityuk.myspecies.info
This extremely helpful website is the home of the mite recording scheme, and includes many useful links.

Facebook group

Soil Biodiversity UK
This friendly group is a great place to learn about this difficult group of invertebrates. Be warned that, even with access to the national experts, many specimens will not be identifiable to species level.

Ticks (Ixodida): 24 species on the UKSI

| Scale | Small | | Effort | Moderate | | Accessibility | 6 |

If you get a sweep net you will soon realise how common ticks are, and if you start looking at them under a microscope you will find that the vast majority of them are Deer Tick/Castor Bean Tick *Ixodes ricinus*. This species accounts for 85% of my tick records, and my next most frequent tick species is the Red Sheep Tick *Haemaphysalis punctata*, at 15%. In my part of the world, Red Sheep Tick is overwhelmingly restricted to the sheep-grazed areas of chalk grassland on the South Downs.

The best way to look for the more unusual species is to search for them on the host animals. This can be done by looking at road kill and/or collecting them off animals that are handled, such as pets or livestock. You can also try 'drag sampling', which involves dragging a towel through vegetation.

Website
Tick ID: www.bristoluniversitytickid.uk
This is an excellent resource, with a key for identifying ticks to species level.

Water mites (Hydrachnidia): 313 species on the UKSI

| Scale | Medium | | Effort | Very difficult | | Accessibility | 15 |

These water mites are found only in freshwater habitats. The best way to find them is to use a very fine-mesh pond net.

Essential texts
There are no recent English texts on water mites.

Terence Gledhill and Kurt O. Viets (1976) *Checklist of Freshwater Mites – OP01*. Freshwater Biological Association.

Harry Smit (2018) *De Nederlandse Watermijten (Acari: Hydrachnidia) [The Dutch Water Mites]*. European Invertebrate Survey/EIS-Nederland.
This Dutch text (a key to 58 genera) appears to be the most recent publication.

Download
C.L. Hopkins (1961) A key to the water mites (Hydracarina) of the Flatford area. *Field Studies Journal* 1(3): 45–64.
This article can be downloaded from the Resources section on the Field Studies Council website.

6.18 Myriapods (millipedes, centipedes, pauropods and symphylans): 190 available species

| Scale | Medium | Effort | Difficult | Accessibility | 12 |

Collective effort = 143 (75.3%) species **Top lister** = Steve Gregory (122 species)

Vital equipment: sieve, tray, hand lens, dissecting microscope, *compound microscope (sometimes needed), Tullgren funnel.*

Society

British Myriapod and Isopod Group (BMIG): www.bmig.org.uk
This group is responsible for a recording scheme for all of the myriapod species. The website includes detailed species accounts and distribution maps, as well as many useful tips for identification. The benefits of membership (which is free) include newsletters and an annual field meeting.

Facebook group

Isopods and Myriapods of Great Britain and Ireland
This group, which is associated with the BMIG, is a very useful source of help with identification.

The majority of millipedes and centipedes can be found by turning logs and stones, sieving moss, leaf litter and litter piles, picking through dead wood, and looking under bark and under stones. Another way to boost your list is to visit hothouses and botanic gardens. Searching around synanthropic sites can also yield new and interesting non-native species.

Kent Pin-head
Polyzonium germanicum.

The huge millipede *Cylindroiulus londinensis*.

Millipedes (Diplopoda): 85 species on the UKSI

Scale: Small   Effort: Moderate   Accessibility: 8

Essential text
J.G. Blower (1985) *Millipedes*. Synopses of the British Fauna (New Series), No. 35. Linnean Society.
At the time of writing, an update to the above synopses is being put together by Helen Read and Paul Lee and will likely be published in 2026.

Centipedes (Chilopoda): 68 species on the UKSI

Scale: Small   Effort: Moderate   Accessibility: 8

Essential texts
D. Barber (2008) *Key to the Identification of British Centipedes*. Field Studies Council.
D. Barber (2009) *Centipedes*. Synopses of the British Fauna (New Series), No. 58. Field Studies Council.
These are basically two versions of the same book. Barber (2008) is more suitable for beginners and easier to find. Barber (2009) is more comprehensive, so if you only get one version, choose this one.

The large centipede *Henia vesuviana*.

D. Barber (2023) *Atlas of the Centipedes of Britain and Ireland*. Field Studies Council. This recently published atlas includes a large amount of up-to-date information that complements the above keys, and due to the limited amount of literature on this group, it is well worth getting a copy.

Pauropods (Pauropoda): 23 species on the UKSI

| Scale | Small | Effort | Very difficult | Accessibility | 5 |

Vital equipment: compound microscope, *Tullgren funnel*.

Pauropods are tiny animals, less than 1.5 mm long, with distinctively forked antennae (like tiny antlers). They are found in soil and leaf litter, but are hard to spot due to their tiny size. Using a Tullgren funnel is a good way to find them from a leaf litter sample. Only four of the 23 species have been listed on the PSL website, with Finley Hutchinson currently the only person to have seen more than one species, and only three listers having ever recorded pauropods.

Websites

A.D. Barber, J.G. Blower and U. Scheller (1992) Pauropoda, the smallest myriapods. *Bulletin of the British Myriapod Group* 8: 13–24. https://bmig.org.uk/sites/default/files/bulletin_bmg/BullBMG8p13-23_Barber-etal_Pauropoda.pdf

www.bmig.org.uk/page/pauropod-checklist
This key can be downloaded from the BMIG website, and there are also a few species accounts and several photos.

Symphylans (Symphyla): 15 species on the UKSI

Scale Small   Effort Very difficult   Accessibility 5

Symphylans look like small, blind, pallid centipedes with long antennae but with far fewer legs. They are detritivores and an important part of the soil fauna. You are most likely to find them under large stones, especially the kind that are partially embedded in the ground. At the time of writing, only four pan-species listers have listed any symphylans. *Scutigerella palmonii* is the most listed species, while the most common species is thought to be *Scutigerella causeyae*.

Essential text

C.A. Edwards (1958) A revision of the British Symphyla. *Proceedings of the Zoological Society of London* 132: 403–39.
  The above text is available at the BMIG website: www.bmig.org.uk/page/symphylan-checklist.

There are species accounts on the BMIG site for only four of these species, but there is a useful checklist, with some notes on how common they are and which ones have only been found in hothouses.

An unidentified symphylan. (Andy Murray)

6.19 Crustaceans (woodlice, amphipods, crabs, lobsters, crayfish, barnacles, shrimps and copepods, etc.): 3,760 available species

Scale Large   Effort Difficult   Accessibility 16

Collective effort = 337 (9.0%) species   **Top lister** = Graeme Lyons (114 species)

Vital equipment: sieve and tray, pond net, suction sampler, dissecting microscope, *push net for shrimping*.

This is a large group of mostly marine species across four classes, the best-known and familiar belonging to the Malacostraca, which contains around 1,460 species. The largest class is the Maxillopoda, with 1,828 species, the vast majority of which are the microscopic copepods. It is not always easy to see how many species are present on the UKSI for crustacea, so some of the figures here are estimates.

Woodlice: *c.*65 species available

Scale Small   Effort Moderate   Accessibility 6

Essential texts

Stephen Hopkin (1991) *A Key to the Woodlice of Britain and Ireland*. Field Studies Council.

Steve Gregory (2009) *Woodlice and Waterlice (Isopoda: Oniscidea and Asellota) in Britain and Ireland*. Field Studies Council.

Although there are around 65 species on the BMIG website, only about 40 of these are native to the British Isles. Of these, a handful of species will account for the vast majority of your records. It's worth learning these species quickly. The five commonest species make up nearly 95% of all of my records, and *Philoscia muscorum* agg. is one of my ten most frequently recorded 'species' of all time. Many of my records these days come from the suction sampler, but the other obvious way to record woodlice is under logs and stones. These two methods will yield quite different species, the former often tolerating damper conditions and the latter frequently preferring drier ones.

A few species are associated with dung heaps, with *Porcellionides pruinosus* usually being very easy to find, so these are always worth searching. Some of the more interesting species can be found in quarries. Garden centres, hothouses and other synanthropic sites will yield interesting species with a strong likelihood of a first for Britain. Coastal sites are also worth visiting, and here pitfall trapping can be a good way of finding some of the scarcer species.

Although it doesn't fit neatly within this group, the invasive Landhopper *Arcitalitrus dorrieni* deserves a mention here as it is covered by the recording scheme, being common in many places in the south, especially in gardens.

Armadillidium depressum is the commonest woodlouse in Brighton.

The attractive woodlouse *Armadillidium pulchellum*.

Freshwater crustaceans: *c.*35 species available ●

Scale Small   Effort Moderate   Accessibility 6

This group includes around 35 species, the largest number being the freshwater amphipods (the largest genus being *Gammarus*, the freshwater shrimps with *c.*14 species), four species of water louse, at least four crayfish and a range of other freshwater crustaceans, such as the groundwater shrimp *Niphargus aquilex* (which only three pan-species listers have noted). I found one of these by sieving *Sphagnum* moss, and it was quite possibly a once-in-a-lifetime encounter with this subterranean species, which spends most of its time below ground in our aquifers. The genus *Gammarus* is by far the most difficult in this group.

As a child, I once filled an ice cream tub with White-clawed Crayfish *Austropotamobius pallipes* we found under rubbish dumped in the brook by my house, and bet my friends they wouldn't put their hands in the tub. It seems crazy that 35 years later this species is a rarity due to the spread of the invasive Signal Crayfish *Pacifastacus leniusculus*.

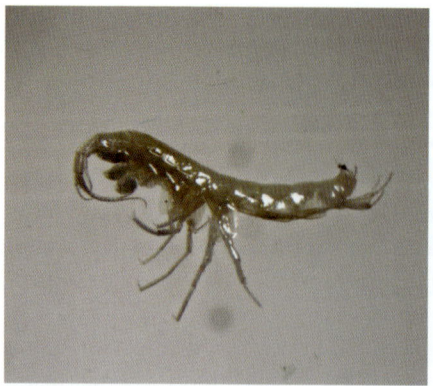

Niphargus aquilex and *Asellus aquaticus*. (Neil Phillips)

Essential text

T. Gledhill, D.W. Sutcliffe and W.D. Williams (1976) *Key to the British Freshwater Crustacea: Malacostraca*. Freshwater Biological Association.

This is an old text and there have been some changes and additions, especially with regard to the freshwater shrimps.

Large marine crustaceans (Malacostraca): *c.*1,350 species available

Scale Large   Effort Difficult   Accessibility 12

The more familiar crabs and lobsters only account for a small proportion of the available species, and once you get away from these larger, more field-identifiable species, crustaceans do become more complicated. However, they have become one of my favourite marine groups as I am more often than not able to ID them to species.

Essential text

Hayward and Ryland (1990), though patchy in its coverage, is very helpful for identifying most larger crustaceans.

Website

APHOTOMARINE is very useful for identifying all of the following marine crustacean groups.

Hairy Crab *Pilumnus hirtellus*.

Crabs, lobsters, shrimps and prawns (Decapoda): c.200 species on the UKSI

| Scale | Medium | Effort | Moderate | Accessibility | 9 |

This group is a good way to increase your list, and crabs and marine shrimps are generally quite easy to identify.

Obviously the best way to find crabs is under rocks in rock pools. Larger rocks will be home to larger crabs, and the less frequently encountered species will be further down the beach on the lowest tides. My most memorable encounter involved looking for Common Goose Barnacles *Lepas anatifera* on the off chance that they might have Columbus Crabs *Planes minutus* among them after a storm on Brighton beach. We found five of these crabs among some flotsam covered in barnacles – a new record for Sussex! This was all about knowing there was a storm coming through and when the next morning high tide was going to be. Most of all, though, it was only possible by seeing other people's sightings of them further west along the coast in Dorset on social media.

Sweep netting through weed, or taking a sample of weed and sieving it in a tray, are good ways to find smaller shrimps along with marine isopods.

Columbus Crab *Planes minutus* among Common Goose Barnacle *Lepas anatifera* (top left). St Piran's Crab *Clibanarius erythropus* in a Painted Top Shell *Calliostoma zizyphinum* – an unforgettable encounter on the Isles of Scilly (top right). Lobster *Homarus gammarus* (bottom left). Common Prawn *Palaemon serratus* (bottom right).

An immature *Stenosoma lancifer*, a local species in the south of England and Wales.

Essential texts

R.W. Ingle (1996) *Shallow-Water Crabs*. Synopses of the British Fauna (New Series), No. 25. Field Studies Council.
   This is a very useful book, but it does not include the hermit crabs. However, Hayward and Ryland (1990) cover these well.

Ray W. Ingle (2005) *Lobsters, Mud Shrimps and Anomuran Crabs*. Synopses of the British Fauna (New Series), No. 55. Field Studies Council (out of print).

G. Smaldon, L.B. Holthius and C.H.J.M. Fransen (1994) *Coastal Shrimps and Prawns*. Synopses of the British Fauna (New Series), No. 15. Field Studies Council (out of print).
   This book covers 44 species.

Hayward and Ryland (1990) cover around 150 species of decapod.

Facebook groups

Crustacea of the NE Atlantic & NW Europe

Porcupine Marine Natural History Society

Rockpooling & Shrimping (British Isles)

Marine isopods: *c.*200 species available on the UKSI

| Scale | Small | Effort | Moderate | Accessibility | 6 |

There is a marine isopod recording scheme that covers around 80 intertidal and brackish species, which sits within the wider BMIG website. You will need to catch most species and look at them closely under a microscope to get an identification. Marine isopods are easily found by netting weed and/or washing it in a bucket or tray.

Essential texts

Ernest Naylor and Angelika Brandt (2015) *Intertidal Marine Isopods*, 2nd edn. Synopses of the British Fauna (New Series), No. 3. Field Studies Council.
   This covers 60 species, and is essential for anyone interested in marine isopods.

Hayward and Ryland (1990) cover around 50 species of marine isopod.

Facebook group

Isopods and Myriapods of Britain and Ireland
    This group covers both marine and terrestrial isopods.

Marine amphipods: *c.*600 species on the UKSI ●

Scale Medium   Effort Difficult   Accessibility 12

This is by far the most difficult category of the larger marine crustaceans.

Essential text

R.J. Lincoln (1979) *British Marine Amphipoda: Gammaridea*. Pisces Conservation (out of print).
    This CD book of an old text covers around 271 marine amphipods.

Barnacles (Cirripedia): 67 species on the UKSI ●

Scale Small   Effort Moderate   Accessibility 6

Hayward and Ryland (1990) cover around 24 species, while Southward (2008) covers 39 species. Barnacles are found on rocky coasts and are common in the intertidal zone. Many will be identifiable from a good macro photo, but you may need to take an adult for microscopic identification.

Acorn Barnacle *Semibalanus balanoides*, a very common species.

Essential text

A.J. Southward (2008) *Barnacles*. Synopses of the British Fauna (New Series), No. 57. Field Studies Council.

Seed shrimps (Ostracoda): 378 species on the UKSI

Scale Medium   Effort Very difficult   Accessibility 15

Most of these bivalve-like crustaceans are microscopic (many species are less than 1 mm long, though some are up to 3 cm). They live on the surface of mud sediment or are planktonic in nature. They pose a significant identification challenge, with just 21 common intertidal species being covered by Hayward and Ryland (1990). A proportion of the 377 species on the list are found in freshwater habitats.

Essential texts

P.A. Henderson (1990) *Freshwater Ostracoda*. Synopses of the British Fauna (New Series), No. 42. Backhuys (out of print).

J. Athersuch, D.J. Horne and J.E. Whittaker (1990) *Marine and Brackish Water Ostracods (Superfamilies Cypridacea and Cytheracea)*. Synopses of the British Fauna (New Series), No. 43. Field Studies Council (out of print).

Martin V. Angel (1993) *Marine Planktonic Ostracods*. Synopses of the British Fauna (New Series), No. 48. Field Studies Council.
Although two of these three volumes are out of print you will need all of them if you want to do the full range of ostracods.

Copepods or sea lice (Copepoda): 1,754 species on the UKSI

Scale Large   Effort Very difficult   Accessibility 20

Nearly half of all the crustaceans on the UKSI are copepods, of which there are 620 genera. There seems to be little literature to help to identify these tiny species, most of which are less than 0.5 mm long. Most resources only get to genera, and then only a fraction of the available genera are covered. Copepods are very much the soil mites of the sea!

Hayward and Ryland (1990) cover over 60 genera/species.

APHOTOMARINE covers around 51 genera/species.

Freshwater species are covered by the following FBA guide.

J.P. Harding and W.A. Smith (1974) *A Key to the British Freshwater Cyclopid and Calanoid Copepods*. Freshwater Biological Association.
This is now out of print, but second-hand copies are widely available.

Branchiopoda: 108 species on the UKSI

Scale Medium   Effort Difficult   Accessibility 12

There are three orders within this class – the Diplostraca (water fleas), the Notostraca and the Anostraca. The rare and protected Tadpole Shrimp *Triops cancriformis* is the only member of the Notostraca. The Anostraca has three species, namely the rare and protected Fairy Shrimp *Chirocephalus diaphanus*, the Brine Shrimp *Artemia salina* (a scarce brackish-water species) and *Tanymastix stagnalis* (another species like the Fairy Shrimp, but which has no records in the NBN Atlas). The vast majority are water fleas, such as the familiar *Daphnia*.

Water fleas (Diplostraca): 104 species on the UKSI

Scale Medium   Effort Difficult   Accessibility 12

Predominantly freshwater in nature, these microscopic animals are a challenge to identify. Traditionally called the Cladocera, they are now known as the Diplostraca. There are some marine species, with two species shown on APHOTOMARINE.

Essential texts

D.J. Scourfield and J.P. Harding (1966) *British Freshwater Cladocera: A Key: SP5*, 3rd edn. Freshwater Biological Association.
   Although out of print, this is still available from the FBA website, though the text is somewhat out of date.

Claude Amoros (1984) *Introduction Pratique à la Systématique des Organismes des Eaux Continentales Françaises. 5. Crustacés Cladocères*. Société Linnéenne De Lyon.
   This key is in French and can be downloaded from the FSC ID Signpost page, which states that it is easy to use. These two keys together will help to get you started with water fleas.

Website

Cladocera Interest Group: www.cladocera.org.uk
   This is a mine of information if you want to take water fleas further. It includes a checklist of the 92 species covered, an atlas with species accounts, and microscopic imagery of many species.

6.20 Entognatha (springtails, proturans and two-tailed bristletails): 427 species available

Scale: Medium | Effort: Very difficult | Accessibility: 15

Collective effort = 168 (39.3%) species

Top lister = James McCulloch (103 species)

Springtails: 406 species on the UKSI

Scale: Medium | Effort: Very difficult | Accessibility: 15

Vital equipment: compound microscope, pooter/tweezers, alcohol, suction sampler.

Springtails are very small, soft-bodied and often highly mobile, and are best collected with a pooter or fine tweezers. They are unusual among invertebrates in that a good dissecting microscope will not help you to identify many individuals to species level – you will need a compound microscope for this.

Essential text

Steve P. Hopkin (2007) *A Key to the Collembola (Springtails) of Britain and Ireland.* Field Studies Council.
This is a useful and very comprehensive key.

A cluster of the coastal springtail *Anurida maritima*. (Finley Hutchinson)

A small number of large springtails are eminently field-identifiable. The species that I record most frequently are *Pogonognathellus longicornis*, *Orchesella villosa* and *Orchesella cincta*, but you'll soon need a more sophisticated approach if you want to increase your springtail list.

Springtails can be found in almost any habitat, and are detritivores, most being in some way associated with decomposing vegetation. I find that the suction sampler is especially useful for finding these animals.

Websites

Frans Janssen's Flickr archive: www.flickr.com/photos/65173625@N00/

Checklist of the Collembola: www.collembola.org

Coneheads or telsontails (Protura): 11 species available ●

Scale Small   Effort Unknown   Accessibility ?

These are distinctively sleek, blind soil invertebrates which only walk on their back two pairs of legs, their over-sized front legs being held up in the air. None of the 11 species on the UKSI have been recorded by any pan-species lister, and I have been unable to find any literature on them.

Two-tailed bristletails (Diplura): 12 species available ●

Scale Small   Effort Very difficult   Accessibility 10

The 12 species on the UKSI could well be an underestimate of the number of species present in the British Isles. These incredibly fast-moving, blind soil invertebrates are rarely seen, poorly understood and often mistaken for symphylans. The two protruding cerci at the rear give these animals a distinctly different appearance to proturans.

Of the 12 species on the UKSI, nine have been recorded by pan-species listers. All species in the UKSI are in the genus *Campodea*, the most frequently listed species being *Campodea staphylinus*.

An unidentified proturan. (Andy Murray)

An unidentified dipluran. (Andy Murray)

Essential texts

M.J. Delaney (1954) *Thysanura and Diplura*. Handbooks for the Identification of British Insects, Volume 1, Part 2. Royal Entomological Society (out of print).
This old RES key can be downloaded from www.royensoc.co.uk and covers 11 species.

Alberto Sendra and Ann Sofia P.S. Reboleira (2020) Euro-Mediterranean fauna of Campodeinae (Campodeidae, Diplura). *European Journal of Taxonomy* 728(1): 1–130.
This French key to Diplura can be found here:
https://europeanjournaloftaxonomy.eu/index.php/ejt/article/view/1181

Website

Chaos of Delight: www.chaosofdelight.org
This site has excellent photos, especially of diplurans and proturans, and is a great introduction to these poorly known invertebrates.

Facebook groups

UK Collembola
This small group is very helpful with identification.

Soil Biodiversity UK
This very friendly group is an invaluable resource.

6.21 Insects: Archaeognatha and Zygentoma (three-tailed bristletails and silverfish): 11 species available

Scale Small   Effort Moderate   Accessibility 6

Collective effort = 10 (90.9%) species   **Top lister** = Mark Telfer (8 species)

Vital equipment: sieve, tray.

Essential text

M.J. Delaney (1954) *Thysanura and Diplura.* Handbooks for the Identification of British Insects, Volume 1, Part 2. Royal Entomological Society (out of print).
This old RES key is downloadable from www.royensoc.co.uk and covers nine species.

There are seven species of three-tailed bristletail and four species of silverfish on the list. As there are so few species, I have provided a few details about them in Table 7.

Dilta hibernica, Ventnor Botanic Gardens.

Table 7. Available species of three-tailed bristletail and silverfish

Order	Species	Common name	Records on NBN	Notes
Archaeognatha	Dilta chateri		4.7%	South Wales
Archaeognatha	Dilta hibernica	Southern Bristletail	10.4%	Western
Archaeognatha	Dilta littoralis		9.3%	Southern
Archaeognatha	Dilta saxicola		–	
Archaeognatha	Petrobius brevistylis		5.4%	Coastal, Wales
Archaeognatha	Petrobius maritimus	Sea Bristletail	38.0%	Coastal, widespread
Archaeognatha	Trigoniophthalmus alternatus		0.9%	Caves and tunnels
Zygentoma	Ctenolepisma longicaudata	Grey Silverfish	1.0%	In houses
Zygentoma	Lepisma saccharina	Silverfish	29.4%	In houses
Zygentoma	Thermobia domestica	Firebrat	0.8%	In houses
Zygentoma	Ateleura cf. formicaria		–	Only on the Isle of Wight to date

6.22 Insects: Odonata (dragonflies and damselflies): 60 species available ●

Scale Small   Effort Easy   Accessibility 4

Collective effort = 54 (90.0%) species   **Top lister** = Philip Rhodes (50 species)

Vital equipment: binoculars (close-focus binoculars are particularly useful for Odonata), butterfly net, *pond net for nymphs, possibly a camera with a good zoom lens.*

Essential texts

Stephen P. Brooks, Steve Cham and Richard Lewington (2014) *Field Guide to the Dragonflies and Damselflies of Great Britain and Ireland*, 5th edn. Bloomsbury Wildlife Guides.

Dave Smallshire and Andy Swash (2018) *Britain's Dragonflies: A Field Guide to the Damselflies and Dragonflies of Great Britain and Ireland*, 4th edn. Princeton University Press.

With only 60 species on the UKSI, these are a fairly easy group to access, as they are large and strongly patterned. However, they are also fast and flighty, so you often only catch a glimpse of them. This means that their behaviour is often as important as their markings for getting an identification, especially when all you can see is a silhouette!

My camera is not suitable for taking shots of dragonflies, and I sometimes have to rely on catching them with a butterfly net. It is important to learn to do this in a way that is not damaging to the insect, but it is often the most efficient way to identify some of the damselfly species during a survey.

Most species fall into one of three categories – common and ubiquitous species, rare and habitat-specific species, or migratory species. Dragonflies and damselflies

A Brilliant Emerald *Somatochlora metallica* that I caught to secure a tentative identification made through binoculars.

The blue-within-blue eyes of the Southern Migrant Hawker *Aeshna affinis*.

often have rather dull common names that are easily confused with each other and/or are potentially misleading if they are named after places. For example, we really should rename the Norfolk Hawker *Aeshna isoceles*, the Green-eyed Hawker (it's really not helpful to name species after places, or people for that matter) due its rapid spread to other counties.

The two-toned pterostigma of the Southern Emerald Damselfly *Lestes barbarus*.

You will very quickly encounter all of the common species by any waterbody, while visits to acidic pools, fens or specific stretches of river will help you to find some of the rarer species. A trip to Scotland will be vital for some of the northern specialists, while twitching or searching sites in the south-east will help you to locate species such as Vagrant Emperor *Anax ephippiger*, Lesser Emperor *Anax parthenope*, Southern Migrant Hawker *Aeshna affinis* or more rare species such as the Scarlet Darter *Crocothemis erythraea*. Brackish grazing marsh around the Thames Gateway has a particularly diverse range of species.

You are often just looking for one small part of the animal to clinch the identification, which is easy enough when you are looking for the bi-coloured pterostigma of the docile Southern Emerald Damselfly *Lestes barbarus*, but a nightmare when you are trying to see a yellow patch on the side of a Brilliant Emerald's face! As soon as you find it again and focus on it through your binoculars the insect will move a couple of metres, and the frustrating cycle will start all over again.

I have to confess they are my least favourite group of insects, perhaps because I am so used to getting close to and interacting with invertebrates in the hand, although close-focus binoculars have really helped me to appreciate them more, as they allow me to focus in on the animals much faster, with a broader field of view. Furthermore, when I'm doing a very active invertebrate survey, looking for dragonflies requires a significant change in my search image, which jars with how I usually survey.

Society

British Dragonfly Society

The benefits of membership include copies of the society's magazine and journal, field trips, meetings and opportunities to take part in the Dragonfly Recording Network.

Facebook group

British Dragonflies and Damselflies

6.23 Insects: Orthopteroids (grasshoppers, bush-crickets, crickets, groundhoppers, stick-insects, cockroaches, earwigs and mantids): 79 species available

Scale Small   Effort Easy   Accessibility 4

Collective effort = 72 (91.1%) species   **Top lister** = John Poland (61 species)

Vital equipment: sweep net, beating stick and tray, hand lens, *head torch*. You will also need good hearing.

Most species reach adulthood late in the season, and there is a plethora of new colonists arriving in the UK, both 'naturally' across the Channel due to climate change and through accidental introductions.

Essential texts

Peter Sutton and Björn Beckmann (2026) *Field Guide to the Grasshoppers of Great Britain and Ireland.* Bloomsbury.
 This new book, illustrated by Richard Lewington, describes all of the species in this section, including all of the non-natives. It covers in detail the more difficult groups, such as the cockroaches, earwigs, stick-insects and recent arrivals. The book contains many useful identification tips, as well as distribution maps, lifecycle charts and sonograms.

Denys Ovenden (1999) *Guide to British Grasshoppers and Allied Insects.* Field Studies Council.
 This is a useful little fold-out guide to the commoner UK species. Many colonists have become established since it was published, including Southern Oak-bush Cricket *Meconema meridionale*, Tree-cricket *Oecanthus pellucens*, Sickle-bearing Bush-cricket *Phaneroptera falcata*, Southern Sickle-bearing Bush-cricket *Phaneroptera nana*, Large Cone-head *Ruspolia nitidula* and several cockroaches, and not all of the stick-insects are illustrated.

Orthoptera (crickets, bush-crickets, grasshoppers and groundhoppers): 54 species on the UKSI

Scale Small   Effort Easy   Accessibility 4

Essential texts

Martin Evans and Roger Edmondson (2007) *A Photographic Guide to the Grasshoppers and Crickets of Britain & Ireland.* WGUK.
 This is a good photographic guide, but many of the species that have become established in the last two decades are missing.

Eric Sardet, Christian Roesti and Yoan Braud (2015) *Grasshoppers of Britain and Western Europe: A Photographic Guide.* Bloomsbury.
This book covers many more species than are found in the UK, but it does include all of the recent colonists that have started to appear along the south coast in recent decades. It is also useful for the Channel Islands species.

At the time of writing there are 54 species of Orthoptera listed on the UKSI. They are often easiest to locate and identify by ear, once you have learned their sounds. Many people start to lose the higher frequencies as they get older (they're still very loud to me), in which case it may no longer be possible to hear species such as Roesel's Bush-cricket *Roeseliana roeselii*. This species is one of my top ten most frequently recorded invertebrates and the species of Orthoptera I record most often, which is remarkable considering that it only colonised our shores in the latter half of the twentieth century.

A sweep net is not particularly useful for species in this group. They are often either buried deep in the undergrowth or jump quickly away as soon as you move the net (or indeed if caught they may jump right out of it), so active searching is often the best method.

Identification of immature grasshoppers is virtually impossible, but most immature crickets and bush-crickets are readily identifiable even as tiny early-instar nymphs.

One day in 2020 I was down at Dungeness with Dave Walker, waiting for nightfall so that we could go and listen for Tree-crickets and see the Sickle-bearing Bush-crickets. An assistant warden showed Dave a photo he had taken of a Large Cone-head several hours earlier. As it was a new record for the site, we immediately

The Jersey Grasshopper *Euchorthippus elegantulus* is a protected species on the island.

A female Roesel's Bush-cricket *Roeseliana roeselii*.

searched the area, but with no success. Several hours later, when it was dark, we witnessed the cacophony of Tree-crickets and spotted a few silent Sickle-bearing Bush-crickets, too. Then I heard a bizarre, crackling, electrical sound behind a moth trap, and asked Dave what kind of battery was making all that noise. He couldn't hear any noise, and I suddenly realised that it might be the sound of a singing Large Cone-head. Then we spotted it! It was so loud to me that it made me feel dizzy when I got really close to it, something I have only ever experienced with cicadas in Australia before. I heard at least three singing adults and we stumbled across a fourth that night. Dave has since discovered that less than 50% of people can hear this species.

The southern counties of England are the best place to get a good list of species in this group.

A Large Cone-head *Ruspolia nitidula* at Old Lodge in Sussex, quite far inland in Ashdown Forest.

An immature Unarmed Stick-insect in Penzance, Cornwall.

Phasmatodea (stick-insects): 10 species on the UKSI

Scale Very small   Effort Easy   Accessibility 2

There are no native stick-insects on the British list. The best way to increase your stick-insect list is to visit the Isles of Scilly and go out at night with a torch. There are at least four species on the islands, and they can still be found even during the migrant season in the autumn. The south-west is the next best place to go looking for this interesting group of non-native insects.

Dictyoptera (cockroaches): 31 species on the UKSI

Scale Small   Effort Moderate   Accessibility 6

We have just three native cockroaches, all listed as Nationally Scarce. The species I record most frequently is the Lesser Cockroach *Capraiellus panzeri*, followed by the Dusky Cockroach *Ectobius lapponicus*. I tend to see our three native species on poor soils (usually with plenty of litter in the case of Dusky Cockroach and Tawny Cockroach *Ectobius pallidus*) in the south-east, with the Lesser Cockroach typically occurring within a few miles of the coast, usually on downland and heaths. Variable Cockroach *Planuncus tingitanus* is spreading rapidly – look for adults near human habitation around October, when our native species are not adult. These additional species can make identification of the whole order challenging.

The rest of the UK species list consists of accidental introductions that are typically restricted to buildings and hothouses. If present in hothouses, naturalised

Tawny Cockroach *Ectobius pallidus*.

cockroaches are usually very obvious and easy to find. Other non-native species can be found in close proximity to human habitation, such as untidy places where food is prepared.

Dermaptera (earwigs): 12 species on the UKSI

Scale Small   Effort Easy   Accessibility 4

We don't have many earwig species in the UK, and it's good to learn the Common Earwig *Forficula auricularia* quickly, as it dominates the records. I have records of four species of earwig, but 95% of my records are of Common Earwig. On the South Downs, especially where Traveller's-joy *Clematis vitalba* is common, Lesne's Earwig *Forficula lesnei* is not particularly scarce, making up 5% of my records. I have only seen Lesser Earwig *Labia minor* twice and Short-winged Earwig *Apterygida media* once.

Website
Mike's Insects Keys are fairly comprehensive for earwigs.

Mantodea (mantis): 2 species on the UKSI

Scale Very small   Effort Easy   Accessibility 2

Praying Mantis *Mantis religiosa* has become established in at least four locations in four different counties in southern England, and is very likely to spread further.

Facebook group
UK Orthoptera – Grasshoppers, Crickets and Allied Insects
This is a very helpful group if you have any questions about identification.

Downloads
Orthoptera and Allied Insects: www.orthoptera.org.uk/node/707
Sound recordings of 25 species can be downloaded here.

A population of Praying Mantis *Mantis religiosa* was discovered on the south coast of the Isle of Wight on 22 August 2025. I took this photo on the 24th!

6.24 Insects: Hemipteroids (true bugs, leafhoppers, aphids, whiteflies, scale insects, psyllids, psocids, thrips and lice, etc.): 2,320 species available

| Scale | Large | Effort | Difficult | Accessibility | 16 |

Collective effort = 1,527 (65.8%) species   **Top lister** = Jonty Denton (956 species)

Vital equipment: sweep net, beating tray, suction sampler, pooter, hand lens, dissecting microscope.

This is a large and mixed group, ranging from species that are easy, such as the shieldbugs, through to the difficult ones, such as aphids and psyllids. When learning true bugs, shieldbugs are often the best place to start as most of them are picture-matchable. After this, the rest of the Heteroptera are mostly manageable, with seed bugs posing little difficulty, plant bugs being a larger family with some challenging genera, and hoppers being more difficult still. Remember that *bugs suck and beetles chew.*

Heteroptera: 670 species on the UKSI

| Scale | Medium | Effort | Moderate | Accessibility | 9 |

This is a perfectly sized group to tackle, especially for a beginner – not too big to be overwhelming, but with plenty of diversity and variety to keep you occupied. It is one of my 'core groups', as I am County Recorder for Heteroptera, and it's also a

"Does my bum look big in this?" The Trapezium Shieldbug *Coptosoma scutellatum* seen from behind.

very useful group to survey because many of its species are phytophagous, so they indicate what food plants are present.

Hard-bodied Heteroptera tend to hibernate over the winter more commonly than soft-bodied species (such as most of the mirids), which are much more abundant by mid- to late summer, when they can be overwhelming in the sweep net. There are so many common mirids that writing them down faster than they can crawl out of your net can be quite challenging.

Many scarce species are associated with warm, broken turf. Suction sampling on sunlit, partially vegetated ground such as that found on chalk grassland, arable land or vegetated shingle can be really productive for bugs, and although many such 'thermophilic' species are classed as scarce, many are also spreading as a result of climate change.

This suborder is constantly throwing up species new to Britain. My first 'first for Britain' was a bizarre-looking shieldbug that I found at Marline Valley, East Sussex in 2019. I christened it the Trapezium Shieldbug *Coptosoma scutellatum*, and it's clear that this species has just hopped the Channel (yes, it can fly!).

Essential texts

T.R.E. Southwood and Dennis Leeston (1959) *Land and Water Bugs of the British Isles*. Pisces Conservation (facsimile edition).
Despite its age, this book contains some very useful information, especially on the food plants of bug species, that is not readily available elsewhere. However, the keys are quite out of date and the illustrations in the facsimile edition are not very well reproduced.

Peter Kirby (2015) *British Heteroptera Keys to Terrestrial Families Other Than Miridae*. Unpublished key.
It's well worth getting hold of this key to all Heteroptera other than the plant bugs and water bugs, as it is the only modern key for some families. It is widely shared among entomologists.

Website

British Bugs: www.britishbugs.org.uk
At the time of writing there is no comprehensive text or useful photographic guide on this group of insects, but this website fulfils those functions and is by far the best resource for beginners. It doesn't cover everything, and for many species you will need a key, but it's a really good start and I still use it on most days in the summer months.

Facebook groups

UK Hemiptera

British Terrestrial True Bugs (Heteroptera) Recording Schemes
Both of these medium-sized groups are worth joining, as there is a great deal of expertise within them.

Shieldbugs, squashbugs and allies: 99 species on the UKSI

Scale: Small   Effort: Very easy   Accessibility: 2

Essential texts

Paul D. Brock (2021) *Britain's Insects: A Field Guide to the Insects of Great Britain and Ireland*. Princeton University Press.

The shieldbugs are extremely well covered in this book, and most of the recent additions are featured. As is often the case with generic invertebrate guides, small families with large, field-identifiable species tend to be covered in their entirety, whereas larger families with smaller, less field-identifiable species are not. When only one or two representatives of a genus are covered, the number of other species in the genus is listed. However, not all genera are covered. The text is brief, but conservation status data and distribution maps are provided.

Richard Jones (2023) *Shieldbugs*. William Collins.

This beautiful book in the New Naturalist series has the most up-to-date key to the shieldbugs and allies, including many species that could turn up in the UK soon. The key is highlighted in green around the edges of the page so that it's easy to find quickly, and the species accounts are very thorough.

Dalman's Leatherbug *Spathocera dalmanii* is encountered on southern acid grassland where Sheep's Sorrel *Rumex acetosella* grows in abundance.

Plant bugs and allies: 472 species on the UKSI

Scale Medium   Effort Difficult   Accessibility 12

This group of around 11 families of terrestrial Heteroptera has its own recording scheme, although it is not very active. The two largest families within this group are the plant bugs or mirids (Miridae) and the seed bugs (Lygaeoidea, in fact a superfamily).

Plant bugs (Miridae): 249 species on the UKSI

Scale Medium   Effort Difficult   Accessibility 12

This is the largest family of the Heteroptera, and there are some problematic groups within it. Small green mirids can be daunting at first, as can the genus *Lygus* and the large genus of arboreal mirids, *Psallus*. Plant bugs are soft-bodied and hard to handle without damaging them, but they are a useful group for surveying, as they indicate what plants are present at a site.

Mirids account for 40% of all of my Hemiptera records, with the Meadow Bug *Leptopterna dolabrata* being my most frequently recorded species, accounting for 6% of my records. Many mirids are only adult during the key summer months, and few are active in winter. *Stenodema* and *Lygus* species are often the sole mirids that I encounter during the winter months, whereas many seed bugs are winter active.

Essential text

B.S. Nau (2012) *Keys to Miridae*. Unpublished keys.
These keys, produced by the late Bernard Nau, are widely shared among entomologists.

Website

Danmarks Blomstertæger: www.larsskipper.dk/miridae/index.htm
This Danish site has photographs of most of the British species.

The mirid *Heterocordylus genistae* is only found on Dyer's Greenweed *Genista tinctoria*, but is often abundant where it occurs.

Beosus maritimus, a mainly coastal species of seed bug.

Lygaeoidea (seed bugs): 115 species on the UKSI

Scale Medium   Effort Moderate   Accessibility 9

Seed bugs (often referred to as ground bugs) consist of a number of bug families defined on the basis of their feeding preference with regard to plant seeds. They include the beet bugs (Piesmatidae), the delicate stilt bugs (Berytidae) and several families that were formerly lumped together in the Lygaeidae. The latter are a group of hard-bodied, ground-dwelling, beetle-like bugs that in many ways are rather similar to ground beetles. Most of them are quite straightforward to identify using Peter Kirby's key (see page 155).

These bugs are most commonly recorded using a suction sampler or when sieving moss or 'tussocking'.

Just 13% of all my Hemiptera records are in this superfamily, and my most frequently recorded species is the Birch Catkin Bug *Kleidocerys resedae*, which as its name suggests is widespread on birches.

Aquatic bugs: 97 species on the UKSI

Scale Small   Effort Moderate   Accessibility 6

The aquatic Heteroptera are a disparate cluster of families. All of these species live under, on or very near water for significant stages of their lifecycle.

Essential texts

A. Savage (1989) *Adults of the British Aquatic Hemiptera Heteroptera*. Freshwater Biological Association.

B.S. Nau (2004) *Draft Keys to Aquatic and Semi-Aquatic Heteroptera*. Unpublished keys. These unpublished keys are extremely useful, and like Bernard Nau's keys to mirids they are often shared among entomologists. They include the most comprehensive key to the difficult shore bugs.

Although this group is a much smaller subset of the Heteroptera, it is a key group when tackling aquatic invertebrates. Due to their aquatic lifestyle, they are rarely recorded unless you have a pond net. The largest family is the Corixidae, with 39 species, but their identification should not be difficult.

Water bugs differ from water beetles in their bizarre array of shapes, sizes and habits. The Water Stick-insect *Ranatra linearis* could not be more different to the Pygmy Backswimmer *Plea minutissima*, and until you realise just how tiny the micro water-crickets (*Microvelia* species) are you could easily overlook them.

Shore bugs can be a tricky family. They are also covered by the Aquatic Heteroptera Recording Scheme (see below), and are best keyed out using Bernard Nau's keys (mentioned earlier) in conjunction with the images on the British Bugs website. An understanding of their often quite distinctly different ecology, and therefore the habitats in which they occur, will aid identification.

Website

The Aquatic Heteroptera Recording Scheme of Britain and Ireland: www.aquaticbugs.com

Auchenorrhyncha (hoppers): 434 species on the UKSI

Scale Medium   Effort Difficult   Accessibility 12

There are nine families here that are always covered together and referred to collectively as hoppers. They include plant, leaf, frog, lace and tree hoppers, as well the New Forest Cicada *Cicadetta montana*.

Vital equipment: pooter, beating tray, sweep net, suction sampler, dissecting microscope.

Essential texts

Robert Biedermann and Rolf Niedringhaus (2009) *The Plant- and Leafhoppers of Germany: Identification Key to All Species*. WABV.

Michael Wilson, Alan Stewart, Robert Biedermann, Herbert Nickel and Rolf Niedringhaus (2015) *The Planthoppers and Leafhoppers of Britain and Ireland: Identification Keys to All Families and Genera and All British and Irish Species not Recorded from Germany*. WABV.

Most hoppers are quite small and few are field-identifiable. However, some common species *are* readily field-identifiable, and you can save yourself a lot of time by

The Pondweed Leafhopper
Erotettix cyane.

learning early on just how variable the common bug of herbaceous vegetation, *Philaenus spumarius*, can be – sometimes it may seem as if every single one you see is unique. This species accounts for nearly 20% of all my hopper records and has a very long season.

A pooter is very useful here, as hoppers will quickly jump out of your tray as you approach them with fingers or tweezers. It's worth taking a range of individuals, as many species can only be identified by examining the male's genitalia.

Website

True Hoppers of the Western Palearctic: www.truehopperswp.com
Although it is very much a work in progress, this new website can be really helpful for identifying hoppers, especially if you find something that doesn't fit.

Psocodea (bark, book, feather and body lice): 136 species in on the UKSI

| Scale | Medium | | Effort | Moderate | | Accessibility | 9 |

Outdoor-living psocids (usually referred to as bark lice) are most often found by beating trees and shrubs in the wild. The indoor-living species (usually referred to as book lice) are mostly synanthropic, found around old books, timber and damp, untidy places, including kitchens. There are around 100 species of bark and book lice. Psocodea now also includes all of the body and feather lice.

Essential text

T.R. New (2005) *Psocids, Psocoptera (Booklice and Barklice)*. Royal Entomological Society.
Most species are covered by this text.

Website

National Barkfly Recording Scheme: https://schemes.brc.ac.uk/barkfly/homepage.htm

This website includes an online key, which uses wing venation patterns as one visual aid to identification, and also has a photo gallery of some species. Using the book and the website it should be possible to get to species level.

Body and feather lice

There are approximately 30 species of body and feather louse on the UKSI, but seemingly very little literature available to identify them to species level. If you had head lice as a child, or your own children have had them, or like me, they found their way into your dreadlocks, *Pediculus humanus* counts. And at least five people have been bold enough to publicly (I'd better check how I spelt that) tick the *other* body louse on the website.

Aphids (Aphididae): 682 species on the UKSI

Scale Medium   Effort Difficult   Accessibility 12

This is a large group of poorly recorded bugs. They are small, soft-bodied, sap-sucking insects, so knowing your plants is vital here. They seem to be more numerous on more substantial plants, and there is great potential here to find species new to the UK, especially by searching exotic plants.

Large Willow Aphid *Tuberolachnus salignus*.

Essential texts

Roger L. Blackman (2010) *Aphids – Aphidinae (Macrosiphini)*. RES Handbooks for the Identification of British Insects, Volume 2, Part 7. Royal Entomological Society.

Roger L. Blackman, Robert D. Dransfield and Robert Brightwell (2019) *Aphids – Anoeciinae, Lachninae, Eriosomatinae, Phloeomyzinae, Thelaxinae, Hormaphidinae, Mindarinae*. RES Handbooks for the Identification of British Insects, Volume 2, Part 8. Royal Entomological Society.

Websites

Influential Points: www.influentialpoints.com
   This cryptically named website can be extremely helpful for identifying aphids.

Aphids on the World's Plants: www.aphidsonworldsplants.info
   This useful online identification guide covers aphids across the globe, so is particularly useful when looking at aphids on non-native plants in gardens and parks.

Facebook group

UK Aphids, Psyllids, Scales and Allies (Hemiptera, Sternorrhyncha)
   This is a small but useful group with a small membership and very little traffic.

Jumping plant lice (Psylloidea): 92 species on the UKSI

Scale: Small   Effort: Difficult   Accessibility: 8

The majority of psyllids are small (under 3 mm long), host-specific, sap-sucking bugs that are not widely recorded. There is a real possibility of finding species new to the British Isles, especially by looking at non-native plants in parks and gardens.

Essential text

D. Hodkinson and I. M White (1979) *Homoptera: Psylloidea*. RES Handbooks for the Identification of British Insects, Volume II, Part 5a. Royal Entomological Society (out of print).
   This book is a little out of date, but it can be downloaded for free from the Royal Entomological Society website.

Websites

Psyl'list – The World Psylloidea Database: www.hemiptera-databases.org/psyllist
   This is a useful resource for helping to identify psyllids.

British Bugs (see page 155) covers psyllids, but only covers around half of the species on the UKSI.

https://norfolknaturalists.org.uk/wp/wp-content/uploads/2024/06/NNNS_Norfolk_Psyllid_guide_v1.1.pdf Here James Emerson is pulling together information on the psyllids of Norfolk.

Scale insects: c.99 species on the UKSI

Scale Small   Effort Very difficult   Accessibility 10

This hugely variable group of hemipterans is possibly poorly represented on the UKSI, and the literature is sparse. Like psyllids and aphids, most scale insects are host specific, and as they are even more likely to be moved around the country with non-native plants there is a strong possibility of finding undiscovered species or recent accidental introductions.

Website

Bedfordshire Natural History Society: https://www.bnhs.co.uk/2019/recorder/big/sternorrhyncha/scales_f1.php
This website contains some useful information on scale insects.

Thrips (Thysanoptera): 182 species on the UKSI

Scale Small   Effort Difficult   Accessibility 8

Most of the UK species are associated with vascular plants, where they are most commonly found on flowers, and are often host specific. There are also a number of species associated with fungi.

Essential texts

W.D.J. Kirk (1996) *Thrips*. Naturalists' Handbooks 25. Richmond Publishing.

L.A. Mound, G.D. Morison, B.R. Pitkin and J.M. Palmer (1976) *Thysanoptera*. Royal Entomological Society (out of print, but can be downloaded from the RES website).

Websites

Thrips of the British Isles: https://keys.lucidcentral.org/keys/v3/british_thrips/
This very useful online key covers 177 species.

Mike's Insect Keys has keys to genera across the three families, and a key to species for the 23 species in the genus *Thrips*.

6.25 Insects: Hymenoptera (bees, ants, wasps and sawflies, etc.): 8,105 species available

Scale: Very large   Effort: Very difficult   Accessibility: 25

Collective effort = 2,358 (29.1%) species   **Top lister** = Julian Small (1,135 species)

This is the largest invertebrate order. The well-known aculeate Hymenoptera account for only a small proportion (8.5%) of the species, and the sawflies and the much larger tranche of poorly known wasps account for all the rest. Beyond the aculeates it would be a daunting task to master this order, and a lifetime's work!

Vital equipment: a butterfly net (an extendable long-handled net can be really useful), or at least a white (and therefore partially transparent) net, hand lens, spy pot (particularly useful for looking at bumblebees in the field), dissecting microscope.

Essential text

H. Goulet and J.T. Huber (1993) *Hymenoptera of the World: An Identification Guide to Families*. Agriculture Canada.
Although this is not a British key, it works well for our fauna.

Aculeate Hymenoptera: 700 species on the UKSI

Scale: Medium   Effort: Difficult   Accessibility: 12

Bees: 283 species on the UKSI

Scale: Medium   Effort: Moderate   Accessibility: 9

All bees are readily identifiable using modern texts. Look for bees at warm sites with plenty of flowers, as well as sites with plenty of bare ground for nesting.

Pantaloon Bee
Dasypoda hirtipes.

The Large Scabious Mining Bee *Andrena hattorfiana* is the biggest species in this diverse genus of beautiful bees. It takes pollen almost exclusively from Field Scabious *Knautia arvensis*.

Essential texts

Steven Falk and Richard Lewington (2015) *Field Guide to the Bees of Great Britain and Ireland*. Bloomsbury.
This text is the size of a field guide, densely packed with information, and the keys are excellent. I think this volume works best when used in conjunction with the following book and Steven Falk's Flickr pages for bees.

George R. Else and Mike Edwards (2018) *Handbook of the Bees of the British Isles*. The Ray Society.
This is an impressive, two-volume set of large A4-size books. Volume 1 consists of the keys, and Volume 2 contains the species accounts and distribution maps. The keys are very comprehensive and there are some very high-quality illustrations, especially of the detail of genitalia, and some very clear microscope images.

Website

Steven Falk's comprehensive Flickr pages are an essential resource for helping you to identify bees.

Aculeate wasps: 329 species on the UKSI

Scale: Medium   Effort: Difficult   Accessibility: 12

These are much more difficult to identify than the bees. They are covered by a number of texts, some of which are fairly recent, though most are in need of revision. A new key by Mike Edwards is in preparation but is unlikely to be published until at least 2026.

Argogorytes mystaceus on a Fly Orchid *Ophrys insectifera*.

Essential texts

Peter F. Yeo and Sarah A. Corbet (2015) *Solitary Wasps*. Naturalists' Handbooks 3. Pelagic Publishing.

M.E. Archer (2014) *The Vespoid Wasps (Tiphiidae, Mutilidae, Sapygidae, Scoliidae and Vespidae) of the British Isles*. Royal Entomological Society.

Website

Steven Falk's Flickr collections are very comprehensive for social, sand and spider-hunting wasps, but less so for other species.

Pompilidae (spider-hunting wasps): 45 species on the UKSI

Scale Small   Effort Difficult   Accessibility 8

These medium to large wasps always seem surprisingly difficult to key out. There is an old RES key that you can obtain from their website.

Essential text

M.C. Day (1988) *Pompilidae*. Royal Entomological Society.

INSECTS: HYMENOPTERA 167

Anoplius viaticus with a large wolf spider. This is the only 'pomp' I feel confident naming in the field.

Website

Steven Falk's Flickr collection is comprehensive and very detailed for this difficult family of wasps.

Download

A New Draft Key to the Pompilidae (Spider-Hunting Wasps) by Graham Collins: https://bwars.com/content/new-draft-key-pompilidae-spider-hunting-wasps-graham-collins-and-also-link-pompilidae

Jewel wasps (Chrysididae): 41 species on the UKSI

| Scale | Small | | Effort | Very difficult | | Accessibility | 10 |

Despite being spectacular insects, some jewel wasps are frustratingly difficult to identify (especially those in the genus *Chrysis*). The following paper is perhaps the best key currently available for identifying them.

I thought *Elampus panzeri* was one of the few field-identifiable species of chrysid, but even this has now been split into two species!

Essential text

Juho Paukkunen, Alexander Berg, Villu Soon, Frode Ødegaard and Paola Rosa (2015) An illustrated key to the cuckoo wasps (Hymenoptera, Chrysididae) of the Nordic and Baltic countries, with description of a new species. *ZooKeys* 548: 1–116.

Ants (Formicidae): 88 species on the UKSI

Scale: Small | Effort: Moderate | Accessibility: 6

Vital equipment: suction sampler, sieve, tray, beating stick, dissecting microscope.

People are often surprised how few ant species there are. In many ways I find them a lot easier than most other Hymenoptera groups, though there are some difficult genera, too. They are easiest to find using a suction sampler.

Essential texts

Gary J. Skinner and Andrew P. Jarman (2025) *Ants*. Naturalists' Handbooks 24. Pelagic Publishing.
This excellent new book covers all 61 species that regularly occur in Britain, Ireland and the Channel Islands, and includes keys to species, distribution maps and a section on hothouse exotica.

Bernhard Seifert (2018) *The Ants of Central and North Europe.* Lutra.
This very comprehensive book contains a vast amount of detail and includes far more species than are present in the UK, making the keys a little difficult to navigate at times, but you can easily annotate them.

Sawflies (Symphyta): 555 species on the UKSI

Scale: Medium | Effort: Difficult | Accessibility: 12

Sawflies are an important group of phytophagous, wasp-like insects, but they are poorly understood and recorded, and there is no modern user-friendly text on the UK fauna. However, the pan-species lister and co-creator of the PSL website, Andy Musgrove has recently written the status review for sawflies for Natural England, which can be viewed here: https://www.sawflies.org.uk/resources/#Status

Essential texts

R.B. Benson (1951) *Hymenoptera: Symphyta*. RES Handbooks for the Identification of British Insects, Volume 6, Section 2a. Royal Entomological Society.

R.B. Benson (1952) *Hymenoptera: Symphyta*. RES Handbooks for the Identification of British Insects, Volume 6, Section 2b. Royal Entomological Society.

R.B. Benson (1958) *Hymenoptera: Symphyta*. RES Handbooks for the Identification of British Insects, Volume 6, Section 2c. Royal Entomological Society.
Despite their age, these three keys are still well worth using. Although they are out of print, all three can be downloaded for free from the RES website.

The sawfly *Sciopteryx soror*.

Jean Lacourt (2020) *Sawflies of Europe*. NAP.
: This book covers considerably more species than are present in the UK, but the key to family is very good. In some respects, though, Benson's keys are better.

Adam Wright (1990) *British Sawflies: A Key to the Adults of Genera Occurring in Britain*. Field Studies Council.
: This small book is now quite dated and the line illustrations are very basic, but the key to family is quite useful.

Website

The Sawflies (Symphyta) of Britain and Ireland: www.sawflies.org.uk
: This is an extremely useful website, and when combined with the above key to genera and the European book, or even the old RES keys, it can often get you to species level. There is also a gallery with photos of both adults and larvae.

Facebook group

British and Irish Sawflies (Symphyta)
: This is your first port of call for any sawfly queries; with it having over 2,300 members, you'll always get a helpful response.

European Sawflies
: It's always worth joining the European groups for a different perspective, or to gain access to some expertise on species that might be very scarce in the UK.

Ichneumonoidea: 4,021 species on the UKSI

Scale Large   Effort Very difficult   Accessibility 20

There are almost as many species in the Ichneumonoidea as there are beetles! There is a very real possibility of finding species new to the British Isles here, perhaps more so than in any other insect group, and most records will be county firsts. The literature is growing all the time, too.

Essential texts
Ichneumonidae subfamilies:

Gavin R. Broad, Mark R. Shaw, Michael G. Fitton, Dawn Painter and Olga Retka (2018) *Ichneumonid Wasps (Hymenoptera: Ichneumonidae): Their Classification and Biology*. Royal Entomological Society.

Some recent keys to certain European species of Ichneumonidae:

N. Johansson (2024) *Review of the Swedish Species of* Lissonota *Gravenhorst (Hymenoptera: Ichneumonidae: Banchinae) with an Illustrated Key to the Females of the Western Palearctic*. Sveriges Entomologiska Förening.

N. Johansson and B. Cederberg (2019) Review of the Swedish species of *Ophion* (Hymenoptera: Ichneumonidae: Ophioninae), with the description of 18 new species and an illustrated key to Swedish species. *European Journal of Taxonomy* 550: 1–136.

N. Johansson (2021) Revision of the Swedish species of *Metopius* Panzer, 1806 (Hymenoptera: Ichneumonidae: Metopiinae), with an illustrated key to the species of Northwestern and Central Europe. *Entomologisk Tidskrift* 142: 37–69.

N. Johansson, J. Hilszczanski and F. Ødegaard (2022) Revision of the Scandinavian species of *Xorides* Latreille, 1809 (Hymenoptera: Ichneumonidae: Xoridinae), with an illustrated key to the species of Northern Europe. *Entomologisk Tidskrift* 143: 183–222.

A rare example of a whole subfamily covered by a good key:

S. Klopfstein (2014) Revision of the Western Palaearctic Diplazontinae (Hymenoptera, Ichneumonidae). *Zootaxa* 3801: 1–143.

Other UK handbooks:

J.F. Perkins (1959) Hymenoptera. Ichneumonoidea. Ichneumonidae, key to subfamilies and Ichneumoninae – 1. *Handbooks for the Identification of British Insects* 7(2ai): 1–116.

J.P. Brock (2017) *The Banchine Wasps (Ichneumonidae: Banchinae) of the British Isles*. Royal Entomological Society.

J.F. Perkins (1960) Hymenoptera. Ichneumonoidea. Ichneumonidae, subfamilies Ichneumoninae II, Alomyinae, Agriotypinae and Lycorininae. *Handbooks for the Identification of British Insects* 7(2aii): 117–213.

I.D. Gauld and P.A. Mitchell (1977) Ichneumonidae. Orthopelmatinae and Anomaloninae. *Handbooks for the Identification of British Insects* 7(2): 1–32.

M.G. Fitton, M.R. Shaw and I.D. Gauld (1988) Pimpline Ichneumon-flies. Hymenoptera, Ichneumonidae (Pimplinae). *Handbooks for the Identification of British Insects* 7(1): 1–110.

Website

The Ichneumon Files: www.bioimages.org.uk/ichneumon_files.htm
Malcolm Storey has presented a huge amount of information here, together with many well-illustrated keys.

Facebook groups

British Ichneumonidae

European Ichneumonidae

Chalcids (Chalcidoidea): 1,778 species on the UKSI

Scale Large   Effort Very difficult   Accessibility 20

Essential texts

Nearctic Chalcidoidea genera:

G.A.P. Gibson, J.T. Huber and J.B.E. Woolley (1997) *Annotated Keys to the Genera of Nearctic Chalcidoidea (Hymenoptera)*. NRC Research Press.

Website

Universal Chalcidoidea Database: https://www.nhm.ac.uk/our-science/data/chalcidoids/database/indexHosts.dsml?index=Hosts&VALGENUS=tetrastichus
This is the old version, which is no longer being updated, but is much more user-friendly.

Ormyrus nitidulus, a chalcid. (Derek Binns)

Downloads

Ian Gauld and Barry Bolton (1988) *The Hymenoptera*. British Museum Natural History/ Oxford University Press. https://www.scribd.com/document/829710724/The-Hymenoptera-Gauld-and-Bolton

Henri Goulet and John Theodore Huber (1993) *Hymenoptera of the World: An Identification Guide to Families*. Agriculture Canada Publication, Ottawa. https://www.researchgate.net/publication/259227143_Hymenoptera_of_the_World_An_Identification_Guide_to_Families

Braconidae: 1,361 species on the UKSI.

| Scale | Large | Effort | Very difficult | Accessibility | 20 |

Essential texts

Braconidae subfamilies:

M.R. Shaw and T. Huddleston (1991) Classification and biology of braconid wasps (Hymenoptera: Braconidae). *Handbooks for the Identification of British Insects* 7(11): 1–126.

C. van Achterberg (1993) Illustrated key to the subfamilies of the Braconidae (Hymenoptera: Ichneumonoidea). *Zoologische Verhandelingen* 283: 1–189.

Nearctic Braconidae genera:

R.A. Wharton, P.M. Marsh and M.J. Sharkey (eds) (1997) *Manual of the New World Genera of the Family Braconidae (Hymenoptera)*. Special Publication No. 1. International Society of Hymenopterists.

Downloads

M.R. Shaw and T. Huddleston (1991) *Classification and Biology of Brachonid Wasps*. Volume 7, Part 11. Royal Entomological Society of London. https://www.royensoc.co.uk/shop/publications/out-of-print-handbooks/vol-7-part-11-classification-biology-of-braconid-wasps-hymenoptera-braconidae/

Society

Bees, Wasps and Ants Recording Scheme (BWARS)
   Members receive newsletters and a handbook on aculeates, and can attend workshops. The recording scheme includes online maps and species accounts. A lot of useful information can be downloaded for free from the BWARS website: https://bwars.com/content/identification-guides-and-downloads

Facebook group

UK Bees, Wasps and Ants
   This is the official BWARS Facebook group.

6.26 Insects: Coleoptera (beetles): 4,301 species available

Scale **Large** Effort **Difficult** Accessibility **16**

Collective effort = 3,735 (86.8%) species   **Top lister** = Peter Hodge (3,091 species)

Vital equipment: net, hand lens, beating tray, sieve, suction sampler, tubes, pooter, dissecting microscope.

This is a large order, with 102 beetle families in the British Isles. I shall cover a selection of the larger and generally more well-known families, which together account for nearly 3,000 species, just shy of 75% of the UK fauna.

Essential texts
If you have a limited budget, the four volumes written by Andrew Duff would be your best option, but will cost around £450 for the whole set. They are highly recommended.

Andrew G. Duff (2012) *Beetles of Britain and Ireland, Volume 1: Sphaeriusidae to Silphidae.* A.G. Duff Publishing (out of print).

Andrew G. Duff (2024) *Beetles of Britain and Ireland, Volume 2: Staphylinidae.* A.G. Duff Publishing.

Mesosa nebulosa, a stonkin' longhorn, which remains the only individual of this species that I have seen.

Andrew G. Duff (2020) *Beetles of Britain and Ireland, Volume 3: Geotrupidae to Scraptiidae*. A.G. Duff Publishing.

Andrew G. Duff (2016) *Beetles of Britain and Ireland, Volume 4: Cerambycidae to Curculionidae*. A.G. Duff Publishing (out of print).

When you are first starting out with beetles a key to family is quite useful (though, as with most keys, the higher up the key is taxonomically, the harder it is to use). With more experience you will rarely need to use keys to family, and an easier way in is to picture match to family using a website such as Kerbtier or UK Beetle Recording (see page 183), only resorting to the key to family when this doesn't work.

D.M. Unwin (1984) *A Key to the Families of British Beetles*. Field Studies Council.

Ladybirds (Coccinellidae): 63 species on the UKSI

| Scale | Small | Effort | Easy | Accessibility | 4 |

This is probably the best-known group of beetles, and my most frequently recorded beetle species is the 7-spot Ladybird *Coccinella septempunctata*. Most ladybirds are large, colourful and field-identifiable (though don't bother trying to count the spots on anything upwards of a 7-spot!). There are also a small number of smaller ladybirds, which require microscopic identification.

Essential texts

Volume 3 of Duff's *Beetles of Britain and Ireland* has a very useful key that works well with the smaller ladybirds, too. The larger species are rarely likely to need keying out, as most of them can be picture matched.

Helen Roy, Peter Brown and Richard Lewington (2018) *Field Guide to the Ladybirds of Great Britain and Ireland*. Bloomsbury.

Scarce 7-spot Ladybird *Coccinella magnifica*. Someone missed a trick by not calling this the Magnificent 7-spot Ladybird!

Agapanthia villosoviridescens, one of the UK's smartest-looking beetles.

Helen Roy, Peter Brown, Robert Frost and Remy Poland (2011) *Ladybirds (Coccinellidae) of Britain and Ireland*. Field Studies Council.

Maria Justamond and David W. Williams (2025) *Micro Ladybirds of Britain and Ireland*. Field Studies Council.
Ladybirds (Coccinellidae) of Britain and Ireland is an atlas, which works very well as an identification guide too, as each species account holds at least one photograph as well as the usual maps and phenology charts.

Longhorns (Cerambycidae): 87 species on the UKSI

Scale: Small   Effort: Easy   Accessibility: 4

This a small family, and almost all of its species are entirely field-identifiable – these distinctive animals are among our most beautiful beetles. The only drawbacks are that there are so few species, and they are mainly very seasonal, with the majority active only briefly in the summer months. You never find many longhorns, so although they are a wonderful gateway family, they are not going to sustain you for long in a pan-species listing (or even a beetle) capacity. Most longhorns are saproxylic as larvae but need flowers to feed on as adults, so look for them in places with plenty of dead and decaying wood *and* with plenty of flower nearby, such as parkland, woodland edges, rides and glades. Beating hawthorn blossom near big old trees is a very effective way to find them.

Essential texts

Volume 4 of Duff's *Beetles of Britain and Ireland* covers longhorns, but the exquisite illustrations by Richard Lewington in the FSC fold-out guide (originally published in 2007 in *British Wildlife*) are also a useful visual guide.

Wil J. Heaney, Kay Potts and Richard Lewington (2018) *Guide to Longhorn Beetles of Britain*. Field Studies Council.

Longhorn beetles account for just 2% of my beetle records, and 29% of these are of the very common *Grammoptera ruficornis* – a small black ubiquitous species that breeds in small dead twigs and branches.

The first time I went out specifically looking for a beetle was in 2009, when I headed to Ebernoe Common in Sussex to search for *Leptura aurulenta*. I found it. Fast forward 16 years, and at the time of writing, I am approaching 2,000 beetle species.

Soldier beetles (Cantharidae): 42 species on the UKSI

| Scale | Small | Effort | Moderate | Accessibility | 6 |

The larger soldier beetles are brightly coloured and mostly field-identifiable, whereas the two smaller genera are better examined at the microscope, and some of them have to be identified by their genitalia.

Essential text

The keys in Volume 3 of Duff's *Beetles of Britain and Ireland* work well, especially for the smaller species.

Website

Weevil Guides: www.tinyurl.com/weevilguides
    This excellent visual guide has been produced by Mark Gurney.

Soldier beetles are very seasonal, found mainly from May through to August, but with the greatest diversity in May and June. Typically the first sign of the very common *Rhagonycha fulva* in late June heralds the end of the exciting part of the beetling season (many of us affectionately call it the 'Herald of Dross' for this reason, which is an improvement on one of its other common names – the 'Hogweed Bonking Beetle'). Soldier beetles account for just 5% of all my beetle records, my most frequently recorded soldier beetle species being *Rhagonycha fulva*.

Click beetles (Elateridae): 73 species on the UKSI

| Scale | Small | Effort | Moderate | Accessibility | 6 |

Many of the species in this family of distinctively shaped beetles are saproxylic and can be found in association with dead wood, but there also plenty of grassland species, which feed on the roots of plants. Click beetles account for just 4% of my beetle records, and my most frequently recorded click beetle species is the ubiquitous (but seasonal) *Athous haemorrhoidalis*.

The key in Volume 3 of Duff's *Beetles of Britain and Ireland* is very good.

Clown beetles (Histeridae), false clown beetles (Sphaeritidae) and sexton beetles (Silphidae): 53 clown beetles, 1 Sphaeritidae beetle and 21 sexton beetles on the UKSI (75 species in total) ●

Scale Small   Effort Difficult   Accessibility 8

Vital equipment: sieve, tray, trowel, gloves, and perhaps a peg for your nose!

Essential text

Steve A. Lane, C.B.H. Lucas and Ashleigh L. Wiffin (2020) *The Histeridae, Sphaeritidae and Silphidae of Britain and Ireland.* Field Studies Council.
   There is a very useful key in this excellent atlas, which is particularly good for the histerids (previously always made far more complicated than necessary in keys). It includes many clear photographs that make these difficult features very easy to navigate.

If you want to find these beetles, any dead animal at the right stage of decay is going to be worth a look. Beat it like a piñata over a tray and watch the treasure fall out, as you gag your way to coleopteran glory. Alternatively, try flipping the carcass over with a stick or boot and if you can stomach the stench, you might find some beetles. Here is a case where my autism works against me, the intensity of my reaction to this sort of smell prevents me from such activities.

Dendroxena quadrimaculata looks completely different from the rest of the family, and is also unique in that it feeds on caterpillars in the canopy, which is probably the reason why I was able to get a photo of this Nationally Scarce species.

A range of histerids are also found under bark and around dead wood, as well as in rotting fungi. Another way to find sexton beetles is by using a light trap, but don't put them in your favourite moth pots as the smell is unbearable. I was at a public moth event at the Lodge years ago, when a sexton beetle flew down an older gentleman's shirt. His wife laughed and stated, "well he does look like a corpse!"

Water beetles (several disparate families): 287 species on the UKSI

| Scale | Medium | Effort | Moderate | Accessibility | 9 |

There are ten families in this group, the largest of which are the Dytiscidae (the predatory diving beetles), with 118 species, and the water scavengers (Hydrophilidae), with 75 species. A small number of hydrophilids are terrestrial.

Vital equipment: a stout, long-handled pond net, large tray or sheet, killing fluid, tweezers.

Pushing the net hard through submerged and emergent vegetation works well, as does running it backwards and forwards through the same patch of vegetation, as this will encourage water beetles out of the vegetation into the water column, from where you can sweep them up.

Perhaps the best way to catch water beetles away from water is to run a light trap near to a wetland on warm summer nights, when the beetles are on the wing. Although this method is unlikely to give you a comprehensive list of beetles, it's

Water beetles, such as this rare species *Graphoderus cinereus*, are difficult to photograph out of the water.

useful for pan-species listers, as it often yields quite unexpected species. If you have a dark car, on a warm summer's day you will often find water beetles landing on it because they mistake it for a pond.

Essential texts

Garth N. Foster and Laurie E. Friday (2011) *Keys to Adults of the Water Beetles of Britain and Ireland (Part 1)*. Royal Entomological Society.
This volume covers the Hydradephaga (Gyrinidae, Haliplidae, Paelobiidae, Noteridae and Dytiscidae).

Garth N. Foster, David T. Bilton and Laurie E. Friday (2014) *Keys to Adults of the Water Beetles of Britain and Ireland (Part 2)*. Royal Entomological Society.
This volume covers the Hydrophiloidea superfamily, which includes both aquatic and terrestrial species.

Both volumes have excellent illustrations, and I can almost always get water beetles to species level with these keys. The plates include photos of *all* species, which is rare among books on beetles.

Leaf beetles (Chrysomelidae): 299 species on the UKSI

| Scale | Medium | Effort | Moderate | Accessibility | 9 |

This family is mostly straightforward, with the exception of the large genus of flea beetles, *Longitarsus*, so don't start with these. As all of the species are phytophagous, it's useful to know your vascular plants before tackling this group (along with the weevils). As they don't have to run after their dinner, they tend to be pretty chilled out, so this, coupled with their large size and often striking metallic colorations and patterns, make them attractive photographic subjects.

Leaf beetles account for 11% of my beetle records, and my most frequently recorded leaf beetle is *Cassida rubiginosa*, a common, field-identifiable tortoise beetle found on thistles.

The key in Volume 4 of Duff's *Beetles of Britain and Ireland* works really well for this family.

The myrmecophile leaf beetle, *Clytra quadripunctata*.

The ground beetle *Agonum sexpunctatum*.

Ground beetles (Carabidae): 377 species on the UKSI

| Scale | Medium | Effort | Moderate | Accessibility | 9 |

A well-known family of classic-looking beetles, mainly predatory ground dwellers, popular with many beginners. They are a great gateway group of beetles, and are well covered in the literature.

Essential texts

Martin L. Luff (2007) *The Carabidae (Ground Beetles) of Britain and Ireland.* Royal Entomological Society.

Volume 1 of Duff's *Beetles of Britain and Ireland*.

Carabids can be readily caught by pitfall trapping, or by using a suction sampler, looking under logs and stones, sieving moss, 'tussocking' or beating. They can be recorded in almost any habitat. There are plenty of rare and scarce species, too, such as the striking metallic and purple green *Agonum sexpunctatum*, which can be seen flashing across black peaty soils on heaths and bogs.

Ground beetles account for around 12% of my beetle records, my most frequently recorded species being *Paradromius linearis*.

Weevils (Curculionidae): 537 species on the UKSI

| Scale | Medium | Effort | Moderate | Accessibility | 9 |

This is the second-largest family of beetles after the rove beetles in the British Isles. They are an important group of mainly phytophagous species, and like the leaf beetles, make excellent photographic subjects for all the same reasons.

Weevils account for 17% of my beetle records (more than any other beetle family). My most frequently recorded species is the ubiquitous pea weevil, *Sitona lineatus*.

INSECTS: COLEOPTERA   181

The bird-dropping mimic weevil, *Platystomos albinus*.

Website

Weevil Guides: www.tinyurl.com/weevilguides

On this website, rather than approaching each species by working through a laborious dichotomous key, Mark Gurney adopts a more visual approach, which I find works best when used in conjunction with a key such as that in Volume 4 of Duff's *Beetles of Britain and Ireland*. The website includes photographs, concise text, salient features for identification and lists of similar species, all of which are very helpful tools for gaining an identification.

Dung beetles and chafers (Scarabaeidae and Geotrupidae):
95 species and 9 species, respectively, on the UKSI

Scale Medium   Effort Difficult   Accessibility 12

Vital equipment: sieve, tray, trowel, gloves, *possibly a hazmat suit*.

If you want to find these beetles, be prepared to spend some time covered in dung and rotting corpses – not everyone's idea of fun. These 'patch' habitat specialists

The Variable Chafer *Gnorimus variabilis* is as rare as rocking-horse droppings. In fact, as it feeds solely on rotten oak timber, its droppings are probably quite similar to those that a rocking horse might produce!

can be difficult to target, as you tend to come across their habitats randomly within the landscape.

Dung specialists often do better on sites that are grazed all the time, and the more unusual soils tend to be home to the more unusual species.

Dung beetles account for just over 1% of my beetle records, and my most frequently recorded dung beetle species is the Minotaur Beetle *Typhaeus typhoeus*.

The key in Volume 3 of Duff's *Beetles of Britain and Ireland* is very useful.

Website

Dung Beetle UK Mapping Project (DUMP): www.dungbeetlemap.wordpress.com/finding-and-recording-dung-beetles/identification

The large genus *Aphodius* has now been split into many smaller genera, and the '*Aphodius*' key on the DUMP website is particularly useful, as these are by far the most difficult species in this group.

Rove beetles (Staphylinidae): 1,155 species on the UKSI

| Scale | Large | Effort | Very difficult | Accessibility | 20 |

This is the largest UK beetle family – there are more rove beetles than there are macro-moths in the British Isles. The recently released second volume of Duff's *Beetles of Britain and Ireland* covers all of the Staphylinidae for the first time, including the large and difficult subfamily Aleocharinae, which accounts for around 471 species. By far my most frequently recorded rove beetle species is the field-identifiable *Tachyporus hypnorum*.

Essential texts

Derek A. Lott (2009) *The Staphylinidae (Rove Beetles) of Britain and Ireland. Part 5: Scaphidiinae, Piestinae, Oxytelinae*. Royal Entomological Society.

Derek A. Lott and Roy Anderson (2011) *The Staphylinidae (Rove Beetles) of Britain and Ireland. Parts 7 and 8: Oxyporinae, Steninae, Euasethetinae, Pseudopsiane, Paederinane, Staphylininae*. Royal Entomological Society.

Websites

Mike's Insect Keys: Keys for the Identification of British Staphylinidae: https://sites.google.com/view/mikes-insect-keys/mikes-insect-keys/keys-for-the-identification-of-british-beetles-coleoptera/keys-for-the-identification-of-british-staphylinidae
These keys are very useful for subfamilies such as the Aleocharinae, Omaliinae, Pselpahinae and Tachyporinae.

UK Beetle Recording: www.coleoptera.org.uk/home
This site provides a good overview of the beetle families, with a link to maps on the NBN Atlas.

Die Käfer Europas: www.coleonet.de
This site is particularly useful if you find something that might be new to the UK. It has very good keys, but the text is in German.

INSECTS: COLEOPTERA

Staphylinus dimidiaticornis is much larger and more striking than most other rove beetles.

Kerbtier (Beetle Fauna of Germany): www.kerbtier.de/enindex.ht
   This is rather like an online collection, and really helpful for comparing species that you might not have seen before. Although not comprehensive, it is a useful guide.

John Walters guides to British Beetles: www.johnwalters.co.uk/publications
   These guides only cover a small range of beetles, but are extremely useful for those species. They are available to download for free from the website as PDF files.

Society

Coleopterists Society of Britain and Ireland (ColSoc): www.colsoc.org
   ColSoc was launched in 2022, and there is no fee to become a supporter of the society. Benefits include regular news and updates, beetle recording support, online networking with experts and enthusiasts, and attendance at field meetings, workshops and lectures.

Facebook group

Beetles of Britain and Ireland
   This large useful group includes many coleopterists who can aid you in getting an identification, or help with general questions about beetles.

6.27 Insects: Diptera (true flies): 7,372 species available

| Scale | Very large | Effort | Very difficult | Accessibility | 25 |

Collective effort = 3,967 (53.8%) species

Top lister = Dave Gibbs (3,184 species before he left pan-species listing), Jonty Denton (1,727 species)

Vital equipment: pooter, long-handled net with a white bag, dissecting microscope, *pan and malaise traps.*

Other than the beetles, this is the only group for which anyone has recorded over 3,000 species, Dave Gibbs's total being the highest in any one taxonomic group. Flies are always likely to be one of the highest possible subtotals that a pan-species lister can realistically achieve, along with the Coleoptera, Hymenoptera, vascular plants and fungi. Therefore, if you're in this for the long haul, you really can't afford to ignore flies.

This is such a huge order, with 107 families, that it can be difficult to know where to start. Some of the easier families have been well covered in recent publications, but the rest of the literature consists of out-of-date RES keys that are available to download for free, and much more recent unpublished literature (in the form of either draft test keys or text that has only been published online). Many of the draft test keys are available on the members' area of the Dipterists Forum (see page 195) or on Mike's Insect Keys, or are shared among entomologists.

Essential text

Gail Ashton, Rory Dimon, Steven J. Falk and Peter Creed (2026) *A Photographic Guide to Flies of Britain & Ireland.* NatureBureau.
   This recent photographic guide covers over 1,200 species and includes distribution maps, conservation status data and plenty of taxonomic information. Although by no means comprehensive, this is the largest number of fly species ever to be included in one photographic guide.

Keys to family

With such a large order, a key to family can be very useful, especially for many of the smaller, less well-known families.

The old AIDGAP key to families can be accessed online here:

D.M. Unwin (1981) *A Key to the Families of British Diptera.* Field Studies Council.
   https://www.gnatwork.ac.uk/sites/gnatwork/files/content/attachments/2022-12-13/Key%20to%20the%20families%20of%20british%20diptera.pdf

A more intermediate key can be downloaded from the Dipterists Forum website here but you'll need to be a member to access that part of the site:

https://dipterists.org.uk/home Stuart Ball, John Ismay and Barbara Ismay (2017) *Introduction to Families of British Diptera.* Dipterists Forum.

For a more detailed approach, with exquisite illustrations, try the following key (the 'quick' key is a great idea):

Pjotr Oosterbroek (2006) *The European Families of the Diptera.* KNNV Uitgeverij.

Website

Steven Falk's collections on Flickr: https://www.flickr.com/photos/63075200@N07/collections/
For almost all of the fly families, Steven Falk's Flickr pages are one of the most important free online resources. For many families all of the species are featured, with detailed photos and informative text on each of them, as well as tips on how to avoid confusing similar species.

The following 13 groups/families have been selected to show some of the more accessible taxa across a range of scales and levels of difficulty. Collectively they cover 2,380 species, which is around a third of the UK's flies.

Bibionids (St Mark's flies): 18 species on the UKSI

There are 14 species in the genus *Bibio* and 4 species in the genus *Dilophus*. All species are often abundant where they occur. The larvae are common soil invertebrates, and diversity is highest in May and June.

Download

Paul Freeman and R.P. Lane (1985) *Bibionid and Scatopsid Flies. Diptera: Bibionidae and Scatopsidae.* Royal Entomological Society.
www.royensoc.co.uk/wp-content/uploads/2022/10/Diptera-Bibionid-and-Scatopsid-flies.pdf
Although the old RES key is somewhat out of date, if you use it in conjunction with Steven Falk's pages for the family you can usually get an identification.

Conopidae (bee-grabbers): 25 species on the UKSI

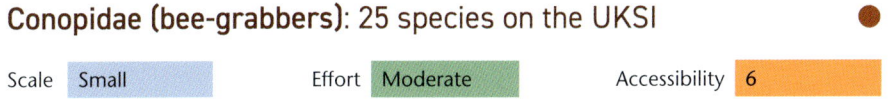

This is a small family of attractive but difficult parasitoids of the adults of various bees and wasps. They are mostly brightly coloured, Batesian mimics. The commonest species by far is *Sicus ferrugineus*, which is easily field-identifiable. There is a good key to species on Mike's Insect Keys (from 2021).

Facebook group

Conopids
This is a large global group, but with enough British and European dipterist members to be relevant.

Scathophagidae (dung flies): 54 species on the UKSI

Scale: Small | Effort: Moderate | Accessibility: 6

The well-known, very common Yellow Dung Fly *Scathophaga sterocoraria* is often seen in huge numbers around cow pats. However, the majority of the species in this family are not in fact dung feeders, but occupy a wide range of ecological niches. Species diversity is higher in the north than it is in the south.

Website

Steven Falk's Flickr collections give very good coverage of some of the most difficult species.

Download

Stuart G. Ball (2014) *Key to the British Scathophagidae (Diptera)*. Unpublished key. This excellent modern key can be downloaded for free from the recording scheme website:
www.scathophagidae.myspecies.info/files/scathophagid_key.pdf

Blow flies (Calliphoridae, Polleniidae and Rhiniidae) and woodlouse flies (Rhinophoridae): 56 species on the UKSI

Scale: Small | Effort: Moderate | Accessibility: 6

There are 39 species in the Calliphoridae, 8 species in the Polleniidae, 1 species in the Rhiniidae and 8 species in the Rhinophoridae. The majority of the Calliphoridae are associated with carrion as larvae, the Rhinophoridae (woodlouse flies) are endoparasites of woodlice, the Polleniidae (cluster flies) are earthworm parasitoids and the single member of the Rhiniidae (the Locust Blowfly *Stomorhina lunata*) is a predator of locust egg pods. The adults are most commonly found in sunny spots and on a wide range of sturdy flowering plants.

Carrion traps are a good way of attracting these flies.

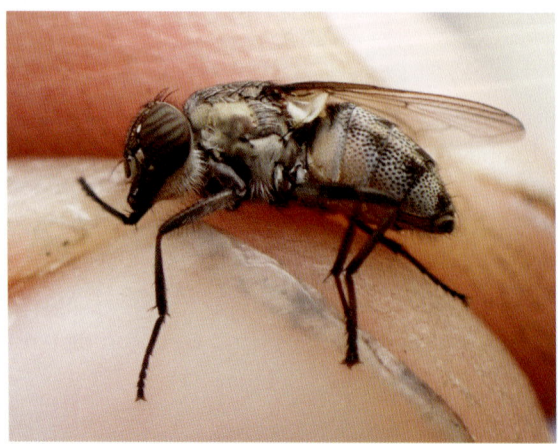

The only member of the Rhiniidae in the UK, the Locust Blowfly *Stomorhina lunata* starts life in the egg pods of locusts.

Essential text

Olga Sivell (2021) *Blow Flies (Diptera: Calliphoridae, Polleniidae, Rhiniidae)*. Royal Entomological Society.
This very good key covers all of the families in this group except the Rhinophoridae. However, the latter is covered by Steven Falk's test key (see below).

Website

www.stevenfalk.co.uk/files/21577/testkeytobritishblowflies132016.pdf
Steven Falk provides fairly comprehensive coverage of the blow flies on his Flickr pages.

Soldier flies and allies: 164 species on the UKSI

| Scale | Medium | Effort | Easy | Accessibility | 6 |

On the UKSI there are 50 species belonging to the Stratiomyidae (soldier flies), 30 species of the Tabanidae (horseflies), 29 species of the Asilidae (robberflies), 14 species of the Therevidae (stiletto flies), 15 species of the Rhagionidae (snipe flies), 11 species of the Bombyliidae (bee flies), 4 species of the Acroceridae (hunchback flies), 3 species of the Athericidae (water snipe flies), 2 species of the Scenopinidae (window flies), 3 species of the Xylophagidae (awl flies) and 3 species of the Xylomyidae (wood soldier flies). On the whole this is a fairly easy group.

Clockwise from bottom left: the glorious bee-mimic *Stratiomys longicornis* (a soldier fly), the first Sussex record of the bee-fly *Villa cingulata*, at Heyshott Down in 2016 (the species has subsequently spread rapidly throughout the county), and the large robberfly *Dysmachus trigonus*.

Essential text

Alan Stubbs and Martin Drake (2001) *British Soldierflies and Their Allies.* British Entomological and Natural History Society.
This book is essential for this group, as it covers all of the above species.

Websites

Soldierflies and Allies Recording Scheme: www.dipterists.org.uk/soldierflies-allies-scheme/home
This website contains much useful information for identification.

Steven Falk's Flickr collection covers all of these families in great detail, and these pages are a very useful way to further consolidate an identification.

Facebook group

British Soldierflies and Allies
This group is very much focused on recording. Campaigns such as Bee-fly Watch, which is part of the Soldierflies and Allies Recording Scheme, are an enjoyable way of seeing how trends in the national emergence and distribution of species can be tracked via iRecord in near real time, and how they change from year to year.

Sarcophagidae (flesh flies): 66 species on the UKSI

Scale: Small   Effort: Difficult   Accessibility: 8

In most cases you really need males to clinch an identification, particularly with species in the large genus *Sarcophaga*, but flesh flies are fairly straightforward if you use the key and match the genitalia. They are particularly easy to record by pan trapping.

Essential text

D. Povolny and J. Verves (1997) *The Flesh-Flies of Central Europe (Insecta, Diptera, Sarcophagidae).* Pfeil Verlag (out of print).

Website

Steven Falk's Flickr collection covers virtually all of the British species.

Download

Key to Adult Flesh Flies (Diptera: Sarcophagidae) of the British Isles: https://osf.io/preprints/osf/vf5r6

Sciomyzidae (snail-killing flies): 72 species on the UKSI

Scale: Small   Effort: Difficult   Accessibility: 8

The larger genera of these important, predatory wetland flies are generally a lot easier to identify if you can find a male and examine the genitalia. Steven Falk's Flickr pages are very useful here, but you will need to get hold of Stuart Ball's key in order to get to species level.

Download

Stuart Ball (2017) *Sciomyzidae (Diptera)*. Unpublished key.
  This key can be found in the members' area of the Dipterists Forum (see page 195).

Picture-winged flies (Tephritidae) and wing-wave flies (Ulidiidae): 107 species on the UKSI

| Scale | Medium | Effort | Moderate | Accessibility | 9 |

Often conflated by beginners, both of these families are phytophagous so have been grouped together here. There are around 87 species in the Tephritidae and 20 species in the Ulidiidae.

Essential texts

M. White (1988) *Tephritid Flies: Diptera: Tephritidae*. Royal Entomological Society.
  The Tephritidae are best covered by this old RES key, which can be downloaded at: www.royensoc.co.uk/wp-content/uploads/2022/01/Vol10_Part05a-White.pdf

John T. Smit (2010) *De Nederlandse boorvliegen (Tephritidae)*. Entomologische Tabellen.
  It is well worth obtaining this Dutch key.

Websites

Mike's Insect Keys are very helpful for most tephritids.

The Ulidiidae are all covered by a visual guide on Steven Falk's Flickr pages:
  www.flickr.com/photos/63075200@N07/albums/72157697220967291/

There is a recent key (produced in 2020) in the members' area of the Dipterists Forum (see page 195).

Empididae (round-headed flies): 205 species on the UKSI

| Scale | Medium | Effort | Moderate | Accessibility | 9 |

Perhaps the best resource for identifying empids is Mike's Insect Keys. Using this you can get roughly 80% to species level. The genera *Empis*, *Rhamphomyia*, *Chelifera*, *Chelipoda* and *Clinocera* are covered in their entirety.

Syrphidae (hoverflies): 303 species on the UKSI

| Scale | Medium | Effort | Moderate | Accessibility | 9 |

Most people probably start with this family of flies, as the relatively small size of the group makes it manageable. There are plenty of field-identifiable species, but many are also only identifiable at the microscope, so there is a range of ways to access the group. Hoverflies are well known and are a useful group when interpreting survey data. Their ecology is mostly quite well known, too.

During a visit to Windsor Great Park in 2010, I stumbled upon this spectacular hoverfly, *Calipobrola speciosa*. It is quite a rare beast and I have not seen it since.

Essential texts

Alan E. Stubbs and Steven J. Falk (2000) *British Hoverflies*. British Entomological and Natural History Society.
This is a great key, and a vital investment for anyone who is taking hoverflies seriously. Steven Falk's illustrations are exquisite.

Stuart Ball and Roger Morris (2024) *Hoverflies of Britain and Ireland*, 3rd edn. Princeton University Press.
This is a useful accompaniment to the above book, but I also find it useful to cross-reference the species I have keyed out with photos of the insects in the field.

Sander Bot and Frank Van de Meutter (2023) *Hoverflies of Britain and North-West Europe: A Photographic Guide.* Bloomsbury Wildlife.
This new field guide adds another dimension, with set specimens, and it also includes plenty of species that might appear in the UK in the future.

Website

Steven Falk's Flickr collection is very useful here, especially for those species in which the sexes can look quite different from one another. I always confirm identifications, especially for the larger genera, such as *Cheilosia*, with Steven's Flickr pages.

Facebook group

UK Hoverflies
This is a lively group, and it has been particularly interesting to watch Roger Morris capture records directly from the group and find many different creative ways to present data on hoverfly recording.

Neurigona quadrifasciata. You can see why they are called long-legged flies.

Dolichopodidae (long-legged flies): 323 species on the UKSI

| Scale | Medium | Effort | Difficult | Accessibility | 12 |

These very handsome flies are rapidly becoming a favourite of mine, they are fairly easy to key out, especially with males.

Essential texts

E.C.M. d'Assis Fonseca (1978) *Diptera: Orthorrhapha: Brachycera: Dolichopodidae.* Royal Entomological Society.

Martin Drake (2025) *British Dolichopodidae (Diptera).* Royal Entomological Society. This new text covers some 327 species on the British and Irish lists, including detailed keys to all species, species accounts, photographs and distributions.

Mike's Insect Keys are very useful for getting to genera.

Muscidae (house flies): 295 species on the UKSI

| Scale | Medium | Effort | Difficult | Accessibility | 12 |

This is a large family of flies and (along with the Sarcophagidae and the Tachinidae) you will need to get to know your way around the bristles of these flies in order to navigate the keys.

Essential texts

E.C.M. d'Assis Fonseca (1968) *Handbooks for the Identification of British Insects. Diptera: Cyclorrhapha: Calyptrata: Section (b) Muscidae.* Royal Entomological Society of London.

Eudasyphora cyanella, a common muscid found throughout the year. (Chris Bentley)

This RES key is very out of date, and a small number of new species have been added to the UK fauna since it was published, but it can be downloaded for free from the RES website:

www.royensoc.co.uk/wp-content/uploads/2022/01/Vol10_Part04b.pdf

Mike's Insect Keys are very useful for identifying muscids.

There are keys available in the members' area of the Dipterists Forum (see page 195).

F. Gregor, R. Rozkosny, M. Bartak and J. Vanhara (2002) *The Muscidae (Diptera) of Central Europe*. Masaryk University (out of print).
This more recent European guide covers all of the recent additions, and it is worth obtaining a second-hand copy.

Website

Steven Falk's Flickr pages provide fairly comprehensive coverage of British species, and are almost complete at the time of writing.

Craneflies (Tipulidae, Limoniidae, Trichoceridae, Cylindrotomidae, Pediciidae and Ptychopteridae): 360 species on the UKSI ● ●

| Scale | Medium | Effort | Difficult | Accessibility | 12 |

This group consists of 229 species in the family Limoniidae, 87 species in the Tipulidae, 13 species in the Trichoceridae, 4 species in the Cylindrotomidae, 20 species in the Pediciidae and 7 species in the Ptychopteridae. Males are often easier to identify than females. Craneflies are difficult to curate in a collection, as their legs get damaged and detach easily. The highest diversity tends to be found in woodlands, bogs and sites near water.

Essential text

Alan Stubbs (2021) *British Craneflies*. British Entomological and Natural History Society.
   This long-awaited text on this important group of flies has good but demanding keys, some very detailed drawings of genitalia, and images of a selection of wings.

Website

Catalogue of Craneflies of the World: https://ccw.naturalis.nl
   This international cranefly website is widely used by dipterists.

Agromyzidae (mainly tiny, leafmining flies): 426 species on the UKSI

| Scale | Medium | Effort | Difficult | Accessibility | 12 |

This large group is the fourth most speciose fly family. It is a good group for pan-species listers, as the vast majority of the species are leafminers, which means that a knowledge of the vascular plants on which the larvae feed is needed to get an identification in most cases. They are much easier to record at this stage than when they are adults, due to the extremely small size of the flies.

Websites

Agromyzidae Recording Scheme: www.agromyzidae.co.uk
   This website has many useful resources to help you to get an identification, including information on host plants, species accounts showing how difficult each species is to identify, and distribution details. There are also test keys to adults for many species, and illustrations of male genitalia for all species.

UK Fly Mines: www.ukflymines.co.uk
   This very useful site is probably the best place to start.

British Leafminers: www.leafmines.co.uk
   This site is useful for fly leafminers (as well as micro-moth leafminers)

Plant Parasites of Europe: www.bladmineerders.nl
   This Dutch site has information on all types of leafminers, not just fly mines.

Phytomyza hellebori, the only leafminer recorded on Stinking Hellebore *Helleborus foetidus*.

Tachinidae: 286 species on the UKSI

Scale: Medium   Effort: Difficult   Accessibility: 12

This is the ninth-largest family of UK flies, and all of its species are parasitoids of other invertebrates, mainly Lepidoptera but also Coleoptera, Hemiptera, Hymenoptera and even other Diptera. The larva literally eats the host alive, killing it.

Essential texts

Robert Belshaw (1993) *Tachinid Flies: Diptera: Tachinidae*. Royal Entomological Society of London.
> This is still the best key for the UK fauna, though it is a little out of date (around 40 species have been added since the key was published). A new key is currently in production.
> The RES key works best when used in conjunction with Steven Falk's Flickr pages for this family, which have photos of about 95% of the UK species, with species accounts for even more. The species are logically arranged like an online museum collection, with hints and tips on similar species, and plenty of information on ecology and distribution.

Hans-Peter Tschorsnig and Benno Herting (1994) *The Tachinids (Diptera: Tachinidae) of Central Europe: Identification Keys for the Species and Data on Distribution and Ecology*. State Museum of Natural Science, Stuttgart.

Even the largest tachinid, *Tachina grossa*, is no match for the UK's largest fly, the predatory Hornet Robberfly *Asilus crabroniformis*.

This European key is also very good, and an English translation can be downloaded from the tachinid recording scheme at www.tachinidae.org.uk/blog/downloads/

Website

The Tachinid Recording Scheme: www.tachinidae.org.uk

This site contains much useful information about identification and the UK species.

Society

Dipterists Forum: www.dipterists.org.uk/home

Membership gives you access to many of the keys mentioned earlier in this chapter, and to a whole mine of information and experience that will be invaluable to anyone who is serious about flies.

Facebook group

UK Flies

This group for flies in general is very helpful for answering queries and assisting with identification of the more difficult fly groups.

Online resources

Although they do not cover all of the fly families, Steven Falk's Flickr pages are one of the very best resources for hoverflies, soldier flies and allies, and tachinids, among others.

www.flickr.com/photos/63075200@N07/collections

6.28 Insects: Lepidoptera: butterflies: 84 species available

| Scale | Small | Effort | Very easy | Accessibility | 2 |

Collective effort = 76 (90.5%) species   **Top lister** = Dave Horton (67 species)

Vital equipment: butterfly net, binoculars (close-focus binoculars are particularly useful for Lepidoptera), camera, large pots.

This highly familiar group of large and impressive lepidopterans have bold, colourful markings that afford them a disproportionate amount of conservation and recording interest. Many of our species have had their habitat greatly restricted, and a large proportion of them are now on the IUCN Red List, with many of these in continuing national decline.

Butterflies are generally very easy to identify and are our best-known insects. A male Brimstone *Gonepteryx rhamni* is perhaps identifiable at a greater distance than any other insect. It is really only the smaller skippers, the female blues and perhaps some fritillaries that are likely to present any real difficulty with identification, along with the time wasters of the invertebrate world – white butterflies in flight!

Essential texts

Richard Lewington (2019) *Pocket Guide to the Butterflies of Great Britain and Ireland*, 2nd edn. Bloomsbury Wildlife.

Barry Henwood and Phil Sterling (2020) *Field Guide to the Caterpillars of Great Britain and Ireland*. Bloomsbury Wildlife.

Some of the best places to see a high diversity of butterflies are the chalk downs of the south-east, mature woodland with significant amounts of open space, and

Adonis Blue *Polyommatus bellargus* is locally common on chalk grassland in the south of England.

A Queen of Spain Fritillary *Issoria lathonia* on the outskirts of Brighton.

wooded heaths. I have yet to see several northern species of butterfly (Mountain Ringlet *Erebia epiphron* and Chequered Skipper *Carterocephalus palaemon*), so I will need to head north at some point if I want to 'complete the set', following on the heels of Patrick Barkham's *The Butterfly Isles*.

Although the best place to see scarce migrant butterflies is the south coast, the east and far west coasts are also well worth exploring, often for quite different species. I have been lucky enough to twitch Long-tailed Blue *Lampides boeticus* and Queen of Spain Fritillary *Issoria lathonia* in Sussex, as well as stumble upon my own Large Tortoiseshell *Nymphalis polychloros* on two occasions.

Larva of the Purple Emperor *Apatura iris*.

You are unlikely to encounter the larva of a butterfly species you have not already seen as an adult, but it does occasionally happen. I saw the larvae of Heath Fritillary *Melitaea athalia* long before I saw the adults, yet in all my years I have never seen the larvae of Silver-washed Fritillary *Argynnis paphia*, Dark Green Fritillary *Speyeria aglaja* or White Admiral *Limenitis camilla*. Generally speaking, most species are either rarely encountered without targeted searches for the larvae, or are so common as to be encountered without any effort (such as the Large White *Pieris brassicae* and the Peacock *Aglais io*).

Websites

It is worth checking to see whether there is a local Butterfly Conservation branch in your area. For example, the Sussex branch has a really active sightings page, which can be a useful tool for connecting with scarce migrant butterflies and rare resident species:

Sussex Butterfly Conservation: www.sussex-butterflies.org.uk/sightings

Facebook groups

Migrant Lepidoptera (UK and Ireland)
> This group mainly covers moths but is also helpful for migrant butterflies, and especially for its updates on suitable weather for migrants (very useful for predicting when to be out looking for butterflies and where).

Caterpillars UK… (Conservation)
> This group provides helpful support with identifying butterfly (and moth) larvae.

6.29 Insects: Lepidoptera: moths: 2,624 species available

Scale: Large   Effort: Moderate   Accessibility: 12

Collective effort = 2,302 (87.7%) species   **Top lister** = Tony Davis (1,783 species)

Vital equipment: light trap, *generator*, butterfly net, head torch, pots, *beating stick and tray*, dissecting microscope, chemicals for genitalia dissection.

Just over 2,600 moths have been recorded in the British Isles, with just over a third of these being macro-moths and just under two-thirds being micro-moths. Running a light trap in your garden is quite possibly the laziest form of natural history there is – you literally press the mains switch, go to bed, and then get up early to identify and record the catch. For most people this will probably be the main way in which they will encounter moths. However, it gets a little harder when you are light trapping on site. Recording larvae and adult moths by day is a little more involved, and is typically how I encounter them when surveying.

The macro-moths are a great group to get into, with the vast majority being field-identifiable by picture matching (there are some first-rate field guides available). Micro-moths are a little more challenging, but there are some groups and families within this disparate collection of usually (but not always) smaller species that are readily field-identifiable, while others can only be identified by examining the genitalia under the microscope.

One of just two Lappets *Gastropacha quercifolia* I have seen in over 30 years of mothing.

Mark Gurney and I stumbled upon this Broad-bordered White Underwing *Anarta melanopa* at the top of the Cairngorms in 2007.

Essential text

Various editors (1983–2014) *The Moths and Butterflies of Great Britain and Ireland.* E.J. Brill.
   The ten-volume set known as 'MBGBI' is very expensive and some volumes are really hard to get hold of, but the genitalia illustrations are excellent, and if anyone wants to tackle the micro-moths in a thorough manner, this series of books is essential.

Websites

UK Moths: www.ukmoths.org.uk
   UK Moths is one of the oldest websites. It's a very useful resource in that it has photos of moths at rest and useful information on food plants and flight times, but the text does not provide any help with identification.

British Lepidoptera: www.britishlepidoptera.weebly.com
   This website has a unique approach in that it uses very large photos of species showing a level of detail not typically available in most guides or online resources. It also includes many micrographs of genitalia.

Facebook groups

Migrant Lepidoptera (UK and Ireland)

Caterpillars UK… (Conservation)

Moths UK Flying Tonight (this group encourages people to update their sightings as soon as possible, so that they can build up a picture of what's on the wing today).

Macro-moths: 1,043 available species on the UKSI

Scale Large   Effort Very easy   Accessibility 4

Essential text

Paul Waring, Martin Townsend and Richard Lewington (2018) *Field Guide to the Moths of Great Britain and Ireland*, 3rd edn. Bloomsbury.
This book contains illustrations of all the macro-moths at rest, and the new edition includes much more up-to-date information. If you can only get one book on macro-moths, this is the one to choose.

Although there are around 15 families within the macro-moths, they are almost always tackled as a group. The largest family is the Noctuidae, with 418 species on the UKSI, followed by the Geometridae, with 333 species.

Micro-moths: 1,581 species available on the UKSI

Scale Large   Effort Moderate   Accessibility 12

Essential text

Phil Sterling, Mark Parsons and Richard Lewington (2023) *Field Guide to the Micro-moths of Great Britain and Ireland*, 2nd edn. Bloomsbury Wildlife.
The new edition of this really useful guide contains plenty of information that was not included in the first edition, and although it does not cover all of the micro-moths, it does include most of the species you are likely to encounter. It is often the only field guide I take with me into the field in summer.

Pyralids and crambids (Pyralidae and Crambidae): 256 species on the UKSI

Scale Medium   Effort Easy   Accessibility 6

Essential texts

Most but not all of these species are covered by Sterling et al. (2023).

Mark Parsons and Sean Clancy (2023) *A guide to the Pyralid and Crambid Moths of Britain and Ireland*. Atropos Publishing.
This book is a natural successor to Barry Goater's excellent *British Pyralid Moths* with photographs of species at rest, images of set specimens and distribution maps. It is essential for any experienced moth-er, especially as it includes the plethora of new pyrales that have been turning up in the UK in recent years. I am still surprised how many species in this book I do not recognise – climate change is making things interesting for entomologists.

Most of the species I encounter in the field are readily field-identifiable. The only ones I struggle with are those in the genera *Eudonia* and *Scoparia*.

The pyralid *Pyrausta ostrinalis* is really localised in Sussex, being found only on several chalk slopes along the Cuckmere Valley.

Tortrix moths (Tortricidae): 409 species on the UKSI ●

Scale Medium   Effort Moderate   Accessibility 9

The majority of the species in this large and distinctive family are field-identifiable. However, it does include some difficult genera, with a significant number of species that need to be dissected to secure identification.

Using MBGBI and Sterling et al. (2023) it should be fairly straightforward to clinch an identification in most cases. UK Moths is always a useful additional aid. By far the most frequent species I see is *Celypha lacunana*, which is readily disturbed from rest and has a long flight period, so it is worth learning this species early on.

Essential texts

J.D. Bradley, W.G. Tremewan and A. Smith (1970s) *British Tortricoid Moths: Vols I & II*. Pisces Conservation Ltd/Ray Society.
  A CD version of this Ray Society Monograph is available. I printed off the plates about 20 years ago and still use them today.

Jon Clifton and Jim Wheeler (2016) *Bird-Dropping Tortrix Moths of the British Isles*, 2nd edn. Clifton and Wheeler.
  The bird-dropping tortrixes are a disparate cluster of unrelated species that have all adopted the same strategy – if you resemble a bird-dropping, no one wants to eat you! This is a helpful guide to these species, which can be quite confusing.

The scarce tortrix *Commophila aeneana*, a bright orange tortrix with wing markings that resemble boiling tar.

Plume moths (Pterophoridae): 46 species on the UKSI

Scale: Small   Effort: Moderate   Accessibility: 6

This is a small family and most of the species are fairly easy to identify, especially if you know the identity of the food plant on which you found the moth. A small number are really common, but the majority of plume moths are scarce. The very common *Emmelina monodactyla* is ubiquitous, and is my most frequently recorded species (accounting for around a third of all of my plume records).

Essential texts

Most species are covered in Sterling et al. (2023), but a more comprehensive account of the family can be found here:

Colin Hart (2014) *British Plume Moths*. British Entomological and Natural History Society.

Moth dissection

Genitalia dissection for moths is a little more involved than that, say, for spiders (in which all of the genitalia are external) and beetles (where you simply have to hook out their aedeagi), especially if you are attempting it on a dried specimen. A description of the chemicals needed and the procedures involved is beyond the scope of this book, so the interested reader is encouraged to visit the following website:

www.mothdissection.co.uk

Light trapping

As a child I ran a Heath trap using an actinic bulb in my garden, and in my small yard in Brighton I now run the same kind of trap. However, if you have a larger garden or want to trap out in the wild, I recommend that you use a Robinson trap, ideally with a mercury vapour (MV) bulb – though it's worth noting that MV bulbs are being phased out and replaced with more energy-efficient alternatives (e.g. 20W actinic bulbs). This is very unfortunate for the study of moths in Europe, as there have been no exemptions for scientific study.

An MV Robinson is the best trap to run if you are trapping on site, but it will need a generator. Cheap but reliable generators can be purchased from suppliers of trade tools and hardware, while actinic bulbs can be run off batteries. Two possible options are:

Anglian Lepidopterist Supplies: www.angleps.com

NHBS: https://www.nhbs.com

By-catch

For the keen pan-species lister one of the best things about moth trapping is the by-catch – all of the other insect life that comes to light. Recently, I have been carrying out a biodiversity survey of the playing field next to my house. In August 2023 I ran a light trap and found two lifers in the by-catch – the mirid *Alloeotomus germanicus*, which was new to West Sussex, and the dermestid beetle, *Dermestes haemorrhoidalis*. I also found the Nationally Rare carabid *Ophonus melletii*. Most moth-ers would have ignored these seemingly insignificant invertebrates, but as Jersey Tiger *Euplagia quadripunctaria* was the moth highlight that night, the most significant records on that occasion were definitely the by-catch.

Facebook group

Moth Trap Intruders UK

This group is great for sharing photos of by-catch and getting help with identification. It gives you an insight into just how many other insects (and more) come to light.

Macro-moth early stages

These days I personally record many of my moths as larvae. The way to approach them is to pick your fights. There are plenty of dull green or brown larvae with no

Some Light Crimson Underwing larvae are amazing lichen mimics.

The Dusky Sallow *Eremobia ochroleuca* is one of my most frequently recorded moth larvae in the sweep net on base-rich grasslands.

discernible features, many of which you are unlikely to get to species, especially dull noctuids and plain green geometrids. However, the ones that have distinctive patterns and protrusions and striking colours or hairs are usually pretty easy. Things to note that will help you with an identification include the number of prolegs, colour of the spiracles, food plant, markings on the head capsule, and more subtle details (e.g. leg colour, small markings on the anal claspers, etc.).

Larvae are not easy to collect. You can put one in a large container with a good supply of its food plant and try to rear it. This is time-consuming and by no means guaranteed to work, but can be very rewarding if you have the time. Alternatively, you can get a good photo in the field. If you can get a shot of the larva from many different angles this is much more likely to help you to reach an identification.

When trying to confirm an identification from the field guides I find it helpful to search for suitable larval images online. If you are looking up a macro-moth, do try searching for the scientific name, as you'll get far more photos from European moth-ers who probably have no idea, for instance, what a Setaceous Hebrew Character is.

During the 2024 survey season I recorded two noctuids new to my list as larvae – the Wormwood *Cucullia absinthii* and Light Crimson Underwing *Catocala promissa*. Both of these are highly distinctive and attractive insects, even as larvae. I would be happy to see the adult but I have no urge to go and look for them now, as I have seen the species already.

This is a great way to add to a survey list, but it might not be the easiest way to add to your PSL list. During a standardised survey, especially in the spring on trees and shrubs, the number of moth larvae I encounter far outweighs the number of adults. Later in the year when there are more day-flying and 'disturbed-from-rest' moth encounters, that's not the case. How else are you going to get the winter-active Mottled Umber *Erannis defoliaria* (which is now listed as Vulnerable) on a site list without light trapping in the winter? As I mentioned earlier, records of larvae (or any other stage) are just as valid as records of adults, and count just as much in PSL. There are a few species where I have seen the larva but not the adult, and there is one species, Kentish Glory *Endromis versicolora*, that I have ticked after seeing neither the larva nor the adult, but the distinctive eggs.

A dozen Kentish Glory eggs.

Beating for caterpillars requires more effort than, say, beating for beetles, as many of them cling very persistently to leaves, so give two or three very hard whacks in rapid succession with your chosen implement. You can also find many larvae by sweeping or with a suction sampler, though I tend to find that the proportion of these that I can identify is much smaller compared with those beaten out of trees. Pedunculate Oak *Quercus robur* is perhaps the best species for obtaining many different species by beating, while Hawthorn *Crataegus monogyna* is often extremely productive, too.

Recording adult day-flying moths
Day-flying moths are typically recorded in a similar way to butterflies. If a closer view is needed, catch them with a butterfly net and transfer them to a container. Many moths that you will see in the daytime are not technically day flyers, but are readily disturbed from rest and will fly a short distance before settling. Netting them is usually quite straightforward in grasslands, but less so among trees and shrubs, where you often have just one chance to try and catch them before they reach cover, so you have to be reactive. Some micro-moths will 'tuck in' and drop when they land, and not fly a second time.

I have found a significant number of migrants in this way, including Bordered Straw *Heliothis peltigera*, Gem *Nycterosea obstipata*, Dewick's Plusia *Macdunnoughia confusa*, *Palpita vitrealis*, *Loxostege sticticalis*, *Tebenna micalis*, *Antigastra catalaunalis*, Small Marbled *Eublemma parva* and Purple Marbled *Eublemma ostrina*. Nowadays I see more Vestal *Rhodometra sacraria* this way than I do at light.

The great thing about recording moths in this way is that any larvae you find, as well as most adults, are likely to be breeding on the site, or at least to have got there naturally. Light traps attract moths across long distances, especially on warm nights when these insects are dispersing or migrating.

Suction samplers are a surprisingly effective way to record micro-moths, especially plume moths and scarce micro-moths such as *Scythris* and *Ochsenheimeria* species, which rarely come to light. For example, using my suction sampler I have found every single specimen of *Scythris potentillella* that has been recorded in Sussex.

Latticed Heath *Chiasmia clathrata* belongs to the family Geometridae, members of which are often readily disturbed from rest by day.

Micro-moth early stages: mines, case bearers and other larval signs

Late in the season this is another useful way to build up a pan-species list, and for many of the leafminers it is really the only way to identify them.

Looking for the larvae of case-bearing moths, such as *Coleophora* species on their food plants, can also be extremely rewarding. Many feed on only one species of plant and can be very rare. If you want to find a rare moth that only eats one food plant, try searching on large stands of that plant, as you are unlikely to find them if there are only one or two individual plants present. It's also worth looking for larval spinnings and other signs, but you will need to research these, and also the food plants, beforehand.

Coleophora pennella (left) feeds solely on Viper's Bugloss *Echium vulgare* but is hard to spot, as its case is constructed out of bits of the same plant that it feeds on. The mine of *Ectoedemia septembrella* (right) showing the lines of frass inside a leaf of Perforate St John's-wort *Hypericum perforatum*. At least it has ventilation!

Websites

British Leafminers: www.leafmines.co.uk
   This site has useful keys to species based on the host plant.

Plant Parasites of Europe (leafminers, galls and fungi): www.bladmineerders.nl
   This is an incredibly detailed site for any plant parasite you might find, it's especially useful when you're not sure where to start with an ID.

Societies

County moth groups and websites
   Joining a local moth group is a great way to learn more about moths. Many of these groups have regular indoor talks and some have regular field meetings. Most counties now have their own online atlases, which provide local information, maps and phenology charts, and are really useful for consolidating an identification.

6.30 Insects: Remaining small orders (mayflies, stoneflies, caddisflies, lacewings, scorpionflies, snakeflies, alderflies, stylops, web-spinners, fleas): 457 species available

Scale Medium   Effort Moderate   Accessibility 9

Collective effort = 329 (72.0%) species   **Top lister** = Jonty Denton (196 species)

Vital equipment: butterfly net, beating tray, beating stick, steel-rimmed pond net, *light trap, suction sampler.*

A significant proportion of these orders have aquatic larval stages, which are usually much more difficult to identify than the adults.

Mayflies (Ephemeroptera): 53 species on the UKSI

Scale Small   Effort Difficult   Accessibility 8

Mayflies can be identified fairly easily both as larvae and as adults. As with many of the invertebrate orders in this section, species-richness is often much higher in the north and west of the UK, where there are so many more fast-flowing, stony-bottomed streams than in the south-east.

The large, impressive Green Drake *Ephemera danica* is my most frequently recorded mayfly species, followed by the small, plain *Cloeon dipterum*.

Essential texts

Craig Macadam and Cyril Bennett (2010) *A Pictorial Guide to British Ephemeroptera*. Field Studies Council.

J.M. Elliot and U.H. Humpesh (1983) *A Key to the Adults of the British Ephemeroptera*. Freshwater Biological Association.

Stoneflies (Plecoptera): 35 species on the UKSI

Scale Small   Effort Moderate   Accessibility 6

I find stoneflies by beating trees and shrubs around waterways. Very few of them are field-identifiable, and most species will need to be identified by genitalia examination at the microscope.

I tend to encounter the same few species in the south-east (over two-thirds of all my stonefly records are of the common species *Nemoura cinerea*). Species diversity always seems to be higher in upland areas and in the west.

Essential text

Craig R. Macadam, Hugh B. Feeley and Jason Doe (2022) *British and Irish Stoneflies (Plecoptera): Keys to the Adults and the Larvae.* Freshwater Biological Association. This new guide replaces the old Freshwater Biological Association key. Many of the line drawings are retained, but the keys have been updated and the species accounts are much more detailed, with colour photos and distribution maps.

Caddisflies (Trichoptera): 205 species available on the UKSI

Scale	Medium	Effort	Moderate	Accessibility	9

This is the largest order in this section. As the vast majority of these species have aquatic larvae, sites such as streams, rivers, lakes or ponds are likely to have the most caddisflies. These insects are often found by sweeping waterside vegetation, beating trees and shrubs near waterways or, as happens very often, disturbing them from rest in the daytime (and then mistaking them for moths). Another really effective way of recording caddisflies is at light, as they are often the most frequent by-catch in moth traps. Very few species are field-identifiable, apart from a handful of very common ones. Most large caddisfly species and all smaller ones will require genitalia determination at the microscope, but some larger species can be examined with a hand lens in the field. Males tend to be slightly easier to identify than females.

My two most frequently recorded species, *Limnephilus auricula* and *Glyphotaelius pellucidus*, are both field-identifiable and account for 40% of all of my caddisfly records. Both are often found a long way from water, usually by beating trees and shrubs.

Glyphotaelius pellucidus, a common field-identifiable caddisfly.

Essential texts

Peter Barnard and Emma Ross (2012) *The Adult Trichoptera (Caddisflies) of Britain and Ireland*. Royal Entomological Society.

Using this book, which has very helpful keys, I can usually get almost all of the specimens I collect to species level. The only drawback is the over-simplification of some of the male genitalia diagrams, especially for the larger species.

Ian Wallace, Sharon and Peter Flint (2022) *Adult Caddis (Trichoptera) of Britain and Ireland: A Practical Guide*. Field Studies Council.

This much more recent book is best used in conjunction with the RES key. It has a different and much more visual approach to keying out species. There are plenty of photos and some good-quality diagrams of genitalia. However, if I had to choose just one text I would opt for the RES key.

In contrast to the adults, larvae are very difficult to identify, and I would suggest that you only tackle these if you have to identify them as part of an aquatic invertebrate survey. If you want to add caddisflies to your personal pan-species list, record the adults, as they are so much easier.

Website

www.trichoptera.senckenberg.science/Trichoptera%20fennoscandinavica-aktuell/introduction.htm?fbclid

This useful website includes photographs of the genitalia of most British species.

Lacewings, alderflies, scorpionflies, waxflies, snakeflies and ant-lions: 85 species on the UKSI

Scale: Small   Effort: Moderate   Accessibility: 6

Essential text

Colin W. Plant (1997) *A Key to the Adults of British Lacewings and Their Allies*. Field Studies Council.

This book is essential if you want to take the above invertebrate orders seriously, but some of the illustrations are a little oversimplified and the keys could be clearer and more concise. It is in need of an update.

Lacewings (Neuroptera): 73 species on the UKSI

Scale: Small   Effort: Moderate   Accessibility: 6

Beating is by far the best way to find lacewings (as well as scorpionflies and snakeflies). Areas with pines always seem to have more of the smaller, brown lacewings. Many species readily come to light.

It's worth learning to identify the species complex *Chrysoperla carnea* agg. early on, as it accounts for the vast majority of the individuals you will find. It makes up over 63% of all my Neuroptera records, and my second most frequently recorded species is *Micromus variegatus*, at 15%. A trip to the café at Minsmere Nature Reserve in Suffolk used to be a guaranteed way of seeing our only ant-lion, *Euroleon nostras*.

Scorpionflies (Mecoptera): 4 species on the UKSI

| Scale | Very small | Effort | Easy | Accessibility | 2 |

This small order consists of three scorpionflies and the Snow Flea *Boreus hyemalis*, and the males of all of them are readily identifiable with a hand lens.

Brock (2021) (see page 156) is a good supplement to Plant (1997), and is especially useful for the three scorpionfly species, which can be identified using their wing pattern.

Two of the three species of scorpionfly are extremely common in the summer. *Panorpa germanica* accounts for 62% of my records, and *Panorpa communis* makes up 34%, so it's worth learning these two species early on.

Snakeflies (Raphidioptera): 4 species on the UKSI

| Scale | Very small | Effort | Moderate | Accessibility | 3 |

These unusually shaped insects live behind bark as larvae, moving up to the canopy as adults. Although I find the larvae quite frequently in this manner, I only encounter the adults a couple of times a year. In fact I only have 27 records of snakeflies, 85% of which are of *Phaeostigma notata*.

The unusual Snow Flea *Boreus hyemalis* is in fact a species of wingless scorpionfly!

I used to find snakeflies quite tricky to identify, as the key in Plant (1997) is unnecessarily complicated. Brock (2021) has photographs of all four species, as well as a page showing the wing venation, so I now use this guide alone for this order.

Alderflies (Megaloptera): 3 species on the UKSI

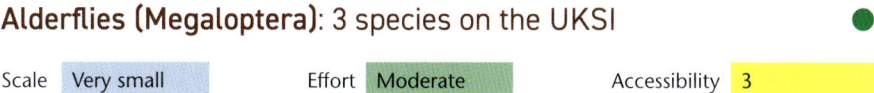

All three UK species are in the genus *Sialis*. They are found around water, and the adults are active between April and June, when they can be either netted in flight or swept or beaten from the surrounding vegetation. For all of them genitalia examination is needed to identify them to species level. Over 95% of my records are of the commonest species, *Sialis lutaria*.

Website
Lacewings and Allies Recording Scheme https://www.laars.jamesjepson.com/
This is a really useful resource, with keys to most species and images of set specimens.

Stylops (Strepsiptera): 16 species on the UKSI

Although 16 species of these strange, grotesque-looking parasites of bees are listed on the UKSI, the true number is currently the subject of debate among taxonomists, and ongoing DNA analysis is likely to reveal exactly how many species there are, and precisely which host bees they are using. The situation is not helped by how infrequently they are found. I occasionally find the adult females in samples of bees, but I have only seen the males a few times, in pitfalls collected in Scotland when I was working for the RSPB.

Essential texts
Due to the taxonomic uncertainty about species numbers there is currently no up-to-date key. Stylops were historically grouped with beetles, and are covered in some older texts, such as the following:

H. Freude, K.W. Harde and G.A. Lohse (eds) (undated) *Die Käfer Mitteleuropas, Volume 8*. Goecke & Evers.

Web-spinners (Embioptera): 2 species on the UKSI

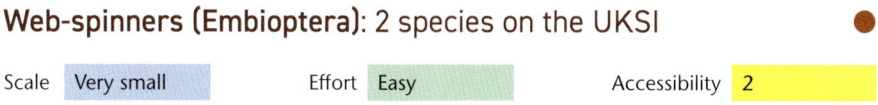

Both species listed on the UKSI are rare accidental imports. They are most probably moved around with materials as they attach themselves via webs to various substrates.

Fleas (Siphonaptera): 59 species on the UKSI

Scale: Small   Effort: Difficult   Accessibility: 8

You are most likely to find a range of fleas if you handle wild birds, bats or other mammals, as many of the species are host specific. Fleas can also be found on recently dead animals, or if you look in old bird and mammal nests. If you have a cat, put a piece of paper under prey items they bring in, and see what drops off!

Essential texts

Amoret P. Whitaker (2001) *Fleas (Siphonaptera)*. RES Handbooks for the Identification of British Insects, Volume 1, Part 16. Royal Entomological Society.

R.S. George (2008) *Atlas of the Fleas (Siphonaptera) of Britain and Ireland*. Field Studies Council.

6.31 Echinoderms (sea urchins, sea stars and sea cucumbers, etc.): 160 species available

| Scale | Medium | Effort | Easy | Accessibility | 6 |

Collective effort = 31 (19.4%) species

Top lister = Kian Hayles-Cotton (18 species)

The best way to find echinoderms is to turn over rocks in rock pools; many, but not all, species will be attached to the underside of the rocks (but do always remember to put them back how you found them). The 160 species listed on the website include 94 species of sea stars and feather stars, 22 species of sea urchins and 44 species of sea cucumbers.

Vital equipment: waterproof camera, *snorkelling/scuba diving gear, dissecting microscope.*

Essential texts

E.C. Southward and A.C. Campbell (2005) *Echinoderms*. Synopses of the British Fauna (New Series), No. 56. Field Studies Council (out of print).
This covers around 94 species (three feather stars, 21 sea stars, 17 sea urchins, 20 brittlestars and 33 sea cucumbers).

Hayward and Ryland (1990) also provide very useful coverage of echinoderms.

The Green Sea Urchin *Psammechinus miliaris* looks like someone electrocuted a Snakelocks Anemone *Anemonia viridis*.

All kinds of wrong, an adult Sea Gherkin *Pawsonia saxicola*.

Sea urchins (Echinodea): 22 species on the UKSI

Scale Small   Effort Easy   Accessibility 4

Of the 22 species listed on the UKSI, 16 species are covered by Hayward and Ryland (1990). My most frequently recorded species is the Green Sea Urchin *Psammechinus miliaris*.

Sea cucumbers (Holothurioidea): 44 species on the UKSI

Scale Small   Effort Moderate   Accessibility 6

Of the 44 species listed on the UKSI, 20 species are covered by Hayward and Ryland (1990). The Sea Gherkin *Pawsonia saxicola* is the species most frequently recorded by pan-species listers.

Spiny Starfish *Marthasterias glacialis*, a large and spectacular species found in the west, in this case on the Isles of Scilly.

Sea stars (previously known as starfish) (Asteroidea): 40 species on the UKSI

Scale Small   Effort Easy   Accessibility 4

Hayward and Ryland (1990) cover the 17 species that you are most likely to encounter, and most should be identifiable from a good photo.

Brittlestars (Ophiuroidea): 44 species on the UKSI

Scale Small   Effort Moderate   Accessibility 6

Keep a sharp lookout for brittlestars, as they can be quite small and easy to overlook. Hayward and Ryland (1990) cover around 17 species. Some species may need to be examined under the microscope to secure an identification.

Feather stars (Crinoidea): 10 species available

Scale Very small   Effort Moderate   Accessibility 3

Hayward and Ryland (1990) cover just two species.

A red Rosy Feather-star *Antedon bifida* swimming through a rock pool on the Isles of Scilly – a highlight of the trip.

6.32 Tunicates (sea squirts and salps): 134 species available ●

Scale Medium   Effort Difficult   Accessibility 12

Collective effort = 51 (38.1%) species

Top lister = Nathan Jackson (30 species)

Vital equipment: waterproof camera, *snorkelling/scuba diving gear.*

Essential texts

Sarah Bowe, Claire Goodwin, David Kipling and Bernard Picton (2018) *Sea Squirts and Sponges of Britain and Ireland.* Wild Nature Press.
  This is a really useful guide, with plenty of colour photos and useful identification tips. As the photos were taken underwater, the species will obviously look rather different from those found on the underside of a rock and removed briefly from the water, which is how most people will encounter them. The book covers around 60 species of sea squirt.

Hayward and Ryland (1990) cover around 64 species but adopt a more technical approach, including microscopic details, leading to a different collection of species being described.

Anne Bay-Nouailhat and Wilfried Bay-Nouailhat (2020) *Guide des Tuniciers de l'Europe de l'Ouest.* M&L Éditions.
  Although the text of this book is in French, it is an excellent resource, covering around 113 species, including many of the more difficult species not included in the above texts. In addition, there is a very useful photographic guide that also covers the salps.

Lightbulb Sea Squirt
Clavelina lepadiformis.

San Diego Sea Squirt *Botrylloides diegensis*, a colonial species which like all Botryllinae share a common anus – house sharing sucks! (left). Neptune's Heart Sea Squirt *Phallusia mammillata*, the largest species of sea squirt (right).

Sea squirts (Ascidiacea): 114 species on the UKSI

Scale | Medium   Effort | Difficult   Accessibility | 12

Without going under water, turning over rocks is the easiest way to find sea squirts. They can also be found under overhangs of rocks and gullies, groynes, breakwaters, bases of piers and on the underside of pontoons, so it's well worth searching a wide range of places. They are firmly attached to the substrate, which means that detaching them for a closer look is not an option. Underwater photos are the best way to see all of the details clearly. Out of the water, much of that detail is lost with sea squirts, but less so with the more brightly coloured colonial species such as the San Diego Sea Squirt.

Salps, etc. (pelagic tunicates): 23 species available

Scale | Small   Effort | Very difficult   Accessibility | 10

Salps and pyrosomes are free-floating chains of pelagic tunicates, with 12 and two species on the UKSI respectively. There are a further six species of larvacean, which are solitary pelagic species, and three species of the tiny, planktonic doliolids. There are very few records of any of these on the NBN gateway.

Essential texts

J.H. Fraser (1982) *British Pelagic Tunicates*. Linnean Society of London.
   This text is out of print but second-hand copies are available.

Bay-Nouailhat and Bay-Nouailhat (2020) also covers salps and has photographs of several species.

6.33 Fish (a paraphyletic group): 718 species available

| Scale | Medium | Effort | Easy | Accessibility | 6 |

Collective effort = 180 (25.1%) species   **Top lister** = Graeme Lyons (101 species)

Vital equipment: metal-rimmed pond net, clear container, waterproof camera, fishing rod, *shrimp net, boat, snorkelling/scuba diving gear, bathyscope.*

Freshwater fish: *c.*80 species available

| Scale | Small | Effort | Very easy | Accessibility | 2 |

Essential text

Peter S. Maitland (2004) *Keys to the Freshwater Fish of Britain and Ireland.* Freshwater Biological Association.

As a child I always had my hands in the local sandy stream, San's Brook. Bullhead *Cottus gobio* was common there under stones, and we actually tickled a Brown Trout *Salmo trutta* once, too. Later we learned that Stone Loach *Barbatula barbatula* and Brook Lamprey *Lampetra planeri* were also present (a bathyscope would come in handy here). As we got older, I added quite a list from rod and line fishing, including species such as Gudgeon *Gobio gobio,* Perch *Perca fluviatilis,* Roach *Rutilus rutilus,*

Bullhead *Cottus gobio* in a stream in Hampshire.

The only Ruffe I have ever seen, electrofished from Strumpshaw Fen. (Matt Self)

Minnow *Phoxinus phoxinus*, Dace *Leuciscus leuciscus* and Crucian Carp *Carassius carassius*.

However, the most fun you can have with freshwater fish is probably electrofishing. You might think that 240 Volts and water wouldn't mix well, but in fact, it's remarkably good fun! I did five winters of this with the RSPB as part of the EU Life project, mainly to see what food was available in reedbeds for Bittern *Botaurus stellaris* to feed on. There wasn't a huge amount of variety in this habitat but occasionally you would see something different, like the introduced Sunbleak *Leucaspius delineatus* on the Somerset Levels or a Ruffe *Gymnocephalus cernuus* in the Norfolk Broads.

Marine fish: *c.*650 species available

| Scale | Medium | Effort | Easy | Accessibility | 6 |

The vast majority of fish species in the region are marine.

Essential texts

Lin Baldock and Frances Dipper (2023) *Inshore Fishes of Britain and Ireland*. Princeton University Press.
   This new book in the Seasearch series is particularly good for divers, as most of the photos are taken under water. It has excellent coverage of complex groups, such as the gobies.

Peter Henderson (2014) *Identification Guide to the Inshore Fish of the British Isles*. Pisces Conservation.
   This is a very useful book for those who are handling fish out of the water (when they look quite different), and it also has many identification tips.

Paul Kay and Frances Dipper (2009) *A Field Guide to the Marine Fishes of Wales and Adjacent Waters*. Marine Wildlife.
   This book has some good simple line illustrations and great photos, and is useful even if you're not based in Wales.

Rock pooling is an exciting way to get a good list of intertidal fish species, especially if you target the lowest (spring) tides. There are three ways of finding fish.

1. Turning stones. This can be immensely frustrating as fish dart from one rock to the next, and in many places the same fish species, such as Shanny *Lipophrys pholis* or Rock Goby *Gobius paganellus*, are overwhelmingly dominant. At lower tides, more unusual species such as Tompot Blenny *Parablennius gattorugine* can also be found. Always remember to put any rocks (or weed) gently back in the same position as you found them.

2. Targeting fish in open water. Shoals of species such as young Sea Bass *Dicentrarchus labrax*, Sand Smelt *Atherina presbyter*, Two-spotted Goby *Gobiusculus flavescens* and various species of mullet fry will use deeper, open water in larger rock pools. Targeting these with a metal-rimmed pond net is challenging but often successful. While using this approach with a huge shoal of silvery fish in a very deep pool on Worm's Head, a tidal island just off the coast of Wales, I secured my only record of Pilchard *Sardina pilchardus*.

3. Running a net through vegetation. Deep pools and channels with plenty of long seaweed can yield some interesting finds, but don't be too rough or you will damage the weed and rocks. This method is particularly good for species of clingfish, pipefish and wrasse. In Pembrokeshire, in a deep pool that was quite high up the beach, I found three very different species in this way – Horse Mackerel *Trachurus trachurus*, Fifteen-spined Stickleback *Spinachia spinachia* and Long-spined Sea-scorpion *Taurulus bubalis*.

While rock pooling in Cornwall in 2022 I found a very live egg case of a Nursehound *Scyliorhinus stellaris*, so was pleased to add this to my list.

The beautiful Connemara Clingfish *Lepadogaster candolii* is common under rocks in rock pools on Jersey.

The Fifteen-spined Stickleback is a thing of beauty in the water.

Light rock fishing
A relatively recent trend in fishing involves targeting small fish (often overlapping with the species that can be found in rock pools) with a small rod and line from piers, sheltered coasts or rocky harbours.

Shrimp netting, dragging and raking
Pushing a shrimp net (a large, wide net) in front of you on sandy beaches can be surprisingly productive, as can dragging a metal-rimmed pond net behind you. Raking can produce sand-eel species close to a retreating tide.

Fishing further out to sea
Whether you fish directly from the shore or go out on a boat, this is a great way to add to your list. I have seen species such as Mackerel *Scomber scombrus*, Garfish *Belone belone* and Atlantic Cod *Gadus morhua* by staying close to fishermen on the shoreline, usually by being in the right place at the right time.

I have spent some memorable days volunteering with Sussex Inshore Fisheries and Conservation Authority (IFCA). We used both seine nets from the coast and small nets deployed off the back of boats. These capture small fish that you are unlikely to catch while sea fishing, as well as species that you are unlikely to find in intertidal habitats.

These species included Tub Gurnard *Chelidonichthys lucerna*, Reticulated Dragonet *Callionymus reticulatus*, Lesser Weever *Echiichthys vipera*, Hook-nose *Agonus cataphractus*, Black Goby *Gobius niger*, Dover Sole *Solea solea*, Witch *Glyptocephalus cynoglossus* and Brill *Scophthalmus rhombus* – none of which I have recorded since. There are a number of IFCAs around the coast of England, so there might be similar volunteering opportunities available in your area.

Tub Gurnard *Chelidonichthys lucerna* and a net full of other small fish from a Sussex IFCA survey.

Larger fish, such as Basking Shark *Cetorhinus maximus*, Blue-fin Tuna *Thunnus thynnus* and Sun-fish *Mola mola*, can be seen from ferries or clifftops.

Sea fishing is another way to see yet more different species. In late 2023 I paid to go as an observer on a sea-fishing boat off Brighton beach. I was very impressed by the huge specimens of Conger Eel *Conger conger* that were being pulled out, but my main reason for going on this trip was to see the Black Sea-bream *Spondyliosoma cantharus*.

Line-caught Black Sea-bream off Brighton beach.

The Undulate Ray is like a piece of Indigenous Australian art that has come to life and leapt right out of the sea!

A second trip in late 2024 was even better, yielding dozens of Common Squid, a few Common Cuttlefish *Sepia officinalis*, a Starry Smooth-hound *Mustelus asterias* and two specimens of Undulate Ray *Raja undulata*, as well as a pod of Bottle-nosed Dolphin *Tursiops truncatus*.

Facebook groups

Inshore Fishes of Britain and Ireland

Porcupine Marine Natural History Society

Rockpooling & Shrimping (British Isles)

6.34 Reptiles (a paraphyletic group): 16 species available

Scale Small   Effort Very easy   Accessibility 2

Collective effort = 13 (81.3%) species

Top listers = Chris Griffin, James Harding-Morris, Robert Jacques, John Poland, Philip Rhodes, Will Soar and Simon West (9 species each)

Vital equipment: reptile refugia, close-focus binoculars, camera with a good zoom lens.

Essential text

Howard Inns (2009) *Britain's Reptiles and Amphibians: A Guide to the Reptiles and Amphibians of Great Britain, Ireland and the Channel Islands*. WILDGuides Ltd.

We have very few reptiles, so you are never going to get a big list with this group. In fact, terrestrial natives are limited to just six species in the UK – three snakes (Adder *Vipera berus*, Grass Snake *Natrix helvetica* and Smooth Snake *Coronella austriaca*) and three lizards (Viviparous Lizard *Zootoca vivipara*, Sand Lizard *Lacerta agilis* and

Immature Grass Snake.

Western Green Lizard on the south coast of Jersey.

Slow-worm *Anguis fragilis*). This rises to seven species for the whole of the British Isles, as the Western Green Lizard *Lacerta bilineata* is native to the Channel Islands.

Sandy heaths in the south of England are the best places for our rare native species, especially Dorset, Hampshire and Surrey.

The naturalised Wall Lizard *Podarcis muralis* is now quite widespread. The large Aesculapian Snake *Zamenis longissimus* is naturalised in North Wales at Conway Bay, and also in and near the grounds of London Zoo, and along the nearby Regent's Canal.

There are small naturalised populations of the Mediterranean House Gecko *Hemidactylus turcicus* at the University of Hull and the University of Nottingham in England, and an introduced population of *Natrix natrix* persists (as our native Grass Snake was reclassified as *Natrix helvetica* when the grass snakes were split). You might be lucky enough to encounter a Leatherback Turtle *Dermochelys coriacea* on a boat trip out at sea (with a further four more marine turtle species on the UKSI), but beyond this the options are limited. The introduced population of Western Green Lizard in Dorset seems to have been collected to extinction by dealers.

Red-eared Terrapin *Trachemys scripta* does not count. This is a long-lived animal and it cannot breed in the UK – if you see one it will have been released by someone. They will not be added to the website unless breeding is proven, but by all means record them. You can also submit records to the Turtle Tally project at www.turtletally.co.uk.

6.35 Amphibians: 20 species available

Scale: Small   Effort: Easy   Accessibility: 6

Collective effort = 17 (85.0%) species   **Top lister** = John Poland (15 species)

Vital equipment: metal-rimmed pond net, clear container for viewing, *bottle traps*, torch.

Essential text

Howard Inns (2009) *Britain's Reptiles and Amphibians: A Guide to the Reptiles and Amphibians of Great Britain, Ireland and the Channel Islands.* WILDGuides Ltd.

With three native newts, two native toads and two native frogs (albeit Pool Frog *Pelophylax lessonae* is a reintroduction), this is another group with limited speciosity. Long established aliens such as Midwife Toad *Alytes obstetricans* and Marsh Frog *Pelophylax ridibundus*, along with rapidly spreading species such as Alpine Newt *Ichthyosaura alpestris*, can add to the numbers, but as with reptiles you're never going to get a huge list. On Jersey, the large Spiny Toad *Bufo spinosus* and the gracile Agile Frog *Rana dalmatina* are both native. Bottle trapping is a harmless way to monitor and record newts.

All amphibians are confined to waterbodies for breeding, but they can also be found a considerable distance away from them. Going out at night with a torch can

Midwife Toadlet.

AMPHIBIANS

Great Crested Newt caught under licence on a training course that I attended.

be a good way to spot newts in ponds, but finding rare native species (and some of the scarce introductions) will involve going to known sites at specific times of year.

You will need a licence to trap and/or handle the Great Crested Newt *Triturus cristatus*, despite the fact that it is a widespread and relatively common species.

A Ferrero Rocher plastic container is useful for amphibians, not just fish in rock pools. Here, showing the markings and colour of a male Alpine Newt.

6.36 Birds: 649 species available

Scale: Medium   Effort: Very easy   Accessibility: 2

Collective effort = 624 (96.1%) species

Top lister = Matthew Deans (565 species)

Vital equipment: binoculars, telescope, tripod.

Almost all pan-species listers have a birding background, which means that birds are the most significant 'gateway' group of all. With many of us starting as children, birding can open the door to all other natural history groups. Birds are generally very easy to identify, with just a few problematic groups, such as warblers and gulls – though these are some of the favourite species groups of many birders precisely *because* they are difficult!

Essential text

There is really only one book to recommend here, and nearly every reader probably already has it:

Lars Svensson, Killian Mullarney and Dan Zetterstrom (2023) *Collins Bird Guide*, 3rd edn. William Collins.

Twitching revisited

I made my first biological record on 23 October 1988. It was a Goldeneye *Bucephala clangula* at Blithfield Reservoir in Staffordshire. I know I started birdwatching a little before this, but it wasn't until the winter of 1988/89 that I began keeping records, this being some 13 years before I would ever hear the term 'biological recording' – I instinctively knew I needed to be doing it.

The top bird lister for Britain and Ireland seems to be Steve Gantlett, at 590 species. As there are conflicting systems for life-listing birds, some people have claimed that the 600 species mark has already been reached, but according to the British Ornithologists' Union (BOU) list that BUBO (and therefore the new PSL website) uses it has not quite been reached.

So, we can assume that if you threw everything at birding for a lifetime, the upper limit is going to be close to 600 species. I researched that figure to show birders how many more lifers you can get by getting into PSL – here you can join the 5,000 club, and not just the 500 club!

As I have hopefully already demonstrated, pan-species listing has already made significant conservation gains to the natural history of these islands, and all of that started for me with birding and twitching.

Much of the excitement of birding is about finding rare migrants oneself and going to see rare migrants that others have found – twitching is great fun and gets you to distant parts of the British Isles. Without some twitching you are unlikely to get beyond 250 species of bird. Yet some naturalists have a very negative view of twitchers and twitching, which I believe is totally unwarranted. The mean number of birds seen by the top 15 pan-species listers (some of the most accomplished naturalists I know) is 421 species. There is no way you can reach that number of birds without a significant amount of twitching. It is also clear that many great naturalists embark on their lifelong taxonomic journey by catching the twitching bug when they are young.

As I explained earlier (see page 9), pan-species listing has already brought significant conservations benefits to the natural history of these islands, and for me all of that started with birding and twitching.

Looking again at the top 15 people on the PSL rankings, all of us have made a significant contribution to conservation in the UK. Most people who care enough to keep a pan-species list are already dedicated to wildlife and its conservation. There is a need to be more understanding of different approaches. How is the listing that we do any different to writing a list for the RSPB's Big Garden Birdwatch, or writing a list of species in a survey, or managing a database of over 350,000 records? How does a list that represents a lifetime of personal effort and achievement really differ from a survey or site list? It only appears different if there is assumed to be a negative underlying motive for it.

It is my hope, that among other things, this book will demonstrate most twitchers are nothing other than passionate naturalists, who genuinely care for the natural world. Given that most pan-species listers seem to follow a route from birding, to twitching, to pan-species listing, I'd say it should never be frowned upon – encouraged even, and even better, harnessed, so that some of that energy and drive can be gently diverted towards PSL instead.

Corncrake *Crex crex* at Beachy Head in 2013. This was my last regularly breeding UK bird that I hadn't seen (excluding Scottish Crossbill *Loxia scotica*, which personally I do not believe is an actual species).

Patch listing and surveying

Patch listing is both rewarding and enjoyable, and considerably less intense than twitching as it removes the need for long journeys and the fear of dipping out. I still do a four-visit Common Birds Census at the Butcherlands rewilding project for Sussex Wildlife Trust. It is mainly about charting the changes in the territories of birds such as Skylark *Alauda arvensis* in the early days, passing to Linnet *Linaria cannabina* and Whitethroat *Curruca communis*, with Nightingale *Luscinia megarhynchos* encroaching from the edges, reaching 20 territories in 2023. Turtle Dove *Streptopelia turtur* has been doing well there, too. However, it's the unexpected finds that keep me engaged – a beautiful adult Honey-buzzard *Pernis apivorus*, a family party of Hawfinch *Coccothraustes coccothraustes* and an adult Lesser Spotted Woodpecker *Dryobates minor* feeding its young. Even a relatively bland area will over 10 years yield some great records if you visit it often enough. Then, towards the end of my time on the staff, I picked up a calling Dartford Warbler *Curruca undata*, which I initially assumed was a winter visitor on its way to the heaths to breed. Five years later that pair are still breeding, hidden away in a sea of bramble scrub, the male never feeling inclined to sing.

On my last day of the survey I picked up a warbler singing that I did not recognise. I couldn't even tell what genus it was, and had never heard a bird sing that rapidly before. It was fast, too, moving away from me across the field, hopping from one bramble bush to another. I had to run to keep up with it, but eventually it popped up right in front of me – it was a beautiful golden orb of a Melodious Warbler *Hippolais polyglotta*. That was a cracking find, it shows if you hit any area hard enough, for long enough, you'll eventually strike gold!

Sea watching

My favourite kind of birding is sea-watching. There is nothing more exciting than having a brief and tantalising view of a bird you rarely see – flying distantly by (as long as it is good enough for an ID). The excitement of trying to get your friends on to it (you've got to have a good system for this – just pointing and shouting repeatedly "it's there" isn't going to cut it, and will make you very unpopular) adds to the excitement. So, agree on a system (such as a clock face for direction, above or below the horizon, flying east or west, decide on some distance bands and have your markers agreed like buoys, fishing boats or distant structures on the horizon). If you don't have a good system, maybe *you'll* be the only one that doesn't get onto something rare flying by. Believe me, that sucks. I've been known to cling to a sea wall like a barnacle for hour-upon-hour, staring through my scope so much I get face cramp while sea-watching. It's addictive!

Getting a big bird list

I rarely twitch more than two or three times a year for birds, so the idea of suddenly diverting all my energy back to this away from my most recent obsessions of spiders, fungi, nudibranchs and sea-fishing, is just not going to happen, as I can do all these things much more locally within my counties, which also has the effect of reducing my carbon footprint. It's also the exact opposite reason why I have pushed the conservation benefits of PSL so hard and for so long, but in the interests of completion, getting a big bird list is going to involve some, or all, of the following:

- keeping a very close eye on BirdGuides and other sources of news;
- potentially moving to a birding hot spot;
- having a bunch of people that can work together to share driving and fuel costs, which will mean being quite organised as a group (I think they're known as 'friends');
- having flexible working hours;
- spending a lot of time in places such as the Isles of Scilly, Shetland and the west coast of Ireland, among others;
- a healthy disposable income.

Personally, I would much rather that people spent more time pan-species listing closer to home, finding their own species and submitting thousands of records of under-recorded groups each year.

Website

Bird Guides: www.birdguides.com

Facebook groups

Rare Birds of Britain & Ireland

UK Bird Identification

6.37 Mammals: 109 species available

Scale: Medium Effort: Easy Accessibility: 6

Collective effort = 94 (86.2%) species **Top lister** = Simon West (71 species)

Essential text

Dominic Couzens, Andy Swash, Robert Still and Jon Dunn (2017) *Britain's Mammals: A Field Guide to the Mammals of Britain and Ireland*. Princeton University Press.

Sea mammals: 37 species available

Scale: Small Effort: Easy Accessibility: 4

Vital equipment: binoculars, scope, tripod. Image-stabilised binoculars (14× magnification) are claimed to be a real game changer at sea.

Essential text

Robert Still, Hugh Harrop, Tim Stenton and Luis Dias (2019) *Europe's Sea Mammals Including the Azores, Madeira, the Canary Islands and Cape Verde: A Field Guide to the Whales, Dolphins, Porpoises and Seals*. Princeton University Press.

This male Walrus *Odobenus rosmarus* stopped briefly in Hampshire, where he behaved better than he did a few days later on New Year's Eve in Scarborough!

The best way to get a good list of cetaceans is by spending a lot of time sea watching, especially on the north and west coasts. However, they do appear elsewhere from time to time – for example, a Humpback Whale *Megaptera novaeangliae* was filmed just off the Sussex coast in December 2022 and was regularly seen in January 2025, when I was lucky enough to twitch it at Beachy Head!

Travelling on ferries is another good way to see cetaceans, though success is not guaranteed. Pelagic boat trips that are looking specifically for cetaceans and sea birds are another great way to see them.

Another increasingly popular way to see large sea mammals is to twitch them (see page 20 in Chapter 3 where I describe my experience of twitching a Walrus).

Websites

MARINElife: www.marine-life.org.uk
   Volunteering for MARINElife on ferries or smaller boats is a great way to see cetaceans, sea birds and other marine organisms while helping to contribute to the conservation of our sea life. Recent sightings from regularly watched routes are posted in the blog section of the website.

Sea Watch: www.seawatchfoundation.org.uk/recentsighting
   Details of casual sightings of marine mammals can be found on this site.

BirdGuides: www.birdguides.com
   BirdGuides often releases news on cetacean sightings.

Facebook group

Shetland Orca & Cetacean Sightings
   Although only relevant to Shetland, this is a useful group for the islands.

Bats: 24 species available

Scale: Small   Effort: Difficult   Accessibility: 8

Vital equipment: bat detector, head torch, *harp traps and acoustic lures if working under licence.*

Essential texts

Kate Jones and Allyson Wash (2001) *A Guide to British Bats*. Field Studies Council.
   This very helpful little fold-out guide is ideal for more casual recording.

Christian Dietz and Andreas Kiefer (2018) *Bats of Britain and Europe*. Bloomsbury Wildlife.
   This excellent photographic guide also contains distribution maps and sonograms. The inclusion of all the European species is very useful if you find something unusual.

Bearing in mind that all bat species are protected, and licences are required for handling, trapping and disturbing these animals, there are two very different ways in which you can record bats.

A Bechstein's Bat after it had been trapped and released.

Passive bat recording

Some bat detectors – including both hand-held ones that simply produce a sound that you then have to interpret, and static detectors that are left in place overnight but can produce sonograms back in the lab – now give a percentage score of what each species might be. I did many bat transects using hand-held detectors for the Bat Conservation Trust at Sussex Wildlife Trust, but found the sounds frustratingly difficult to interpret (in stark contrast to learning bird song and the sounds of crickets and grasshoppers). Even fairly obvious-sounding species such as Barbastelle Bat *Barbastella barbastellus* took me about five years to figure out.

Active bat recording

This is a much more involved approach, and will require a licence, so is generally only undertaken by professional bat surveyors, or by people who work for nature conservation charities or are volunteers for relevant bat groups.

While working for Sussex Wildlife Trust, one of my responsibilities was to contract bat specialists to find roosts of the rare bats on the reserves and in the wider landscape. This involved highly experienced bat specialists catching and radio-tagging bats using an acoustic lure and a harp trap. As a result, I have seen Barbastelle Bat, Bechstein's Bat *Myotis bechsteinii* and Alcathoe Bats *Myotis alcathoe* up close. I was even fortunate enough to see a Noctule Bat *Nyctalus noctula* in the hand, though it produced possibly one of the most unpleasant sounds I have ever heard made by an animal.

Another way to see bats is to help with a roost count by a local bat group. Various old tunnels and buildings around the UK are accessed during the winter months by specialised volunteers who organise standardised counts of roosting bats. During one such count I was lucky enough to see the only Greater Mouse-eared Bat *Myotis myotis* in the UK at the time.

Small mammals: 23 species available

Small mammal traps are a very useful way to record smaller rodents and shrews, though you need to get a licence to run them for catching shrews. Always make sure the impact is justified by being careful to submit your records. Squirrels, Mole *Talpa europaea* and Hedgehog *Erinaceus europaeus* are also classed as small mammals. Depending on where you live, a trip to the Lake District, Northumberland or Scotland, or to an island such as Anglesey or the Isle of Wight, is a good way to see the Red Squirrel *Sciurus vulgaris*.

Website

www.gov.uk/government/publications/shrews-licence-to-take-them

Larger mammals: 25 species available

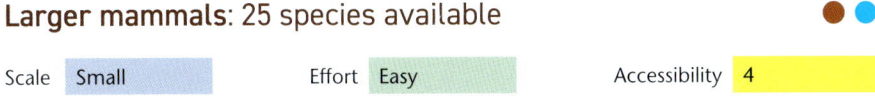

We have so few large mammals that only deer are likely to present problems with identification. In the UK we have two native species and several non-natives that are now causing significant damage to our woodlands. If you want to find some of the rarities such as Wild Cat *Felis silvestris* and Pine Marten *Martes martes* you will need to head to Scotland (but increasingly less so in the case of Pine Marten). Three feral

Hazel Dormouse *Muscardinus avellanarius* (left) is a little harder to see, as you will need to either obtain a licence to disturb them or join a licensed surveyor as a volunteer, as I did on the only occasion when I have seen this dormouse and Red Squirrel (right) on the Isle of Wight. Look at their little feets!

livestock species are also on the list – Feral Goat *Capra hircus*, Feral Sheep *Ovis aries* (such as those on St Kilda) and Feral Cattle *Bos taurus* (on Swona).

Great apes (Hominidae): 1 species available

| Scale | Very small | Effort | Very easy | Accessibility | 1 |

Don't forget to count Human *Homo sapiens*! So much of the devastation wrought by us Humans on the natural world has been a result of people perceiving themselves as better than or separate from nature. Clearly, we need to start seeing ourselves as a part of our ecosystem. Therefore, we all come with a pan-species list of one (despite only 65% of pan-species listers having seen this species).

Society

Mammal Society: www.mammal.org.uk
> This website has photos and species accounts of most native species. The benefits of membership include journals, magazines, access to discounted courses and involvement with local groups.

6.38 Other animals (nematodes, tardigrades, rotifers, other worms, hemichordates, etc.): 1,949 species available

Scale **Large** Effort **Very difficult** Accessibility **20**

Collective effort = 64 (3.3%) species   **Top lister** = Jonty Denton (25 species)

This is a 'polyphyletic ragbag group', and probably the most poorly understood group in PSL, consisting of all the animals we couldn't fit in anywhere else. They can be loosely grouped into microscopic invertebrates found in marine and freshwater habitats (such as tardigrades and rotifers) and small, obscure groups of marine invertebrates (such as lampshells and mud dragons). The largest groups are presented first, and only the groups shown in bold type in Table 8 have been described in detail.

Essential text
Hayward and Ryland (1990) are very useful for the marine phyla, but coverage is patchy.

Table 8. Groups included in the 'other animals' category.

Rank	Phylum	Common name	Number of species
1	Nematoda	Nematodes	804
2	Rotifera	Rotifers	574
3	Gastrotricha	Hairybellies	209
4	Nemertea	Ribbon worms	99
5	Tardigrada	Tardigrades or water bears	55
6	Entoprocta	Goblet worms	52
7	Kinorhyncha	Mud dragons	37
8	Chaetognatha	Arrow worms	20
9	**Brachiopoda**	Lampshells	18
10	**Hemichordata**	Acorn worms	13
11	Gnathostomulida	Gnathostomulids (marine)	12
12	Orthonectida	Mesozoans (marine)	12
13	Dicyemida	Mesozoans (marine)	7
14	Nematomorpha	Hairworms (mix)	7
15	Phoronida	Horseshoe worms (marine)	5
16	Priapulida	Priapulids (marine)	4
17	Loricifera	Loriciferans (marine)	3
18	Acanthocephala	Thorny-headed worms (marine)	3
19	Cycliophora	Cycliphorans (marine)	1

C.I. Morgan, and P.E. King (1976) *British Tardigrades*. Synopses of the British Fauna (New Series), No. 9. Academic Press (out of print).
Unfortunately, but perhaps not surprisingly, second-hand copies of this book are hard to find.

The following European keys cover most British species:

R. Bertolani (1982) *Tardigradi (Tardigrada) Guide per il riconoscimento delle specie animali delle acque interne italiane* 15 (in Italian).

H. Dastych (1988) *The Tardigrades of Poland*. Monographie Fauny Polski, 16 (in English).

Anonymous (1928) *Bärtierchen (Tardigrada)*. Tierwelt Deutschlands, 12. Gustav Fischer (in German).

Y. Séméria (2003) Tardigrades. *Faune de France*, 87 (in French).

The following key is less up to date, but free to download:

L. Cuenot (1932) Tardigrades. *Faune de France*, 24 (in French)
https://faunedefrance.org/bibliotheque/docs/L.CUENOT(FdeFr24)Tardigrades.pdf

Goblet worms (Entoprocta): 52 species on the UKSI

Scale Small   Effort Very difficult   Accessibility 10

These small, often colonial, filter-feeding animals are anchored to their substrate by a stalk.
Hayward and Ryland (1990) have a fairly limited key that covers eight species in four genera.

Mud dragons (Kinorhyncha): 52 species on the UKSI

Scale Small   Effort Unknown   Accessibility ?

These are tiny (less than 1 mm long) worm-like creatures of the marine benthic layer. They are not covered in Hayward and Ryland (1990).

Arrow worms (Chaetognatha): 20 species on the UKSI

Scale Small   Effort Very difficult   Accessibility 10

The majority of these tiny predatory worms are planktonic, although a number of species also occur in the benthic layer.

Essential text

A.C. Pierrot-Bults and K.C. Chidgey (1988) *Chaetognatha*. Synopses of the British Fauna (New Series), No. 39. Linnean Society of London (out of print).

Lampshells (Brachiopoda): 18 species on the UKSI

Scale Small   Effort Difficult   Accessibility 8

Not to be confused with the Branchiopoda, these are remarkably bivalve-like creatures of the benthic layer. Many species seem to live beyond the intertidal region, so are unlikely to be encountered when rock pooling.

Hayward and Ryland (1990) cover around half of the species on the UKSI.

APHOTOMARINE features eight species.

Acorn worms (Hemichordata): 13 species on the UKSI

Scale Small   Effort Difficult   Accessibility 8

These soft-bodied worm-like creatures typically live in soft substrates. Hayward and Ryland (1990) cover 11 species with a key.

Chapter 7

Pan-species listing in unusual habitats and specific situations

This chapter will discuss a range of scenarios, guilds and habitats that all require a slightly different approach to pan-species listing (PSL).

Galls and mines

Galls are an ideal group for PSL. Not only do you need to know the host plant species, but also galls are caused by a whole range of taxa, including fungi, beetles, mites, flies, wasps and psyllids, among others. Furthermore, many of the species that cause galls themselves fall victim to parasitoids. Other species (such as the weevil

The fly gall *Harmandiola globuli* on Aspen *Populus tremula* (top left). *Diastrophus rubi* (a wasp) on Bramble *Rubus fruticosus* agg. stems (top right). The mite *Aceria squalida* on Small Scabious *Scabiosa columbaria* (bottom left). The weevil *Gymnetron villosulum* on Pink Water-speedwell *Veronica catenata* (bottom right).

Curculio villosus) start life in a specific gall that they did not cause (in this case Oak Apple *Biorhiza pallida*).

Galls and especially mines are often much more evident later in the summer I often take large plastic bags with me for galls and mines when I'm surveying at this time of year, labelled by location. The best approach is to do your research first and then go out purposefully looking for them; for me, this is quite time-consuming, and doesn't really fit with the way I survey invertebrates (you don't get many galls in the sweep net or beating tray). Your best chance at finding any species that feeds on a specific host is to search large stands of that plant.

Essential texts

Margaret Redfern, Peter Shirley and Michael Bloxham (2023) *British Plant Galls: Identification of Galls on Plants and Fungi,* 3rd edn. Field Studies Council.

M. Chinery (2011) *Britain's Plant Galls: A Photographic Guide.* WILDGuides Ltd.

Society

British Plant Gall Society: www.britishplantgallsociety.org
 The benefits of membership include a twice-yearly magazine, access to all meetings and advice from specialists.

Facebook group

British Plant Galls (in association with BPGS)
 This is the Facebook group of the British Plant Gall Society.

Mines

There are some excellent websites to help you to identify leafminers.

Websites

British Leafminers: www.leafmines.co.uk
 This website is most useful if you are already fairly certain that you have a moth mine.

The leaf and stem mines of British flies and other insects: www.ukflymines.co.uk
 This is more fly focused, but it is also useful if you have no idea what insect is causing your mine.

Plant Parasites of Europe: www.bladmineerders.nl
 This site covers most species that might parasitise plants in some way, ranging from mites to flies to fungi.

Rock pooling

Essential text

P.J. Hayward and J.S. Ryland (1990) *The Marine Fauna of the British Isles and North-West Europe.* Oxford University Press.

If you want to see a wide range of species and also the most unusual species, you really need to target the best tides of the month, or better still of the year. The spring tides in March and September are particularly good for rock pooling. The most productive sessions occur on calm, warm, dry days. A very strong offshore wind can prevent a low tide from going out as far as predicted, and the strong ripples on the water surface can make it almost impossible to view anything in the pools. Typically, on such days, you are limited to what you can see on the underside of rocks and structures that are already emerging from the water.

Always try to work on the intertidal areas that are rarely exposed, which will be as close to the water's edge as you can get. Work on the pools you can reach that are closest to the sea (but not connected to it), or even turn stones that are still in the sea (though obviously you'll be unlikely to catch anything that swims off). If you rock pool after heavy rain, the water may well be very turbid. This can make rock pooling a real chore, as you can't even see the base of the rocks that you're turning over. Avoid turning over embedded rocks, as they will have nothing underneath them and you will just cloud the pool with black anoxic mud.

The best places for rock pooling are typically those that have some shelter from the prevailing winds. It's also worth thinking about substrate. The deep channels in the chalk at Rottingdean in Sussex, which are covered in seaweed, are quite good but very slippery. A few miles down the road in the sheltered lagoon at Holywell there is a relatively flat and firm-bottomed expanse that is a joy to work, especially as it is peppered with large stones of a size that can be turned easily.

If you wear thigh waders (I use steel toe-capped ones, so I am less worried about dropping heavy rocks on my toes) you will be able to work in deeper pools without getting wet, and you can also kneel down in shallow water, mud and weed, which is very useful if you want to take macro shots. An alternative approach I have also adopted recently is to wear a wetsuit; not only does this allow for greater flexibility but you can switch easily between snorkelling and you don't need to worry much about getting cut off on the incoming tide. A sturdy metal-rimmed net with a long stout handle can also serve as a useful balancing aid and support for when you slip, allowing you to move faster and with more confidence over slippery rocks.

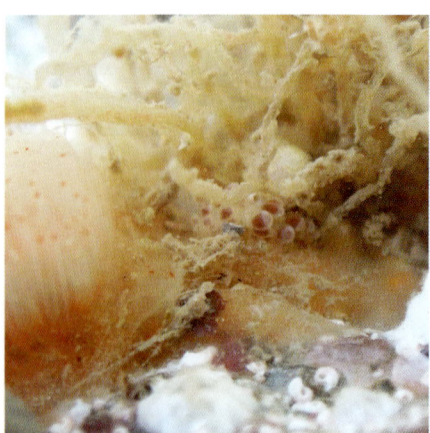 

The nudibranch *Catriona aurantia* at Menai Bridge. I searched the bases of the food source (the hydroid *Ectopleura larynx*) using close-focus binoculars until I found one. The left image is the animal tucked in under the hydroid, the right image after I encouraged it out into the open.

When you crack open the Ferrero Rocher, only to find a very rotund Tompot Blenny inside instead.

Close-focus binoculars (6.5× or 8.5× magnification) are very useful, allowing you to kneel down right next to a pool or tray of netted material and scan through the water with them, spotting things that are barely visible to the naked eye. They are also good for scanning the underside of rocks for nudibranchs and other tiny creatures. In Jersey I spotted my first ever Candy-striped Flatworm *Prostheceraeus vittatus* using this method.

Another great little tip for observing species as they appear underwater is to get hold of a 16-pack of Ferrero Rocher. First, eat the chocolates (or throw them at a rival pan-species lister). This leaves you a flexible, clear container with rounded edges, which helps to view fish submerged in water from the sides and front, such as in this image of a Tompot Blenny (as opposed to an opaque tray, viewed from above). The shallower lid does the same thing for tiny nudibranchs.

Do remember to gently put all rocks back exactly as you found them, as many plants and animals will not survive if you leave rocks upturned. Ideally you want to spend around an hour before and an hour after low tide, pushing out until you get to slack water and then working back with the tide. I always allow for three hours to be on the safe side.

Barnacle-covered rocks can be extremely sharp, so it is worth wearing workman's gloves to protect your fingertips from getting lacerated. Most important of all, don't get stranded – always keep an eye on the tide. I put my phone in a waterproof case and set an alarm for when low tide is, this allows me to relax and go into full hyper-focus mode up to that point.

Edmundsella pedata as a tiny, brightly coloured ball of jelly on the underside of a rock and underwater in fully unfurled "I can't believe that's a British species" mode.

Rock pooling at night can yield a different set of species all together, we recently found a couple of Topknots *Zeugopterus punctatus* creeping actively around in rock pools at Eastbourne, a species known to be at least partly nocturnal. Head torches are essential, while a bright waterproof torch with a UV function can illuminate even more species that you might otherwise have missed. There is more risk here, though, so I would avoid rock pooling at night on your own.

You'll need to plan your rock pooling days or hours in advance to take advantage of the next workable low tides, and weeks or months in advance to align with the best tides of the year or month. A low tide in Cornwall will be at a completely different time of day to low tide in Sussex on a given date.

Essential text

Liverpool and Irish Sea Tide Table (Year) by Laver's. Available to purchase from www.laverpublishing.com. You can adjust the tides for your region using the neat map. This will allow you to plan around the best tides of the year up to 12 months in advance.

Website

Tide Forecast: www.tide-forecast.com
   For more short-term planning, websites like this can be useful but will only allow you to look a few weeks ahead.

Aquatic invertebrates

Vital equipment: waterproof notebook, wellies/thigh waders, waterproof clothing, tweezers, alcohol, large tray or plastic sheet for sorting through samples, bucket.

Searching for aquatic invertebrates is very different to searching for terrestrial ones. A strong, long-handled, metal-framed net is essential, whether you are pond-netting or kick-sampling. These net bags don't last long if used a lot. Much of the effort from vigorous pond netting comes from your core and lower back, so try not to over-reach – using a push–pull motion is better for your back than a sweeping, side-to-side motion. Target emergent and submerged vegetation, and work backwards and forward through each patch of vegetation several times, as it often takes a few passes before you start to dislodge animals into the water column. You'll find different species in different types of vegetation, so sample a whole range of different plants in different growth forms to get the maximum number of species.

There is no point in sampling the water column for macro-invertebrates, or in sampling the mud (unless you are targeting bivalves and are equipped with a bucket and sieve to clean your sample).

Many people use a large plastic sheet for picking through large samples of vegetation, but I just use a large tray. I simply flip it over for any 'wet work', so that the inside of the tray remains dry for terrestrial invertebrate work.

A waterproof notebook and pencil are essential, and you will need fine forceps to collect invertebrates quickly before they move off the sheet or tray (they are also almost impossible to pick up with your fingers). It is important to adopt biosecurity measures to minimise the risk of spreading disease or invasive non-natives during aquatic sampling. Clean and disinfect your wellies/waders, pond net and sheet/tray before each use. This is especially important when moving between sites, as it's very easy to spread invasive species such as New Zealand Pygmyweed *Crassula helmsii*.

Microscopic invertebrates (BY BRIAN EVERSHAM)

Microscopic invertebrates are extremely diverse and abundant in freshwater (including very small waterbodies such as in tree hollows, as well as temporary pools), soil and moss. Large numbers of genera and species may be found in almost any moist or wet habitat. One fairly species-rich starting point is the specialist fauna to be found in *Sphagnum* mosses – it is possible to spend hours at the microscope studying a single drop of water from a tuft of *Sphagnum*.

If you're familiar with terrestrial invertebrates or mosses, compound microscopy may feel very different, and the literature ranges from simple illustrated introductory guides that often stop at genus level to highly technical monographs, some of which assume you have access to high-quality optics and phase-contrast illumination. However, there are a number of very useful publications that may help pan-species listers to get started.

R. Fitter and R. Manuel (1986) *Field Guide to Freshwater Life of Britain and North-West Europe.* Collins.
   This is an excellent introduction to freshwater organisms, from algae and fungi to fish, with keys and tables to major groups, and to families, genera or species. It is sometimes frustrating when it stops at genus level and fails to mention that there is only one British species in the genus, and it would be more useful if it mentioned the number of species in each group. It is illustrated with line drawings throughout, as well as many small colour photos.

W.J. Garnett (1965) *Freshwater Microscopy,* 2nd edn. Constable.
   This introduction to the biology of freshwater organisms covers both compound (slide) microscopy and organisms that can be studied at 20–40× magnification under a stereo dissecting microscope. The coverage of field techniques and microscopes is somewhat dated. The rest of the book covers algae, fungi, vascular plants, protozoans, sponges, coelenterates, platyhelminths, nematodes, annelids, rotifers, polyzoa, molluscs, crustaceans, mites, tardigrades and insects (mainly

aquatic larvae). It is illustrated throughout with black-and-white photos and a few drawings. However, there are no keys – the book mainly provides example species rather than diagnostics.

Marjorie Hingley (1993) *Microscopic life in* Sphagnum. Naturalists' Handbooks 20. Pelagic Publishing.

This very helpful introductory guide to the *Sphagnum* assemblage is also helpful for identifying unknown micro-organisms from any habitat. It includes a brief overview of a very wide range of organisms, from bacteria and single-celled algae to mites and spiders, illustrating a few of the species most often found in *Sphagnum*, including Cyanobacteria, algae (desmids, diatoms and green algae), Protozoa (flagellates, ciliates, naked and testate amoebae, heliozoans), rotifers, microturbellaria, nematodes, enchytraeids, tardigrades and gastrotrichs. There is a key to the major groups, followed by genus-level keys to a selection of flagellate protozoa, desmids, diatoms, testate amoebae and rotifers. There is also a short descriptive guide to common *Sphagnum* species, and checklists of the species in the above major groups as well as crustaceans and oribatid and other mites that have been recorded from *Sphagnum* in Britain. Usefully, it also illustrates a few other micro-objects which may cause confusion, such as *Sphagnum* spores, fungal spores, pollen and moth scales. It is well illustrated with line drawings and four colour plates.

Saline lagoons

This unusual habitat often has a very limited fauna, and the few species that are present are typically rather rare. Lagoons with widely fluctuating salinity tend to be the most uninteresting ones, as very few species can tolerate such sudden and dramatic changes in their environment. Stable lagoons are generally much more speciose, whatever their salinity level.

Submerged vegetation is typically very limited, often including only the tasselweeds, a few pondweeds and Brackish Water-crowfoot *Ranunculus baudotii*, etc., while Sea Club-rush *Bolboschoenus maritimus* is often the only emergent species. Pond netting these plants will generally yield the brackish water bugs, beetles and crustaceans. Brackish water is often quite shallow, and the benthos is usually easy to sample – species such as *Corophium volutator* can be found by taking core samples of the mud and sieving it. There are often other surprises hiding in the mud, too, such as worms, snails and crustaceans, but species diversity here is typically quite low.

R.S.K. Barnes (1994) *The Brackish-Water Fauna of Northwestern Europe*. Cambridge University Press.

This book is like an abridged version of Hayward and Ryland (1990) (see page 88), targeted at saline lagoons and brackish waterbodies, though the illustrations are very basic.

Hot houses, botanical gardens and garden centres

Essential equipment: sieve, tray, head torch (for darker corners), paint brush.

This is an area of natural history that I am not particularly excited by, as it feels too much like going to the zoo (it literally is that in some cases). Yet for the entomologist there are many valid ticks to be had – for example, naturalised non-native cockroaches, spiders, beetles, molluscs, ants and myriapods can all be found in such places. It's the entomological equivalent of the kind of botany I call 'grotany', yet some pan-species listers absolutely love it. As part of my research for this book, I took a trip to Flowerland in Buckinghamshire with Esmond Brown and Gen Popovici. I was not expecting to come away with five new species of spider, all naturalised on one site, with multiple adults of both sexes found, as well as egg sacks and immatures. I also found a stonking large jumping spider outside on a bay tree. We found only one of these, and the others hadn't seen it there before, so it's unlikely that it has naturalised yet. Therefore it won't be added to the website until evidence of a breeding population is collected.

Garden centres are a little easier to visit than hot houses, and they can be a great place for spiders. I have yet to find a garden centre that *doesn't* have the spider *Uloborus plumipes*. The messiest garden centres are the best, as they allow populations of spiders to build up over time. A very productive way to find them is to run a small paint brush round the underside of the lip of plant pots (both indoors and outside the buildings) so that you scoop up spiders in their webs.

Some interesting hot houses and garden centres in the south of England that are worth visiting include the Eden Project, RHS Garden Wisley, Whipsnade Zoo, Ventnor Botanic Garden and Flowerland but most regions will have their equivalents, so it's worth doing some research and see if you can find a site near you. There is a good chance that no one will have looked at it yet. You will always need to get permission to record in such places and be prepared to be asked a lot of questions by members of the public. In biomes, dress like you are going on holiday (those places are warm!).

Sieving the leaf litter in hot houses is generally the best way to find invertebrates, but looking under stones and logs or searching foliage can also be productive. Of course, biomes and hot houses are typically sealed environments, so the species that

Esmond Brown using a small paint brush to sample a plant pot in a garden centre (left). The large jumping spider, *Helpis oxidentalis*, that Gen Popovici found using exactly the same technique.

Uloborus plumipes looks more like a gardening tool than a spider, and seems to be present in every garden centre (left). The adult male *Icius hamatus* that I found outside on a bay tree at Flowerland (right).

find their way there are unlikely to survive and breed outside. This is in contrast to many garden centres and plant nurseries, where invertebrates can readily escape from the indoor areas or are simply introduced directly to the outdoor areas, from where it is easy to 'jump the fence' to the wider environment.

This casual approach to biosecurity is playing a significant role in spreading non-native species into the wild in the UK. I recently surveyed a large garden site in Sussex where thousands of individuals of non-native species have been planted. During that survey I recorded three invertebrates new to Sussex that had been imported with the plants – a bug, a snail and a problematic ant – all of which are now free to enter the wild, just over the garden wall. There is clearly a need for garden centres, plant nurseries, and the parks and gardens that buy these plants to take more precautions to improve biosecurity.

For the pan-species lister, such scenarios present both an opportunity to find species new to the UK, and the chance to record them and inform the relevant staff, organisations or statutory bodies about them.

If PSL did not treat all species on our lists equally, there would be very few people recording these species. Of course we value the native species above all else, and would prefer that the casual introduction of non-native species could be stopped right now, but until that time we can do our best to help to catalogue the situation.

Caves, tunnels and other subterranean habitats

Exploring these habitats is a great way to find a surprising number of species in addition to bats (but make sure you have the necessary licences for the latter). They include spiders, bristletails, woodlice, slugs, moths, mosses, butterflies and flies, among other species.

Wellies, waders and/or waterproofs are recommended, and a light but powerful head torch is the most useful bit of kit you can have in subterranean habitats, especially when you come across unexpected ones such as pill boxes, accessible hollow trees or abandoned buildings. Always keep a torch in your day bag, as you never know what you might find.

The cuddly looking *Meta menardi*, illuminated by a head torch.

There is a section of underpass in Pembrokeshire, illuminated from the ground, which I spotted from the rock pools below. I was fairly certain that a few spider species would be present, and with the aid of my head torch was able to find *Metellina merianae*, Cave Spider *Meta menardi* and *Nesticus cellulanus*, as well as Herald *Scoliopteryx libatrix* moths. People who walked along the footpath every day were completely unaware of the large Cave Spiders and their eggs sacks just a metre or so above their heads, which I had great fun showing them.

Quarries and scree

Vital equipment: knee pads or kneeling pad, tray, sieve, suction sampler, gloves (for turning rocks).

Quarries are another unique habitat and have their own specific assemblage of wildlife. They are also very under-recorded, as they are often in private ownership and with restricted access.

They are particularly good for spiders, but are also home to a range of plants and invertebrates. A suction sampler is very useful for the latter. Rocks in contact with the ground are always worth turning over, and you are more likely to find some subterranean spiders if the rocks are at least partially embedded in the soil. The most exciting part of chalk pits is the scree. Rather than just turning the top stones over, dig down further and further and eventually you will come across some of the more interesting species, including those that are typically known from caves, or rarely seen above ground. Quite a few spiders fall into this category.

Many of these sites are dangerous, and wet chalk is particularly slippery, so care is needed. Avoid working directly underneath someone else, and try not to go to these places alone (or if you do, tell someone where you are going).

Cliffs

Soft rock cliffs are especially good for a wide range of taxa. However, avoid visiting them after significant periods of heavy rain, as landslides are a real risk. There is always a risk of getting stuck in soft mud in such places too. Areas of very active erosion that have not had chance to vegetate over will not yield much, but partly vegetated areas will be more productive.

Coastal, grassy cliffs, such as those around the Lizard in Cornwall, are home to many rare species. The undercliff can often be very different from the clifftop, and again it is well worth turning over stones – those that are in contact with the soil will have different species beneath them to stones that are in contact with other rocks.

A suction sampler will be invaluable in these habitats, and does very well at acting like a third leg for support, especially when working straight up or down hill!

Safety first – looking for spiders at Kynance Cove, Cornwall. (Louis Parkerson)

Urban settings

I do most of my recording on rural sites, whether they are nature reserves, farms or rewilding projects, so my work rarely includes a significant urban element. When recording in my spare time I also tend to travel to such sites, as I like being in the wilderness, and dislike crowds and traffic. However, if you don't look for the plethora of life that's right under your nose, in urban fringes, road verges and street pavements, and in parks and town centres, you will miss out on a lot of lifers.

This is a very different kind of natural history. A lot of the plant species are aliens, with many species that don't even appear in Stace (see page 82), and rarely feature in other field guides.

You have to work a bit harder to find new things in such places, especially if you've worked them well already, but this is pan-listing at its best – forcing you to scrape the barrel. You might be surprised at just how much is in the bottom of that barrel! One of the best ways to be successful here, though, is simply to go outside and look in the first place! The '1,000 for 1KSQ' – to find this number of species within a square kilometre that includes your home – is well worth doing at least once in your life, and it might help you to find that evasive rust, leafmine or book louse.

London

Who would have thought that a trip to London would reveal itself to be such an exciting place for invertebrates, but whenever I get a gig in London, I'm always really excited, as there is a good chance I'll get some lifers. Many species new to Britain turn up in the capital, from accidental introductions due to the presence of so many people, and from natural colonisation events via the Thames Gateway from the Continent (seemingly as many species enter the UK via this route as they do from Channel hopping to the south coast).

Mating pair of *Cryptocephalus rufipes*, seen in the car park at the entrance to the London Wetland Centre in 2019.

I missed out on a first for the UK by less than two weeks back in 2019 in the car park of the London Wetland Centre (of course, Mark Telfer got it ten days before me). I recognised this smart *Cryptocephalus rufipes* as I had seen it in Jersey two years before. Crazily, there was another recent colonist in the car park that day. Two unexpected lifers, both of which were county firsts, before we had even signed in at reception!

The Channel Islands

Although they are not part of the UK, the Channel Islands are part of the British Isles and so they do count on your pan-species list! The great thing about these islands is how much they tell you about what species will be coming to the south of England in a few years' time. I only recognised *Cryptocephalus rufipes* (see previous page) at the London Wetland Centre having already picked it up in Jersey two years earlier. I have visited Jersey three times now (in 2017, 2024 and 2025), and was surprised how many bugs were present on these islands that are either not yet in England, or have since arrived.

However, by far the best thing about Jersey is the rock pooling on the south coast, particularly the area between Green Island and La Rocque. It is out of this world. In September 2024, after turning a large stone on a spring tide and seeing *five* Green Ormers *Haliotis tuberculata* for the first time, I stopped and squeezed my fists really tightly and just shouted "AGGGHHH!" due to being so overstimulated and excited! It was my highlight of the year and the best rock pooling experience I have ever had. The tantalising glimpse of an immature Spiny Squat Lobster *Galathea strigosa* just added to it.

The most important thing to note here is that spring tides can be dangerous. These tides go out for miles and then come in very quickly, so you are at risk of becoming cut off if the gullies back-fill behind you. Therefore, rather than going out as far as you can, look around you for the *lowest area* that you can see. It's counter-intuitive if you are used to rock pooling, but it works.

The tank-like Broad-shouldered Shieldbug *Cydnus aterrimus* at Les Blanches Banques, Jersey in 2017.

The spectacularly strange and unexpectedly fast Green Ormer *Haliotis tuberculata*.

Immature Spiny Squat Lobster *Galathea strigosa*. (Nicolas Jouault)

Mountains

There can be few places more exciting than being in the mountains – the extreme environment, with its thin crisp air, the challenging landscape and the thrill of perhaps seeing something few people have ever seen. Alpine flora is just glorious in every way, if you time it right too. Richard Gallon taught me that the most interesting invertebrates are to be found once you get over 750 metres, and this seems to apply to the mountain flora too, so pick a mountain taller than this to find the really good stuff. You also need to time your visit carefully to see many of these species, and have very precise grid references for them.

Although you will find some invertebrates with a suction sampler, this equipment is quite cumbersome to carry up a mountain, and on the whole you will be better off looking under rocks.

I have always been struck by the way alpine plants grow in distinct clumps, this being largely due to flushes, changes in soil chemistry or the rocks themselves. It was astonishing to see the tiny slither of accessible, available habitat on which the Snowdon Lily *Gagea serotina* grows on Yr Wyddfa (Snowdon). I never would have found it without Richard.

You have to time it right if you want to see Snowdon Lily in flower.

Chapter 8

How to become a super-naturalist: hints and tips

In this chapter I shall describe some aspects of being a naturalist and the things that help me to do it well. Although these hints and tips are very relevant to PSL, they will also help anyone who has an interest in natural history. I shall cover the following broad categories.

- biological recording
- fieldcraft
- health and well-being
- knowledge and the natural history community
- species names
- memory training and brain hacks
- entomology
- digital resources
- social media and natural history.

Biological recording

The basics

Biological recording is at the heart of PSL, and all those taking part are strongly encouraged to make biological records, to store them digitally and ideally to share them with online recording platforms, local record centres and/or recording schemes – preferably using iRecord if you are a more casual recorder. However, you will almost certainly need to do something a little different if you are carrying out formal surveys (for example, I manage my own offline Recorder 6 database).

When making a biological record, at the very minimum you need to collate the following information:

- the name of the *species* you have recorded
- the name of the *location* where you found the species, and an associated *grid reference*; this is a spatial reference, ideally using British National Grid in Britain and the Isle of Man, but using the Irish Grid in Ireland and the UTM

- (Universal Transverse Mercator) in the Channel Islands, as close to where you found it as possible
- the *date* that you recorded the species
- the name of the *person* who found and recorded it.

Other information that is not essential but improves the quality of the record includes:

- *measure of abundance* (e.g. 1 Adult Male, *c.*30 Present or Present Singing)
- the *determiner* (the person who identified it, often the same person who found it, but it is always worth clarifying this, especially when someone else has identified something for you, even if it was online)
- the *sampling technique* (e.g. sweep net, suction sampling, light trap, field observation)
- whether the records are associated with a particular *survey*
- *notes* (I use this to state clearly if my grid reference is a site centroid or an exact grid reference, etc. as there is nowhere else to put this information in Recorder 6 – you might want to have a separate field for this, to keep your 'notes' free for other things).

The following book provides much more comprehensive coverage of the world of biological recording:

Sarah Whild (2026) *The Biological Recording Handbook*. Pelagic Publishing.

Why PSL prefers iRecord to iNaturalist

As a group, we actively encourage anyone who is not managing their own database or who doesn't already have another means of submitting their records to use iRecord (unless they are based in Ireland or in a UK county that specifically promotes a different platform). iRecord is designed for the UK's recording community, with most counties having locally based volunteer experts for the majority of species groups. They verify records before they are fed back to local record centres and recording schemes. It's a great system that works well most of the time – it can occasionally be a little glitchy and appear not especially user friendly. You can sign up here: https://irecord.org.uk

iNaturalist is a global version of iRecord originating from the USA, the founders of which state that their priority is not the quality of the data but the engagement it generates. It is slicker and more user friendly, but it is not designed for the UK's natural history community. If you are using iNaturalist, it's worth taking the following steps to make your records as useful as possible:

- Change the default setting on your account from CC-BY-NC licence (non-commercial) to the standard CC-BY or CC0 licence, so that your records can be used commercially by local record centres. For example, if you want your records to be used to inform a planning application, you will need to change this default setting.
- Always set your real name as your display name. Random usernames associated with records cheapens biological recording. Setting your real name shows you

are serious about your records and helps County Recorders to recognise you, aiding the verification process. Perhaps most importantly, it fosters a feeling of pride and ownership of your records. Some people claim that usernames are safer for people who wish to be anonymous online, but this surely does not warrant usernames being the default option. In such cases, would it not be better to employ a pseudonym?

There are several other reasons why iNaturalist is not as suitable as iRecord for rigorous biological recording:

- It has default image-recognition AI at the point you add photos, which is not easy to opt out of. There is little incentive here to encourage people new to natural history and recording to identify wildlife for themselves.

- All it takes for a record to be considered correct or 'research grade' is for one other user (who may not be an expert and is likely not local) to agree with your record. With iRecord, on the other hand, the verifiers are usually County Recorders with a great deal of experience, especially within their counties. Nonetheless, iNaturalist's 'verification by democracy' – although flawed – is at least rapid, allowing for the mobilisation of records in some ways far more quickly than iRecord. Please note, though, that before records can find their way to your local record centre from iNaturalist, they do have to go through iRecord first. However, when there are no County Recorders or local experts set up to verify records in iRecord for a certain species group (such as many of the marine groups), using iNaturalist is far better than records simply sitting in limbo.

- It doesn't facilitate recorders to take agency over their own records. iNaturalist 'holds' you at a beginner level where you will be submitting photographs of common species every time you encounter them. Furthermore, I could never take, label and upload ~600 images a day when surveying, but I could easily add those records into iRecord via a spreadsheet, where they would find their way into wider use. If I had started using iNaturalist instead of my own database, there's a chance I might not have developed the skills to record as much as I do, my recording limited by how many photos I could take and the time it takes to upload and label them.

- If a record is entered into iNaturalist without a photo, it will usually not be confirmed as 'research grade'. This can give new recorders the impression that only large, field-identifiable species are worth recording. Such 'casual' observations will not find their way to iRecord, nor to County Recorders and ultimately your local records centre.

- Although this varies across – and even within – different taxonomic groups, for invertebrates especially, the use of non-standardised common names can cause a great deal of confusion for beginners, especially when they communicate with specialists.

Despite all this, iNaturalist is a very slick, well-organised and user-friendly website, able to pull together species lists in a way that really engages with global projects such as City Nature Challenge (see page 325), and it has created a real sense of community among some naturalists. iNaturalist is drawing lots of new people into observing wildlife, with many universities promoting it to students over iRecord. But I fear some people who would be better suited to iRecord are using it due to how much faster records are seemingly 'verified' or simply because they have been exposed to it first. Research-grade records do make their way to the Global Biodiversity Information Facility (GBIF), where they can be used by anyone. However, due to the issues I have outlined, and the fact that you can submit *all* records to iRecord, whether or not they have a photo, from a recording point of view I do not see what iNaturalist provides that iRecord does not do better.

Structured surveys, casual recording and incidental records

Surveys are nice and structured – you have a brief and a method, and all you have to do is stick to them. For instance, at work if I'm carrying out an invertebrate survey I stick to invertebrates (unless I hear a rare bird or find an interesting plant). However, if I go out recording for my own enjoyment (casual recording), things are a little more relaxed. I have my favourite groups (for example, I always record every spider) but I do also record quite a few other invertebrates, too.

Of course, it's not possible to record everything all of the time. Even in my big year of recording in 2024, when I made over 62,500 records, there were vast numbers of things I didn't get around to recording. I'd end up recording *Pholcus phalangioides* every time I made a coffee, Herring Gull *Larus argentatus* every time I opened the curtains and Human every time I looked in the mirror! We all have many such encounters each and every day – the reality is, even the most ardent biological recorder really only records a tiny fraction of the wildlife they identify (let alone encounter), and that is a good thing. We'd soon all be drowning in overwhelmingly repetitive and meaningless data if we recorded everything, all of the time.

Yet how *do* you decide what to record from incidental encounters? For example, on your way to the shops each day you might walk past the same patch of Hairy Bitter-cress *Cardamine hirsuta* and the same Great Tit *Parus major* singing from the same tree, after the first time you record them, they're very quickly going to get tedious and a bit pointless (quite unlike with structured surveying when these repetitive records are often really important). Yet every day you go out the door there will usually also be one or two standout records – species that you don't usually see on your daily walk to the shops, for example – and these are the ones to record. The Bloxworth Snout *Hypena obsitalis* that flies at you when you open the front door, the Lime Hawk-moth *Mimas tiliae* larva spotted running across the road, the Brambling *Fringilla montifringilla* calling nasally overhead, the Spitting Spider *Scytodes thoracica* on a neighbour's wall or the unusually late Field Grasshopper *Chorthippus brunneus* – all of these are useful records that are easy to forget if you don't capture them quickly.

On my computer I have a spreadsheet where I enter these casual records from such encounters. They include records from my back garden, records from my local patch, species I find in my house or on the way to the shops, and so on. These records soon add up and they can be surprisingly important in a local sense. If I make two or three such records a day, that's about 1,000 records over the course of a year.

Patch listing

I recently became involved with the 'friends of' group for the playing field next to my house. The area is about five hectares and consists mainly of amenity grassland, with an interesting hedge to the north and some small scattered pockets of richer habitat. About 20% of the field is fenced off as a community area. I decided to start walking the site in the mornings before working on this book. I set the site up as one big recording compartment with two sub-compartments. I can take a gentle walk around the field and be back home within 20 minutes. I wish I had started doing this when I first moved to the area, as I love having a patch to list!

On my first day there I picked up a couple of Firecrest. A few days later came the most random record of all, a male Stonechat *Saxicola rubicola* right in the middle of Brighton. Just the other day I walked around the top of the field where there is large patch of Red Dead-nettle *Lamium purpureum* in flower. I was hoping to find some bees or a bee-fly, but in fact saw three *Anthophora plumipes* males, as well as Brown Rat *Rattus norvegicus* and *Zilla diodia*, my first spider record for the site!

Something wonderful happens with patch listing. The goalposts shift and you become very excited by new things, even when they're not in fact all that unusual – what is key here is that it's the *first encounter for that area*, which is almost certainly all about dopamine. It seems that we can subconsciously 'hack' the threshold required for a given dopamine hit from a wildlife encounter. It's clearly all relative – if I was in the wider countryside I would not normally look twice at *Zilla diodia*, but the fact that it was my first spider on this list was really quite exciting.

If you can develop an emotional connection to the species that you record and/ or the location where you record them, this will help you to remember those species better. And repetition is always key to building a long-lasting knowledge base.

Sharing your data with the relevant record centres and recording schemes

If you already submit your data directly to iRecord, everything is already done for you and you can skip this section. However, if like me you manage an offline database or, for some other reason, don't or can't use iRecord, this section will be relevant.

One of the most important things to avoid is a backlog of records. During the course of writing this book I also got on top of a backlog of about 20 years of notebooks. These spanned the time period from when I first started recording up to the point when I set up my database. Although these historic records are a much lower priority for me (they are mainly birds and common garden moths, in contrast to the under-recorded groups on under-recorded sites that are now my main focus) they are still useful, and they will be lost for ever if I don't digitise them. So if you do end up with a backlog, try to enter those records as soon as you can, especially if you think they might be records that no one else has made.

I regularly share my data with my local record centre, Sussex Biodiversity Record Centre, as 72% of my records are from the two Sussex counties. However, I do have many small bundles of records from other counties that I am often slower to submit.

Some recording schemes link directly to the record centres, in which case you might not need to submit your records to them directly. I have just sent 14,500 records from two years of spider recording to the Spider Recording Scheme, but this certainly does not link to the local record centres. As a verifier for bugs, beetles and spiders in two counties, I would also make a plea for patience with your iRecord verifiers. We provide this service for free, and it can be a daunting task to keep on top of all these records as well as our own data.

Data flow is clearly an issue in the UK, and if we were building a system from scratch it would probably not look like the one we have now. However, we need to work within the one that we have, in order to avoid alienating the recording community.

How to process a backlog from old notebooks

The act of writing this book has changed my approach not only to PSL but also to biological recording. I have limited myself to about an hour a day (around 150–350 records) for working on my backlog – prioritising my Sussex records and generally making birds the lowest priority. I would recommend processing the most recent part of your backlog first, as these are generally the most useful records (though this will depend on what you have recorded and how you have recorded it) and then working *backwards* in time.

My earliest records go back to 1989, mainly birds at that stage. Moth recording commenced with me in late 1992 and plants and other things are scattered but started in the early 1990s (although I never kept such rigorous lists of plants as I did of birds and moths). I kept fastidious notebooks, right up to about the point I started blogging and keeping a database. Although I use notebooks in the field still now, they are a merely a means of getting data out of the field and into my database. So, sadly, I don't have notebooks from about 2010 onwards that I can look through and reminisce. But keeping my blog is like a notebook that everyone can see and learn from, and it is considerably more enjoyable to look through my photos than handwritten lists of species.

I have a similar approach to processing my verification backlog. I try to verify 100–200 records a day in the winter months, and in this way have verified over 25,000 records over the years. However, my best advice on managing a backlog is to never let it develop in the first place.

How to capture metadata when making hundreds of records quickly

A technique that I have recently adopted when recording in my notebooks is to use a four-coloured pen to capture data on collection methods. When writing records down quickly I find it hard to make additional notes on what method I have used, and this system really does work. I use the following colour code (which is itself a mnemonic so that I don't forget my own system):

- red – suction sampler (red is exciting and suction samplers are exciting!)
- green – beating (trees are green)
- blue – sweeping (the only colour left, following the above logic)
- black – all other methods, from direct observation to sieving (black is generic, so is saved for more generic methods and observations).

The three colours are reserved for the three large pieces of kit that I carry (suction sampler, beating tray and sweep net). This system has become second nature to me – I even click the pen to the next colour before I have changed my equipment and sampling technique!

Of course you don't have to use this system for the same collection methods, or indeed for any collection method. It could be used for any metadata that you could potentially lose if you don't put such a system in place.

Recording codes, abbreviations and other shorthand

Another way of making recording easier is to use species codes – either a bespoke system or something a bit more standardised. Mick Crawley, a prolific recorder (and quite probably the most prolific biological recorder of all time in the UK) who has submitted over 1.6 million plant records, says that when using a bespoke system he starts to think of the species in terms of their shorthand code, rather than their actual names.

The recording cards of the Botanical Society of Britain & Ireland (BSBI) adopt a standardised approach, using the first five letters of the genus and first three letters of the species (for example, *Ranunculus repens* becomes *Ranun rep*). iRecord uses a similar approach but with two letters instead of five for the genus (so *Ra rep* in this example).

As I record across many different taxonomic groups in my pan-species surveys it's not that easy for me to use these abbreviations, but I do regularly use my own. For example, I tend to use my own codes for many of the syllable-heavy vascular plants (especially the grasses and other very common plants), as this saves a lot of time when trying to write species down quickly. For example:

- AMG: Annual Meadow-grass
- SVG: Sweet Vernal Grass
- GBFT: Greater Bird's-foot-trefoil
- HWDW: Hemlock Water-dropwort.

I also draw a line down the centre of each page, effectively doubling how much I can write in each notebook. Despite my best efforts, I always tend to come home with a few species names I can barely read. To try and stop myself drifting into the illegible scrawling that happens when I get tired or overstimulated, I tell myself the following: "if you can't read it now Graeme, you won't be able to read it in a week."

The British Trust for Ornithology (BTO) has produced a set of two-digit species codes for birds. These are invaluable as they save much time and space, and are well worth learning. They can be found on the BTO website:
www.bto.org/sites/default/files/u16/downloads/forms_instructions/bto_bird_species_codes.pdf

I come up with my own 'BTO-style' codes for ubiquitous inverts with common names, such as the grasshoppers, butterflies and dragonflies, as this can save lots of time.

Vice-counties

Hewett Watson's vice-counties were originally defined in 1852, and many recording societies still use them. They were designed as a more stable alternative to the ever-shifting administrative boundaries of our actual counties, and as such they do not align with the present-day county boundaries. They are particularly useful for setting up your locations hierarchically in, for example, a Recorder 6 database.

If you want to know what vice-county your record was in – for instance, when entering historical data or when you're not on your home turf – just enter a grid reference on the following page of the BSBI website and it will tell you where you are: Botanical Society of Britain and Ireland: https://database.bsbi.org/gridref.php Whild (2026) covers vice-counties in far more detail.

Fieldcraft

Go out at odd times of the year

One of the best things about spidering is that there is so much to be found in the winter months. Sieving *Sphagnum* in the winter can be extremely productive. The City Nature Challenge (CNC) is an international initiative to get people out recording wildlife in cities, but it takes place from late April to early May, which unfortunately in the UK is not the best time for getting the longest species lists for most taxa. However, March and April are my favourite months for surveying and casual recording, mainly because hardly anyone goes out surveying for invertebrates in early spring, so it's very easy to find species that people have not found before. It's quite possibly the best time of year for spiders, too, and because there is very little foliage around to beat or sweep, a suction sampler, sieve and tray are essential.

I associate the month of March with some of my most enjoyable natural history experiences too, mainly because it's one of the few times of the year I have some free time – there is always plenty to be found on a dry, mild day at this time of year.

How to know exactly where you are (or were)

A decent hand-held Global Positioning System (GPS) is an essential item for any naturalist. This is a great way of getting spot grid references when surveying and recording, but you can also get very close to this for free on your smartphone. Free apps can sometimes be less accurate and can drain your battery, and you're unlikely to be able to put down waypoints, but they are a convenient and inexpensive way of generating spot grid references. I clip my GPS to the left shoulder of my rucksack with two carabiners so that I can read it without unclipping it. This is much quicker than repeatedly having to get my phone out and unlock the screen, which can really eat into surveying time.

Another very useful tool that I use all the time is 'Grab a Grid Reference', which is designed for biological recorders. I have set up every new compartment in my Recorder 6 database using this website, which is hosted by the Bedfordshire Natural History Society:

www.bnhs.co.uk/2019/technology/grabagridref/gagr.php

Not only can you use the aerial photographs to pinpoint where you are to the nearest tree or $10\,m^2$, but also you can see what the associated 100 and $1{,}000\,m^2$ areas look like.

I don't use Gilbert21, but some pan-species listers use it regularly. This app is basically a Dictaphone, but for every sound file that you create it saves the date and grid reference in the file name. You can then sync it straight to the cloud. There is also a desktop application that you can pull these back into, which gives you a table of dates and places. You click on each, listen to what you've said, and type it in. This is almost certainly more useful for casual recording when you don't have predetermined compartments to work with. It would work well if you need your hands free and/or you struggle to write down records at high speed.

Gilbert21: https://github.com/burkmarr/g21u

'Lyons' razor'

When it comes to 'difficult' species pairs I have come up with my own principle, which is a slight variation on 'Occam's razor'. With difficult species pairs, where one species is frequently encountered (e.g. Sparrowhawk *Accipiter nisus*) and the other species is much scarcer (e.g. Goshawk *Accipiter gentilis*), my principle is this: 'If you're not sure, it's the common one'.

Although this isn't completely foolproof, generally it has served me well over the years. Also there is a tendency towards a degree of wishful thinking, especially when you are starting out, that is somewhat reined in by this approach.

Use Lyons' razor with caution. I would hate to stop someone discovering something new to Britain by making them doubt themselves too much. This principle really does only apply when you're not sure.

The irony is, I don't even own a razor!

Tubes

I never go anywhere without plenty of containers. I tend to buy a fresh batch of small glass tubes at the start of the year (they need to be small enough for your hand lens to function), mainly for getting a close up view of tiny invertebrates, for micro-moths I can't do in the field and species I want to photograph later. While I'm surveying, I put almost anything else that I can't identify in the field into a carefully labelled airtight push-top container (I buy them in bulk from Sarstedt) with some 70% alcohol, leaving my small glass tubes for the inevitable micro-moths that need a closer look and can't be put in alcohol as they would then lose all of their scales and become virtually unidentifiable. As I use these same pots for many different functions, it's the moth scales that become problematic by late summer. Do try to keep those tubes clean if you can (or consider having separate tubes just for moths), as having to peer into a dirty tube with a hand lens can slow you down in the field.

Microscopes

Always get the best microscope you can afford, as soon as you can. A top-quality microscope is a worthwhile investment if you are in PSL for the long haul and have the funds available, but my advice would be to go for something that lets in as much light as possible, even at high magnifications. A top-quality microscope will likely cost you a few thousand pounds, rather than several hundred but is an important investment.

I grew up below the poverty line and expendable income didn't really arrive for me until I started freelancing, so it wasn't the first thing that came to mind. When I was about 30, I was kindly given my first microscope by Mark Telfer. Then a year or so later I upgraded to one that cost me about £600. That lasted me about a decade, and then I bought a second-hand Leica (again from Mark) and with an additional objective lens and some powerful eye pieces, I am now pretty much able to do anything (that a dissecting microscope can do). That lot second hand came to around £2,000 with all the eye pieces and additional lighting. I couldn't have afforded that back in 2010 but I could have a few years later; instead I struggled on, not knowing what I was missing until I made the jump.

Have a practical field system that works for you

I would rather take too much equipment with me than travel all the way to a site and then realise I had forgotten a key item (there is nothing more frustrating than losing your last writing implement when you are miles from a stationary shop – you only do that once I can tell you!). Having a system that works for you then is vital. My day pack is pretty much always packed with the same items, including now a tiny branch of Ryman stationers.

The only book I occasionally carry with me in summer is the *Field Guide to the Micro-moths of Great Britain and Ireland*. You can't put micro-moths into a tube of alcohol, so I can end up filling all my glass tubes really quickly in the summer. If I carry the guide with me, I can process a few species as I go.

Typical contents of my day bag: Camelbak®, binoculars, GPS, six Sarstedt tubes filled with 70% bioethanol, spare pens and pencils, hand lens, large emergency pot, compass, head torch, dibber, hand sanitiser, plasters, spare contact lenses, pen knife/multi-tool, antihistamines, sun block, spare camera battery, notebook, card reader, three spare lithium-ion batteries for suction sampler.

HOW TO BECOME A SUPER-NATURALIST: HINTS AND TIPS 269

Entomological utility belt (aka the bug bumbag) (left to right): drinks holder with whatever sample tube I am working on for rapid access, camera, main bag. In the main compartment of the bag are lots of small glass tubes (plus a few bigger ones and a bee spy pot). The small front compartment has pre-cut card strips, a small pencil and a sharpener for my data labels. Anything taken home live in a tube goes in one of the two small side compartments.

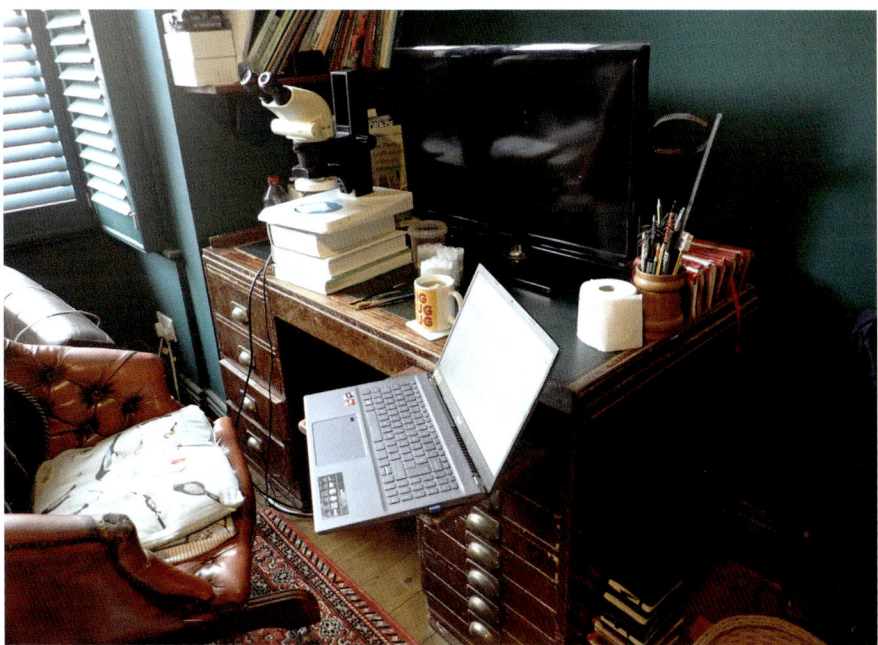

My old desk.

Army surplus equipment is often very good quality and value for money for fieldwork, especially utility belts with many different pockets, which are very useful for entomologists.

Office, desk and library

Any pan-species lister is going to need a good work space. A bright, clean study works for most. While mine is neither bright nor that clean, it works for me and I am rather fond of the lovely desk my late friend Tony Gowland gave me. Book shelves that are easy to get to and big enough to cope with a pan-species lister's ever-growing epic library are worth thinking about, as is space for the inevitable specimen collections.

Weather

As an entomologist who does not specialise only in bees, and has a busy summer of surveying, all I really need is for the weather to be fairly warm (t-shirt weather days), dry, not too windy and with limited dew.

There are many days when the weather is just right, and others when the weather forecast is atrocious and those predictions almost always prove to be correct. But what do you do on days when the forecast is borderline? I have a simple rule for this: 'If it's borderline, *always* go out'. Nine times out of ten this pays off. There is nothing worse than sitting at my desk looking at the glorious sunny day outside and regretting that I had not gone out.

There are many different weather apps and websites available now, but one of the useful things you can do in showery weather is to track storms, almost in real time, using a weather radar site. These are a really useful way to stay dry on such days. They are often much better at predicting whether a storm is going to hit your specific location than a general forecast. However, they are less good at predicting low cloud and sea fogs!

For recording insects, a warm day after rain is often ideal, but do watch out for rain overnight, as this can cause real problems the following day, especially in the field layer. Like dew, it can make the suction sampler completely ineffective, and it can sometimes take until lunchtime or longer to dry out.

Search image

Developing a search image (that is, a highly specific image in your mind's eye of what you are looking for) is almost entirely subconscious, but you can sometimes fall into a trap if your search image shifts from being an 'open' one (where you are on the alert for anything and everything) to something very specific (where you can filter out almost everything except the one species you are searching for). When this happens, you can often miss things. To me this is the natural history equivalent of driving to work and not remembering the journey.

In addition, you can sometimes start to filter out ubiquitous species. I have experienced this with the song of the Wren *Troglodytes troglodytes* on bird surveys. I almost stopped hearing these birds on several occasions. The way to overcome this is to take a few minutes to forget about what you are doing (and clear your brain's 'temporary files') and then start again. I find this works exactly the same way as when you are trying to remember a name but you keep drawing a blank, even though you know you know it, the harder you try, the further the name is from reach. You stop trying to remember it and then recall it near instantly.

One of the strangest search image experiences I have had occurred on the island of Ynys Feurig off the village of Rhosneigr on Anglesey. To get out to the small shed that was the base of operations for helping to protect the terns from predation, I had to navigate my way round hundreds of neat little nests of two or three tern's eggs. It was very stressful, as terns were constantly dive-bombing me, but I managed to be very quick and didn't stand on a single nest. After about a week of this I started seeing nests of two or three eggs when I closed my eyes at night. Later on, after all of the eggs had hatched, when I walked on to the island all I could see was nests of chicks.

This is where it got really interesting; there was a lag of about a week where, although all the eggs were all hatched, and all I was seeing were nests of chicks as I walked onto the islands, when I closed my eyes to sleep at night, *I still saw clutches of eggs*. When you make changes (even temporary ones) to your brain, they are never instant. But it flows the other way too, with search image often springing back into action years later when the relevant triggers are observed. Natural history is like sport in this sense; there is a kind of 'muscle memory' element to learning, with repetition a vitally important part of this process. I can't emphasise this point enough, repetition is everything in this game.

Notebooks in the field

To avoid losing data, never take a notebook into the field if it contains unentered data from another survey. If you are doing back-to-back fieldwork this may be impossible, as you might have to empty hundreds of records a day on returning home. I get around this by buying a pack of ten notebooks at the start of the year and use them on a rotation. If I end up with a pile of eight or nine unemptied notebooks, then I know I need to catch up on my records.

I would love it if there was voice recognition software that would turn species lists of scientific names in to an Excel file but there is currently not, however this could change soon with advances in Artificial Intelligence. Entering data directly into a database or iRecord via a smart phone, is simply not going to work for me – I am processing too many records too quickly for this to work, especially the big farm bioblitzes using the PSL approach where I can make a 1,000 records in a day. If you are just casually recording, though, you might find entering records straight into iRecord on a phone works for you.

Turning over stones and logs

Stones and logs that are in close contact with the ground but not fused to or buried in it tend to have the greatest diversity of species under them (however, you may find some subterranean species this way, especially spiders). Logs or stones that are only in partial contact with the ground are not usually worth investigating.

Sawn cross-sections of logs can be a very good place to search (especially for saproxylic species) when the sawn sections are stacked on top of each other, as many species will squeeze into the cracks between the discs of timber. You might consider arranging them like this to form a casual 'trap' if you are planning to return to the site at a later date.

Always look at the underside of the log or stone that you have turned first, not the space underneath it, as animals can fall off the underside of the log pretty quickly. You usually get a few seconds before they drop or run for cover, however, and most interesting spiders will be on the underside of the log. Finally, wherever you are, always remember to return logs and stones carefully to their original position.

Be prepared for the specimen that doesn't quite fit

Always be on the lookout for something a little different. Every first for the British Isles starts off as something that doesn't look quite right, or doesn't quite key out. It's this pool of encounters that you really have to scrutinise and it's worth giving extra time to such specimens, otherwise you could miss something new.

Health and well-being

Keep hydrated

It is vital to stay hydrated in the summer months. I use a three-litre Camelbak®, basically a large, flexible insulated plastic bladder that fits in your backpack, with a long flexible drinking tube that you can open with your mouth. It allows you to stay hydrated while on the move, without ever having to remove your back pack and stop for a drink.

They are difficult to keep clean, so are best stored in the freezer when not in use, but there is a great benefit to doing this. Fill it up the night before and put in the freezer, it will half-freeze and give you ice-cold water all day the next day, even during a heatwave. The other huge bonus here is the fact that the cold will keep your back cool on the hottest days of the year. A word of advice, though: don't freeze it for two nights or more – you'll be sucking on a block of solid ice until home time!

There is nothing worse than finding that you have drunk all your water by lunchtime on a hot day, so leave more bottles of water in your car, if you have one. I tend to fill a five-litre bottle up at the start of the week and dip into this as needed. Rehydration salts/electrolytes are also worth thinking about if you have to spend a lot of time outside in really hot weather.

Sunblock

I'm with Baz Luhrmann on this. If you are outside several times a week, every week, in the summer months you will be at higher risk of skin cancer. Always keep some factor 50 sunblock in your bag, and apply it at least twice a day, binocular straps are particularly bad at rubbing it off the back of the neck.

Ticks and Lyme disease

Tick bites can spread the bacterial infection known as Lyme disease. On sites where there will be many ticks, especially grassland or woodland with a lot of bracken, always tuck your trousers into your socks and/or use gators, and tuck your top into your trousers. I've only ever had 16 ticks attached to me in my entire life and most of these got me on my midriff or my arms; I've never had them on my legs. Take particular care when walking through bracken, especially in places like the New Forest.

Cattle

Cattle are possibly the biggest hazard you will encounter on site. Always ask the livestock managers beforehand about the temperament, age, sex, numbers and location of any livestock at the site you plan to visit. My advice would be to avoid livestock altogether if you can, and never turn your back on them. Avoid being in a field with a bull, but also be very careful around cows with calves, and never walk between them. I was once charged by a cow that was reacting to my suction sampler at a site I had thought was completely safe, with livestock I had known for years.

If you have to enter a field with cattle and they then start to approach you, don't run away, as they will undoubtedly start running after you. Most cattle, especially young bullocks, are just being inquisitive, and if you can get out of their line of site they will often quickly lose interest.

Dogs

With the growing numbers of dog walkers, partly due to an increase in dog ownership during the pandemic, the number of irresponsible dog owners (and badly behaved pets) has shot up, and this poses a significant risk to field naturalists. If you are on the ground working or taking photos, some dogs may jump all over you. Often they are just being friendly, but I have had dogs jump in my tray, or bark in my face at eye level, and I have been bitten twice. I have many times had people say "it's because you have unusual equipment" when a dog harasses me, I would argue that it's because their dog isn't under control, but if you have a lot of equipment it does seem to make dogs nervous and go on the offensive.

Simply being on the ground doing something different often makes you a target for an angry or frightened dog, and many dog walkers seem to act like this is fair, too! So if you are working on or near a footpath and you see or hear a dog walker coming, it's worth keeping them in line of sight, bracing yourself for such an encounter or moving away if you can.

If necessary your net or beating stick is a good way to physically keep an aggressive dog at a distance, while you wait for the owner to secure them.

Look after your eyes and ears

I am quite severely short-sighted, and a huge benefit to me has been the use of contact lenses for fieldwork in all weathers – I never wear my glasses for this. I update my prescription regularly so that my vision is as crisp as it possibly can be, at both short and long range. People are often surprised at how good my eyesight in the field is, but this is entirely down to how I use my eyes and a total reliance on corrective lenses.

If you go to a lot of loud gigs, even one bout of excessive noise can affect your hearing permanently. I always use ear plugs now to prevent this, as I need to have owl-like hearing for my work.

Stay physically fit

Many of my surveys regularly require walking ten miles with 15 kg of gear. To some people with underlying health conditions this might not be at all achievable, but it should be ok for most with a little training. Yet I recognise I am in a minority for even attempting this in the first place, let alone making it a core part of my working day, but it's a hugely rewarding one and is a very cheap way to stay fit and healthy. Conversely, I now spend some six months over the winter sitting on my butt and when my surveys season arrives again in early April, it's often a shock to the system. The first day is always tough, so it is important to keep active over the winter.

Ankle support is vital if you are working on slopes or rough terrain, and a decent pair of walking boots is a good investment. If you have knee problems and tackle a lot of slopes, as I do on the South Downs, it can be the downhill slopes that cause most difficulty, so keep this in mind when descending a mountain – I am often fine going up but struggle on the descent. If you have had a challenging day in the field, treat it like a workout and stretch afterwards. This will help with conditioning and allow you to get back out as soon as possible.

One-armed sweeping is vital for my method of surveying, and I need a strong right arm for doing this. Callisthenics (especially pull-ups and press-ups) can help to keep you strong, and circuit training, running or swimming help to keep your lungs and cardiovascular system in good condition, too. Bouldering can be useful practice if you ever need to do some serious climbing and maintain upper-body strength.

Look after your mental health

Being driven can be great for achieving things, but it's a double-edged sword. I can point to two particular times in my life when I wasn't enjoying my listing escapades as much as I should have been, driven more at these points by a constant need to be finding new things. This is a very real risk in pan-species listing. On both these occasions, there were actually other stresses in my life leading to this but they also both occurred when I was putting myself under huge pressure during self-imposed challenges. It's important, therefore, to maintain perspective and learn to recognise this.

This is a familiar feeling for me though when I look back. I can remember the first time we went to Norfolk with the YOC and I *didn't* get any lifers. It felt like the bottom had fallen out of my world for a while. All the other kids were playing football after returning to the camp site and I just felt bereft. Of course, I now recognise the work here of dopamine, or more importantly the absence of dopamine after it was expected. My advice on how to deal with this, is to make sure you get your dopamine from a wide range of sources and not just PSL. Keep your other hobbies going. I get dopamine from watching a new film, doing something creative, beating a personal best at callisthenics, for instance, or getting a bingo in Scrabble.

The restorative effects of being in nature are becoming increasingly well known, but perhaps less is known about the mental health benefits at the extreme end of natural history. I find that being in nature has a hugely calming effect on anxiety, and spending time in it day after day, searching for wildlife and making hundreds of records, is an intensely mindful practice, close to what is often called the 'flow state'.

Walking around 50 miles a week carrying heavy equipment is also very physically tiring, so I sleep better in the summer.

My winters could not be more different. Every winter my mental health tends to decline. I am often stuck in my study working on several projects at the same time, and though I really enjoy keying things out and writing reports, my mind has a tendency to wander – something that rarely happens in the field. As the winter progresses, so does my anxiety about my health and future. A great way to help with this is by meditating, even for just ten minutes a day, it brings an element of mindfulness into my winters that I have to work hard to achieve, yet I receive this effortlessly in the summer just by being in nature. Within a few weeks of my field season starting in April I feel restored, calm and happy again and feel less in need of meditation. It's being in nature that does this – having my ears bombarded with bird song, having my hands in the earth, the smell of freshly sieved moss on my fingers and the thrill of the unknown. Most days when I am surveying, I forget I am even a person. This is my church.

Look after your back

It's not the fieldwork that I find the most troubling, but the long winters sitting at the microscope. Good back support is clearly vital for your car seat and desk chair, but all of this goes out of the window when you lean forward to look through a microscope. Make sure that your microscope is high enough or your chair low enough. When I raised my microscope by about 10 cm by putting it on some thick books, all of my back problems stopped within three months.

Hyper-focus is great for getting long sessions of complex things done, but it's rubbish for not making you aware that you might be in discomfort, as it can literally shut out everything else. This can result in things like, not eating, drinking, stretching, or even noticing that you are in pain – so try to be distracted! Except of course, when you are out in the field, when that laser-focus will be the best friend you've ever had.

Pilates or yoga are great ways to prevent back problems from developing in the first place. It's also important to stand up and stretch, or walk around the house occasionally during those long microscope sessions. Set an alarm every hour or so to make sure you do this.

Knowledge and the natural history community

Challenge the status quo

As an example, with regard to the identification of early stages of spiders, much of the literature will tell you that you can't identify species as immatures. However, the reality is that for many families a large number of species *can* be identified in the early stages. If no one takes even a tentative first step towards doing this, the collective knowledge about this won't move forward either. This is an important part of learning, and it's good to push at the boundaries of what we know, while accepting that there are always some limitations to this approach. This approach could even be used in conjunction with DNA or by working with other such naturalists (i.e. checking each other's homework), to further consolidate those field identifications.

Knowledge and ego

We've all met territorial naturalists who have a tendency to be threatened by the next generation, or those that fiercely gatekeep their knowledge in order to maintain their relative or perceived position in the hierarchy. Don't be that person. Cherish the time you spend with people who know more than you, as these encounters will naturally get fewer and far between as you age and acquire more knowledge. There will always be someone, somewhere who knows more than you about a particular group. There is a flow of knowledge that must be maintained so that the next generation can pick up the baton. Never miss an opportunity to help this flow.

If you meet someone who knows more than you do about a particular group, yet they are younger than you, or seem to have been doing it for less time than you, rather than feeling threatened by this, accept it graciously. If they are putting in the hours and sharing their efforts and knowledge, this can only be a good thing. This does happen. I watched my friend Tylan Berry go from knowing very little about money spiders six years ago to being one of the best all-round arachnologists in the UK, becoming considerably better at money spiders than I am. I think this is utterly amazing; it really isn't something I should be threatened by, because Tylan puts the hours in. He shares his efforts and knowledge, which is brilliant. Ego can be a damaging thing in this field; it's important to check in now and again if you find yourself drifting into this unhealthy way of thinking.

The best naturalists and pan-species listers are not motivated by ego, but by a desire to see new species, learn by teaching themselves, as well as learning from and teaching others – it's a two-way flow of knowledge. It's not a good feeling when someone puts you down because they feel threatened by what you have achieved, and such behaviour impedes the flow of knowledge. Conversely, neither is it productive when someone inserts themselves higher up the hierarchy of knowledge than their experience merits on social media, something that can 'turn off' genuinely experienced naturalists from getting involved with such groups. In turn, this ultimately has a negative impact on the wildlife that we are trying to record. Fortunately, gatekeeping of knowledge seems to be happening less and less often, perhaps due to the opening up of natural history online, which has made it harder for such difficult characters to dominate their areas of interest.

Read *British Wildlife*

This publication covers all areas of conservation and natural history. Published eight times a year, it includes cutting-edge conservation and research, habitat management advice, reserve close-ups, occasional identification guides, book reviews and taxonomic summaries (in 2014, I even published an article on PSL in *British Wildlife*).

It is the most widely read natural history magazine in the UK, by amateur naturalists as well as by professional conservationists. It's my favourite publication to leave by the loo and more people in conservation read this than pretty much any other magazine or journal, so if you have something to say, this is the place to say it. It's also remarkably good value, at around £40 a year.

Join societies and groups

Many but not all of these groups have an annual membership fee, but this is more than repaid by the benefits in terms of access to field surveys, courses, online literature and other resources. I regret not having joined the British Arachnological Society much earlier than I did, as the benefits of having access to the Spider Recording Scheme dataset are so great.

Amass a library of natural history books

As a pan-species lister, the more books you have the more species you can identify, so your purchases are a good investment.

Storage and cataloguing present a genuine problem here, though, and moving house is an actual nightmare for me on this front. I have no idea how much I have spent on books over the years but, other than my house, collectively they are the most expensive things I own. Some people are happy to amass huge numbers of texts digitally. Although this might save you money, there will be many books you can't access this way, and I like giving money back to the authors. Given that you likely make less than minimum wage for writing a natural history book, who is going to write them if naturalists stop buying them? I also really enjoy flicking through field guides over and over again; I really don't like to give myself any more additional screentime than I absolutely have to.

You really do need to think about what is going to happen to all of these books when you die too. Having a well-thought-out recipient in a will for them, along with any equipment and reference collections, will likely make sure they all end up going to good use.

Annotate your books

Shock, horror – I write in my books. Make your books really work for you by jotting down in them any notes that you think will rapidly aid identification (hey, it worked for Darwin!). For the few books I am really precious about defacing, I tend to buy a second copy.

I have used some books as tick lists for many years. When I find glitches in keys, or new species are added to the UKSI, it's useful to retrofit these species into the

The sketch I made of a male *Mermessus trilobatus* palp in 2019.

keys. With so many new species turning up in the British Isles, it doesn't take long for a key to become out of date.

If you spread yourself very thinly over a large number of groups it can be difficult to keep up with the bewildering number of new recruits to our fauna. Where there are new species turning up I sometimes find it useful to draw them and insert the drawings in my books. For example, if I find a spider that I do not recognise I find it useful to draw the palp or epigyne, and if necessary post it on the relevant Facebook group in order to get an identification. This is also a great way to memorise detail – after one very intense drawing session I will have memorised that palp forever, but I still keep the sketch in my book as a useful reminder.

Give talks

PSL is all about facilitating the flow of identification knowledge from those who know more to those who know less, and often the best way to spark someone's interest is by running a course or giving a talk.

I try never to miss an opportunity to give a talk in the winter months, when fieldwork is not taking up all of my time.

Here are my tips for giving natural history presentations that allow you to disseminate your knowledge in a contemporary and engaging way:

- *Make them species heavy*. Make wildlife the focus of every presentation you give, whatever the topic. It's perfectly possible to give an engaging talk on a specific conservation subject *and* anchor it with the wildlife species that it's concerned with.

- *Limit text but go big on images*. There is nothing worse than listening to someone read out exactly the same text that is on the screen. Remember what you are going to say so that you can deliver the talk much more spontaneously, and replace the text with images of what you are talking about, with bullet points of text at most.

- *Don't be afraid of graphs and charts*. Few people are likely to be put off by graphs, and a well-prepared graph or chart can often be the highlight of a year's work, and can really help to convey quite complex ideas.

- *Don't underestimate the power of humour*. People will engage more with your talk and remember it better if you can manage to be both informative and funny. I once gave a talk about PSL to a crowded pub in Brighton for an event called 'Nerd Nite'. It's the first time I have had three pints before a presentation – it descended into standup but was probably the best talk I ever gave.

Lead and attend courses

Leading courses is a great way for you to pass on detailed knowledge and also learn yourself. You don't have to be a complete expert – you just have to know more than most of the people you are teaching.

Over the years, I have run courses on spiders, beetles, general entomology, bryophytes, fungi, habitat management for invertebrates, grasses, sedges and rushes, bird survey techniques, saline lagoon invertebrates and vegetation survey techniques. It's also important to reciprocate by attending courses, I have been on beetles, aquatic molluscs, bryophytes, springtails, and grasses, sedges and rushes.

Become a county (or national) recorder

I have been County Recorder for Heteroptera (true bugs) for about 10 years, and for spiders for about half that time (this is around 1,400 species across two counties). My main duties include:

- managing two county lists (and a combined county list) for both groups
- writing articles for the county's annual wildlife recording newsletter and helping with the national articles in *British Wildlife*
- verifying records in iRecord (I also do beetles here, though I am not County Recorder)
- occasionally checking others' specimens
- generally promoting the two groups through social media
- occasionally running field trips.

As with most things, you only get out of it what you put into it. I have never managed a national recording scheme, but essentially it's a scaled-up version of being a County Recorder – managing a greater number of records and possibly even an online mapping package.

An active County Recorder is far better than no recorder at all, so if an opportunity arises, and you have a fairly good knowledge already, even if you are not as expert as you'd like to be, take it. Remember that everyone has to start somewhere with this, and no one is born into the role. Enthusiasm and a willingness to learn, along with other relevant skills such as IT or data handling skills, could be enough to get you started.

You can always take your foot off the gas when life gets difficult with such voluntary roles. With all such roles, be very careful not to over-commit, something easier said than done for the keen naturalist who finds it difficult to say no!

What to do when you find something new

All of the following scenarios involve making and submitting a sound biological record:

- *Finding something new to you*. Put it on your list on www.panspecieslisting. com of course, and if it's interesting enough and you feel inclined, share the news on social media and learn a bit more in the process.
- *Finding something new to a site*. Let the landowner know, and also the County Recorder if it's a significantly unusual or rare species. You can also share the news on social media.
- *Finding something new to the county*. You are likely to have been in touch with the relevant County Recorder already if you know that something is new to

the county, but if you haven't, let them know. As one of their key roles is to keep track of what species the county has, they will be very grateful to receive such records. When you reach 'peak natural history', you might be surprised to find such encounters can happen monthly, if not weekly.

- *Finding something new to Britain.* This will involve confirmation from the national experts on the taxonomic group to which the species belongs, as well as all of the above actions. Such species should be written up in the appropriate national newsletter or journal. This will enable others to learn from your finds and will also prevent other naturalists from incorrectly claiming the species as their own find. Eventually your discovery will need to be added to the UK Species Inventory, too.

- *Finding something new to science.* This will involve finding the relevant taxonomic expert for the species group or family, and helping them to get it described, named and coded, in addition to all of the above actions.

Conservation status

Conservation status is used to quantify and qualify just how rare a species is. It is a great tool for assessing assemblages of species, and is probably most useful to entomologists and arachnologists, as they have so many species to process.

There are several different systems of designation (running in parallel) that are used to assess the conservation status of species. Many invertebrate orders and families have not been assessed for many years and are long overdue an update, whereas some have been assessed much more recently.

The IUCN Red List is essentially a measure of threat. Although it is now mainly a quantitative assessment, it does include a small qualitative element. A species that is not considered likely to be under threat any time soon will be listed as 'Least Concern' (LC). In order of increasing threat level, the conservation status categories for species on the Red List are as follows:

- NE – Not Evaluated (this category includes recent colonists and non-native species, though there is a strong argument that the former should be evaluated as if they were native)
- DD – Data Deficient
- NT – Near Threatened
- VU – Vulnerable
- EN – Endangered
- CR – Critically Endangered (this category can include species that are possibly extinct)
- EW – Extinct in the Wild
- EX – Extinct.

Rarity is calculated using two categories which relate to the quantitative aspect of species distributions:

- Nationally Scarce – species which have been recorded from 16–100 hectads (10 × 10 km squares) in Britain (the old status categories of Nationally Scarce A and Nationally Scarce B have been grouped together in this new format)
- Nationally Rare – species which have been recorded from 15 or fewer hectads in Britain.

For England only, there are also Section 41 (which used to be known as the Biodiversity Action Plan or BAP) species. This can be very useful. For example, Brown-banded Carder Bee *Bombus humilis* is an uncommon bee that falls into this category, and as of 2022 the Dingy Skipper has lost its Vulnerable status (and is clearly too widespread to be Nationally Scarce), leaving it only with its Section 41 status. Such a status is valuable, as this is a good positive indicator species that is easy to lose on a site through inappropriate grazing or management. Both of these species have fallen through the gaps between the IUCN Red List System and the Nationally Rare/Nationally Scarce hectad count system, but both are useful positive indicators that deserve some recognition. The legislation around Section 41 species changes very infrequently, which has resulted in these species still having some formal designation which helps entomologists to classify how good a site is. However, the converse can also apply. A species that is now much commoner than when it was first assessed might lose both its IUCN Red List status and its Nationally Rare/Nationally Scarce status, but it will retain its Section 41 status indefinitely. From a legal point of view, Section 41 species actually carry more weight than the other conservation status systems. Equivalent systems are used in Ireland, Wales and Scotland.

If you are going to disturb any species on Schedule 5 you will need a licence from Natural England in England, or from NatureScot in Scotland, Nature Resources Wales in Wales, and Northern Ireland Environment Agency in Northern Ireland.

If you can memorise the conservation status of species this will be very useful when you're out in the field with others. This might seem like highly pointless information to fill your brain with on top of all the ID stuff but if your mind is like mine, networks of data (such as parallel overlapping systems of information) help cement everything together – like the steel in reinforced concrete. You can instantly show trainees, colleagues or clients exactly which of the species you have found are rare, and also provide some context about just how rare they are, and how up to date the conservation statuses are. It also makes report writing easier, especially when you are compiling species accounts, as well as making blogs and articles more informative.

Learn the ecology of each species

Of course, memorising their conservation status is just one way to cement knowledge about species. Discovering what each species feeds on is a great method to learn how to read the countryside as an entomologist – in much the same way as learning which plant species need acidic soils and which ones need calcareous soils helps you to read it as a botanist. All of this information helps you to become a competent field ecologist.

The more additional information you have about each species, the more you can build networks of knowledge that you can use to interpret the results of survey data analysis. This in turn will equip you to explain far more comprehensively what a landowner can do to make improvements.

Join the network of bird observatories around the UK

Bird observatories are invaluable for wildlife recording and the continuity of natural history knowledge in the UK. As many organisations have taken on additional areas of work in recent decades, wildlife recording has inevitably become a smaller proportion of their output, while the remit of bird observatories has remained fairly static – to record wildlife.

Down here in the south-east of the UK, Dungeness Bird Observatory, Sandwich Bay Bird Observatory and Portland Bird Observatory are all fantastic sites for seeing some very rare wildlife, and they all have a warm and welcoming atmosphere.

Have no fear

Some of the most common mistakes you can make with keys include:

- making an incorrect assumption early on
- inverting a couplet and misreading which way to go
- assuming the key was correct when it wasn't.

One of the best ways to avoid the first two errors is to read through the keys twice, with a brief pause for reset after the first reading. However, this might be easier said than done when you are working through a lot of samples.

One of the most important things to remember is that any key, no matter how difficult, was written by other people, who are also fallible. Sometimes the illustrations need refreshing or updating, sometimes additional information is needed in order to key things out, and sometimes the keys themselves are wrong. Don't despair, though – the PSL community are very approachable and asking for help is strongly encouraged.

Share your knowledge and findings online

One key aspect of the PSL community is sharing what you find. You can do this in several ways.

- Writing a blog or a social media update about something new you have found helps you remember it and its salient features;
- If you are writing in real time, it gives others the opportunity to see what you have seen;
- Or even better, it encourages people to go and find their own records of those species, in different locations;
- By being part of the online natural history community, newcomers can be inspired by your achievements and it's reassuring for your friends to see that you're still active.

Not everyone has time for this, but it's an important way to help and inspire others, so anything you can do to show them what you have been finding is worthwhile. Some people name-drop nearly every new species that they find, which is a great way for others to see what listers are up to.

At the time of writing, very few pan-species listers are actually blogging any more. My own blog (around 1,000 posts over 15 years) is still ongoing, but it has slowed down a lot. My posts are now longer, more complicated and less fun, mainly due to having less time, so any posts I write become massive 'information dumps'.

This is partly why I started the #speciesaday series on Twitter (now moved to Bluesky), which is effectively a microblog. I rarely get time to write a full blog now, so writing about just one species a day (which typically takes me just five minutes) is really enjoyable, and the 280-character limit is a welcome restriction. As my database now operates like a small record centre, I can generate quite meaningful maps within Sussex (for many invertebrates at least), and pairing up a species with a map and information such as the species' conservation status, how many records I have, some phenology, what it eats, etc. is a really great way to engage, and it seems to be popular. I shall never run out of species in the British Isles, as I find them faster than I can post them! I also write weekly field day reports on Instagram that are a bit more involved (but lack something really novel that would warrant featuring on my blog), giving a flavour of what it is like to be a busy entomologist on the south coast.

The new PSL website automatically displays what your last species was for all to see, which is a big advantage over the old site.

Don't forget your garden

I have been thrilled to see the kind of species that Mark Telfer has been finding in his garden recently, from Large Cone-head *Ruspolia nitidula* to Glanville Fritillary *Melitaea cinxia* and a flyover Citrine Wagtail *Motacilla citreola*. Garden PSL is very enjoyable, even if you have a small plot. Running a light trap is a useful way to get a good list over a long time, especially as you will also add to your list through by-catch, but make sure that your neighbours will not be disturbed by it.

Even in the actinic trap in my little garden, which is next to a main road with bright street lights, I have trapped Sombre Brocade *Dryobotodes tenebrosa*, Golden Twin-spot, Plumed Fan-foot, Radford's Flame Shoulder *Ochropleura leucogaster*, Blair's Mocha *Cyclophora puppillaria*, Langmaid's Yellow Underwing *Noctua janthina*, Clancy's Rustic *Caradrina kadenii*, Scarce Bordered Straw *Helicoverpa armigera* and *Palpita vitrealis* – all in 2024 and 2025.

Stay on wildlife-friendly farms

A great way to add to your PSL list, while at the same time contributing to the local community and economy, is to stay in accommodation on wildlife-friendly farms. They might even let you do some casual recording there. I recently stayed at Gorwell Farm in Dorset, where the accommodation was perfectly set up for running a moth trap in the back garden, and this really enhanced the trip, especially as it was very migrant heavy. I caught my first Old World Webworm *Hellula undalis* there, found *Theridiosoma gemmosum* new to the hectad and identified a bird new to the site as a flyover, Crossbill *Loxia curvirostra*.

Species names

Use of common names

I really enjoy using common names. The names of our birds, fish, fungi, plants and macro-moths in particular are wonderful. They are part of our national (and natural) heritage. I am equally happy to use scientific names if there is no accepted or widely used English alternative. Attempts to try to enforce standardised common names are rarely successful – for example, those for bryophytes and spiders have not stuck.

Problems arise when common names that are not widely accepted are published. If these are then picked up by beginners they will find it very difficult to communicate with specialists using these names. It is interesting how quickly people can give up trying to learn scientific names. Yet many children can rattle off dinosaur names completely undaunted and with surprisingly good recall. I think, as we get older, much of the time it is fear of not knowing how to pronounce these names that makes people reach for something more familiar sounding instead.

The best way to learn scientific names is through repeated exposure – and take comfort from the fact that *no one knows how to pronounce them*! They're not Latin. They *contain* a lot of Latin, but also Greek and bits of other languages, new words, proper nouns, herbs and spices and who knows what else! Yes, there are purists who might try to correct you from time to time, but they are wrong to do so. If you hear us at a PSL field meeting it will sometimes seem as if we are talking several different languages!

Some naturalists prefer widely used and standardised common names to scientific names. For the pan-species lister, you will tackle a mixture of groups that have widely used common names, along with groups that have thousands of scientific names. For me, with a head full of scientific names, it's a huge relief to use the widely accepted common names for a few groups when I can. For example, why would someone spend the time learning the scientific names of macro-moths when they have such wonderful common names? The same can be said for vascular plants too, although many botanists would likely disagree. Either way, when it comes to widely used and accepted common names, it's important not to judge those who choose to use them over scientific names.

Creating new common names

I don't believe that a taxonomy-based approach would come up with common names such as Blackbird, Mistle Thrush, Redwing, Ring Ouzel and Fieldfare – it might give us very dull alternatives such as Black Thrush, Great Thrush, Red-flanked Thrush, Mountain Thrush and Greater Winter Thrush. We need a mixture of 'wildcard' names along with a more logical approach. The English names for bryophytes are particularly clunky, with most of the mosses containing the word moss! One example where this has worked and fitted seamlessly with existing names is the fungal nomenclature devised by Liz Holden in 2003. She achieved something remarkably creative here, with wonderful names like Warlock's Butter *Exidia plana* meshing seamlessly with the very old name of Witches' Butter *Exidia glandulosa*.

Warlock's Butter *Exidia nigricans*.

The British Mycological Society website has a very useful set of criteria which could be used to come up with common names for any taxonomic group, and it is well worth looking through these before attempting to do so:

https://www.britmycolsoc.org.uk/english-names.html

We also need to stop naming species after places, as it causes far too much confusion when they are subsequently found elsewhere. Species named after people are also of very little help to aid identification.

Capitalising of common names

The correct way to write common names of species is to use initial capitals for all of the main words in the name (e.g. Golden Twin-spot, Forester, Mediterranean Gull, Greater Streaked Shieldbug). As you will have noticed throughout this book, all species common names have been written this way, followed by the scientific name (in italic type) at the first mention, unless there is no widely used common name, in which case the scientific name only is given.

In recent decades, many institutions (including the Natural History Museum and the BBC, among others) and their associated publications, websites and interpretation materials have moved to writing the common names of species all in lower case (with the exception of parts of names that are proper nouns – people and places). The rise of 'nature writing' has led to more and more species' common names being written in this way. It seems that ease, a weak argument that capitals make text harder to read (something I just don't see as a real issue) and the fact they are not proper nouns have become more important factors than scientific rigour, clarity and a greater command over the language we use to describe species. To me and many other naturalists, though, it's also a measure of how much we *value* our species.

It could be argued that for the purposes of clarity it is far better to write Small Blue than small blue (a small blue what exactly? Names written in this way can often look like the noun is missing entirely). Relatively speaking, all blue butterflies are small, so they could all really be classed as small blue butterflies. However, if you don't capitalise a species name correctly, you cannot tell whether the words are ordinary adjectives or whether you intended to mean one particular species. For instance, writing the name of the Scarce 7-spot Ladybird in lower case causes confusion with the very common 7-spot Ladybird as it is impossible to distinguish whether the word 'scarce' is being used as an adjective or whether it is actually part of the species name. Do you mean to say that the ubiquitous 7-spot Ladybird, is scarce, or do you mean to refer to the more scarce species, the Scarce 7-Spot Ladybird, that lives only in association with Red Wood Ants *Formica rufa*? Only the correct capitalisation can tell the reader what you really intended to convey.

For example, I have also had my descriptive words confused with the name of the species I am talking about. I once described a species of beetle as 'the striking red and black saproxylic click beetle, *Ampedus elongantulus*'. The organisation in question insists on using all lower case letters when writing common names, and so are used to reading and writing names in lower case. Therefore, they assumed that the species' common name was 'red and black saproxylic click beetle' (when in fact it has no common name at all). Worse still, to make their social media post appear as what they perceived as being less intimidating, they removed the scientific name altogether, leaving just my description of the species, written as if it was a species name.

I believe that species names should be considered to be proper nouns in just the same way that people, places and brands are. There is clearly flexibility in our language to allow breeds of animals to be treated as proper nouns (such as British White cattle and Staffordshire Bull Terriers, for example). Wild species are just as deserving of this treatment as our domesticated pets and livestock.

The way I see it, it's the DNA of the animal that's unique, defining the very notion of what a naturally occurring species is. No different to a breed of animal or a make of car, the only difference is one is entirely natural and the others are human-made. There are lots of copies of all of these examples, but the blueprint or DNA is the same or very similar for each one. I think this is one small but truly meaningful way for us to show how much we revere and respect our wildlife.

I started an online campaign that has convinced a number of wildlife charities, institutions and publishers to move towards (or return to) the use of initial capitals in common names. Not only does writing the name in this way make it very clear which part of the sentence is actually the name, thus improving clarity and accuracy, but also we have a standardised system for writing scientific names (they are to be written in italics with the genus capitalised and the species and any subspecies not capitalised), so we really should have a standardised system for the common names, too.

Hyphens

Hyphens are a very important tool when writing common names. They tell you what part of the organism is associated with the adjectives. For instance, many people write Yellow Horned-poppy incorrectly as Yellow-horned Poppy – in fact it is the petals that are yellow, not the long fruits ('horns'), and it is the poppy itself that is horned.

The name Great Spotted Woodpecker tells you that it is the woodpecker that is great, not its spots. This kind of subtlety is slightly lost in the way that the BSBI writes common names of plants. Their rule is that there should be no trinomials. This is why Common Bird's-foot-trefoil is double hyphenated. This system generally works well, except in a few very obvious cases. Perhaps the most annoying of these is Early-purple Orchid, which really should be Early Purple Orchid. The purple is not early. Here we have two systems in conflict with each other, the 'no trinomials' rule should be a guide and not a hard-and-fast rule, and in this case, if Early Purple-orchid is not suitable, we should relax and go for Early Purple Orchid but never Early-purple Orchid! I also prefer the concept of one universal system being used across all taxonomic groups (including marine recorders, who seem to use lower case more than most other naturalists).

In addition, hyphens often tell you when a species is in a different genus. For example, Bastard-toadflax is not a true toadflax in the way that Common Toadflax is. However, hyphens tend to be quickly jettisoned by those who write common names in lower case.

The choice of whether to include the definite article, i.e. the word 'the', as part of the species name for some single-word moth names, is a matter of preference ('The Forester' or 'Forester' being both equally valid). For such single-word names like this, I tend to add 'The' when it's at the start of the sentence but not usually within a sentence, when listing the species or when used as a title for a species account.

The Forester *Adscita statices* is one of the three species of forester moth in the UK.

Common names on social media

I recently sampled the names of British species written in tweets. I took the last 100 posts to see how many had been written using initial capitals and how many were all lower case. I ignored single species names at the start of posts, as it was impossible to say which way they had been written. Remarkably, even when using social media, where accuracy is not always seen as a priority and word space is at a premium, I found that typically 70–80% of people would capitalise (and hyphenate) the common names. Admittedly these were people I followed on Twitter, who are more likely to be expert naturalists with knowledge of how to correctly capitalise species names, so there is probably some bias here.

I believe that this shows a genuine trend among naturalists, with the people who write about species the most (i.e. those that know these species better than anyone) overwhelmingly opting in favour of using initial capitals in the names.

Genus or generic level

If you are writing about species at the generic level – for example, our three species of forester moth (Cistus Forester, Scarce Forester and the Forester) – it is important to switch to all lower case to show that you are not talking about a particular species, just the genus.

What happens when there is only one species in the genus? For instance, if you were writing about Red Fox as a species, where its ecology is important for context, then it's clear that the species is needed, especially if it's your first written reference to it. However, if you were writing an entire book about just this species, it would clearly be sensible to refer to it in the generic sense as a fox where appropriate throughout the book, only referring to it as Red Fox as and when relevant. Yet most articles about wildlife are not typically only about a single species, and instead most species are only mentioned once or a handful of times – for example, in an article about species in a magazine, or a social media post summarising sightings on a nature reserve. Therefore, if you were to mention a species only once, or just a handful of times, in a book or article, those mentions should always include the full name. This distinction is quite possibly the most subtle but confusing part of writing common names accurately, but it is worth getting it right.

Finally, I have included opposite a quick 'style guide' that you can print off and stick behind your desk (or your boss's desk!).

Memory training and brain hacks

Active recall techniques

I imagine most readers of this book will have permanently stored in their memories the difference between a Blue Tit *Cyanistes caeruleus* and a Great Tit. But how many will have stored in their memories the difference between the epigynes of the money spiders *Diplocephalus picinus* and *Diplocephalus latifrons*? It's all about repetition and exposure (I see these two species all the time), and the more times you identify a species, the easier it gets. Many people think that you will reach a point where you simply can't take any more in, but pan-species listers, entomologists, mycologists and

The correct way to write common names of species: some examples

All initial letters of all words in the name are capitalised, unless they appear immediately after a hyphen – for example, Round-headed Rampion. You only need to capitalise actual species and not generic terms like sparrows which are used to refer to either no particular species or to multiple species where more than one might be relevant.

The following are all incorrect:

- round headed rampion
- Round headed rampion
- Round-Headed rampion
- Round Headed Rampion
- Round Headed-rampion (the flower head is round, not the whole plant).

The exception to this rule is if the name after the hyphen is a proper noun, but this is rare – for example, Herb-Robert.

If you are talking about Round-headed Rampion and Spiked Rampion collectively, you should drop the capitals and refer to them as 'rampions' in lower case, and not Rampions.

The rules for scientific names are slightly different. Here you simply need to capitalise the first letter of the first word and to use lower case for any subsequent words. They should always be in italics.

Some examples of sentences containing common names of species

- Common Bird's-foot-trefoil *Lotus corniculatus* and Greater Bird's-foot-trefoil *Lotus pedunculatus* flowered profusely after a hard graze at Butcherlands and then a pause in grazing over the following summer. The author had never seen so much bird's-foot-trefoil growing in any rewilding project.
- The Red Fox *Vulpes vulpes* seems to be much more common in urban parts of Sussex than it is in rural parts. So much so that, in some places in Brighton, foxes can keep you awake all night with their noisy antics.
- All three species of forester moth (Cistus Forester, Scarce Forester and the Forester) have all been recorded at Wilding Waterhall.
- Common Spotted-orchid, Common Twayblade and Early Purple Orchid are three of the commonest orchid species in Sussex.

any naturalist who has thousands of species to deal with will tell you that, strangely, the more you do the easier it starts to get – assuming that you have enough time to dedicate to it, of course.

Some people will have more retentive memories than others (I know I am fortunate in this respect, almost certainly due in part to my neurodivergence), but there are techniques you can use that will really help you to remember facts. I have been applying these techniques my whole life (often subconsciously), so that it is impossible for me to say how much I can attribute having a good memory to any one of: neurodivergence, repetition and exposure, or passive and active memory training techniques. One thing's for sure, they all play a part.

Mnemonics

A mnemonic is a story that you tell yourself, about something you are trying to remember involving (counter intuitively) additional related information, that actually helps you remember that thing better. I became transfixed by this method as a child when watching a documentary about someone who actively memorised long sequences of information using this technique. I think it was around this time that I started doing it both consciously on occasion, but more often completely subconsciously.

The following is an example of a mnemonic I came up with (I discovered later that the arachnologist Michael Roberts came up with it first, although I further embellished my version with some illogical nonsense):

There are three spiders in the genus *Metellina*. Two are very close to one another: *Metellina mengei* is adult much earlier in the year, peaking in May, while *Metellina segmentata* peaks in September. The third species, *Metellina merianae*, is quite different – I typically find it in dark, shady places or buried deep in tussocks (it's very common in cave entrances, for example). Therefore my mnemonic for this genus is:

- *Metellina mengei* = May.

- *Metellina segmentata* = September.

- *Metellina merianae* = Maid Marian has to rescue Robin Hood from the cave.

I am not suggesting that you use exactly the same version, but simply demonstrating how mnemonics work. Having something that works specifically for you is the key factor. It must be something that you came up with and that you therefore have a connection with. I might not have been the first to come up with this particular mnemonic, but I did so independently enough for it to feel as if I did, and as a result I remember it well. A personal, emotional or even humorous connection works even better.

I always used to struggle to remember the difference between Heather Ladybird *Chilocorus bipustulatus* and Pine Ladybird *Exochomus quadripustulatus*. The more I tried, the more confused I got, so I chose the most daft mnemonic I could come up with, and I still regularly use it years later. Pine Ladybird has four red dots in a trapezium shape, whereas Heather Ladybird has four red dots in a line. My mnemonic for remembering the difference between these two species is:

- Pine = not in a line.
- Heather = by a process of elimination it's not Pine.

This mnemonic has become so embedded that I can't undo it and replace it with something more sensible. This is a sign of a very successful mnemonic.

I use other mnemonics to help me to memorise, for example, the specific part of a spider's palps that clinches the identification of that particular species. There is a tiny spiral structure at the end of the palp of the male *Tetragnatha montana*, which is one of the commonest species in the genus.

- *Tetragnatha montana* = the spiral at the end of the palp looks like a path up a mountain, *montana*.

Some things I remember are probably less mnemonic in nature, but still work. I always remember *Sitona hispidulus* for two of its key identification features, which can be reduced to:

- flat eyes, hairy bum = *Sitona hispidulus*.

The two small bird-dropping longhorn beetles, *Pogonocherus hispidulus* and the slightly smaller *Pogonocherus hispidus*, are quite easy to tell apart in the field but harder to match to their names, which are annoyingly similar (If I don't use this mnemonic, without fail my brain will try to match them the wrong way around). My mnemonic for these is:

- smaller beetle = smaller name.

Emotional connections and cultural references

The reason why cultural references work so well is that they have an emotional connection with the things you apply them to. This emotional connection is a vital part of memorising what you see. In fact, I find that if I am not enjoying myself when I see new things, I tend not to remember them so well. The one bird that I have seen (and I know I have, because I made a sketch of it) but of which I have no memory whatsoever was a Buff-breasted Sandpiper *Calidris subruficollis*. I have no memory of the trip, the location or the bird, as I put down no mental markers. Mindfulness is a really key tool here for making sure that you put down markers (in much the same way that if you are present when you lock the front door, you don't panic half an hour later thinking that you've left it unlocked) – so always try to be in the moment.

Anyone who has spent time in the field with me, especially when I found something I have never seen before, will see me losing it through over-excitement. My voice will go several octaves higher and I properly become overjoyed, not in a contrived way; it's totally uncontrollable, and I know it's infectious too. This is a huge part of remembering things in detail for me. Love every minute of it, and you'll put down better memories.

Cultural references help in a big way, but what is most interesting about trying to list these is that although I have access to many of them when I'm out in the field, they are only really accessed or triggered when I see the species in question. They almost seem to disappear when not in use (and are annoyingly difficult to recall when trying to write a book about them, I think this shows how highly efficient our brains are)! Here are a few examples:

1. Orthoptera sounds
   - Roesel's Bush-cricket *Roeseliana roeselii*. If you have ever had a tattoo, you'll know this now common cricket sounds just like a tattoo gun.
   - Common Green Grasshopper *Omocestus viridulus*. This species always sounds to me like a sewing machinist with a tight deadline.
   - Stripe-winged Grasshopper *Stenobothrus lineatus*. The world's tiniest brush cutter, scything away back and forth.
   - Wart-biter *Decticus verrucivorus*. A tiny bicycle building up speed downhill.
   - Large Marsh Grasshopper *Stethophyma grossum*. Someone slowly clicking their fingers to a beat.

2. Amphipod or isopod?
   - Amphipods look as if they have been squashed between two amps. Isopods don't.

3. Blackcaps sound like R2-D2
   - I grew up watching *Star Wars* and one way I learnt to separate Blackcap *Sylvia atricapilla* from Garden Warbler *Sylvia borin* is based on the fact that I find Blackcap song often sounds just like R2-D2 warbling away, yet Garden Warbler song never does. I have never struggled to separate these two species.

4. *Ceratapion gibbirostre*
   - There is a very sharp point on the rostrum of this weevil, which separates it from the closely related *Ceratapion carduorum*. I can remember which species is which because Barry Gibb(irostre) has a very sharp voice.

There are some species that you will always conflate or struggle to recall. For example, I can never match the names of the solider beetle genera *Malthodes* and *Malthinus* correctly, yet in my mind's eye I can tell exactly which is which, so I have simply given up trying to remember them.

Do as much as you can in the field

Field identifications require significant prior microscopic experience of the species that you're identifying, good eyesight and, more often than not, some memorising of keys. The benefits are clear – you don't have to kill as many animals, and you save time later on, so you can spend that time working through species that can only be identified back at the microscope. Field identifications allow you to make a huge number of records using the PSL approach. For me, this has to be efficient – if a field identification is eating into recording time I will collect the species and identify it when I get back home.

The down side of field identification is that you have to be very sure about it, and there is no opportunity to go back. It's a judgement call – and you will get better at this with each such encounter. Recalibrate some of the more difficult species at the start of each field season by taking the odd one to key out, to make sure you've remembered them correctly. I cover this in more detail on page 335.

Entomology and arachnology, etc.

Find a method of dispatching that works for you

Some people would prefer never to take specimens. However, there is a limit to how far you can get with that approach, and I strongly suggest that you reconsider it if you want to take entomology seriously. Find a system that works for you. As a pan-species lister you are probably covering a wide range of taxonomic groups. Some people prefer a pooter, some prefer a jar containing a tissue/and or plaster of Paris soaked in ethyl acetate, and some (like me) use a small container of 70% alcohol or bioethanol. Others use more than one of these methods.

I have never used a pooter, and I don't intend to. I have used a killing jar containing ethyl acetate, and this system does work much better for bees, larger flies and Lepidoptera. Smaller flies, springtails and hoppers (which all have a tendency to bounce away from approaching tweezers) are definitely much better collected with a pooter. However, spiders are much better preserved if they are put straight into alcohol. I use small, airtight, water-tight push-cap tubes from Sarstedt. I have a drinks holder attached to my utility belt, and I place whatever tube I am working on in that. For my field data labels I use thin strips of card cut to size (and kept in the front pocket of the utility belt, together with a small pencil and sharpener). Don't forget that you will have to write in pencil or alcohol-resistant ink (I use a 0.05 mm Staedtler Pigment Liner for my spider collection labels, but you can use this for field labels too) if you are placing a label in alcohol.

For me the big advantage of this system is that I can just put everything straight into storage after a day in the field. No immediate processing is needed the same evening, which is a great relief when I am doing fieldwork all week long in the summer.

Keeping reference collections

You could write an entire book on this. Storage is a genuine issue for me, as is curation of multiple collections across multiple taxa which most pan-species listers may also face – from spiders, to bryophytes, to beetles, to bees and more. I won't go into huge amounts of detail here, as it has been covered to some extent in the taxa breakdowns above, but it is worth planning ahead with collections. In most cases, they take constant work to prevent infestation by pests. Additionally, I unexpectedly lost several store boxes of beetles over a summer to mould, which was utterly heart-breaking.

You can use a chemical suppressant (though these need to be replaced regularly), occasional freezing and visual checking, but prevention is always the best option, and one increasingly popular system involves replacing wooden storage boxes with airtight plastic storage boxes, themselves contained in Ziplock plastic bags, themselves stored on stand-alone metal shelving set well away from walls to help to prevent both damp and pests from damaging your collection.

One particular problem for the pan-species lister is how to display and store all of their material. You will also need to constantly reclassify and rearrange your collections over the years as they grow.

Digital resources

Photography

Entire books have been written on the topic of wildlife photography, but I am not aware of any about wildlife photography specifically for the purposes of identification. Using photographs to attempt to identify species as an alternative to taking specimens is not something I would encourage here. However, for some groups (e.g. many plants, many marine groups, larvae and larger insects) it is often easier to take photos than to collect specimens. For smaller species, you will need to know exactly what part of the animal should be targeted in the photograph. This will be time-consuming and fiddly, and will often be easier to do with a hand lens.

I personally use an Olympus TG-7 because it is rugged, compact (so can be tucked away in a bag or pocket) and has a plethora of features that seem to be aimed specifically at the naturalist, though it is not great at taking photographs of anything distant. The built-in photo-stacking function in the macro mode really does wonders, and the camera also generates a photo without it, allowing you to choose which looks best.

I find that its underwater capacity (it even works in sea water) is ideal for taking photographs in rock pools. Do rinse it with tap water afterwards, though, or the focus mechanism will start to stick. I have used the macro mode with an additional ring-light to take shots of tiny nudibranchs in rock pools in poor light. Capturing images of marine organisms in a black container (painting an open-topped plastic box works well) is a great way of showing detail on pale, translucent animals such as crustaceans and nudibranchs.

I recently lost my TG-7 in rank vegetation in an arable field at Wild Ken Hill; all the summer's photos were on the memory card, a disaster! Some five weeks later, after several storms, I found it! It was completely unscathed; there was no moisture in the camera – it was even still fully charged. It was like I had just dropped it and picked it straight back up. It had even survived the area being topped by a tractor! They really are as tough as they say, but do make sure you back up regularly and consider using a small GPS device attached to your camera to help you locate it if lost.

One of the most terrible things about my photography is the absolute chaos that is my image archive. It is so much easier to use my blog as online storage, and search for a specific picture that way via Google. But my advice is that you treat it like a specimen collection and curate it well – for a great example of this, see Steven Falk's Flickr stream.

Patience is key with macrophotography – as is stability in the absence of a decent tripod. I get down on the ground, and if my elbows or wrists are in contact with the ground I can often make a very secure tripod with my own body. This is usually sufficient to get some good shots with the built-in photo-stacking.

If you are trying to take a photo of something very thin, such as a grass or sedge, and the camera keeps focusing on the background instead, put your finger at a similar distance, and let the camera focus on it. Then move your finger away and you should find that your skinny subject is now in focus.

Play around with your data

If you manage a large database as I do, don't forget that you can use it for fun as well as work! My daily #speciesaday (formerly on Twitter, now on Bluesky) is an example of this. I take one species a day, generating a map, carrying out any data cleaning if necessary, talking briefly about the species and its ecology and then talking through my map of the species. This is a great way of gaining insight into what, where and how you record. You might discover things you had no idea about. It also shows you where your gaps are, and this is a great way to focus your casual recording, if square-bashing is your thing. And of course it inspires others to do the same.

One of my favourite activities as an entomologist is storing summary statistics for every survey I do in a spreadsheet, and then analysing it in one chart. 'The proportion of species with conservation status' is a great metric for measuring a site's quality for invertebrates, but what does a proportion of, say, 7.6% mean to a client without context?

If I state that my rolling mean is 7.2%, you will now know that your site is slightly above average for one of my surveys. If I plot these proportions in 1% intervals in a frequency distribution chart, and then colour-code each survey by broad 'site type', you have even more context. You can see in my chart (shown below) a clear peak around the 6–7% mark, but also a second peak in the 9–10% range, which seems

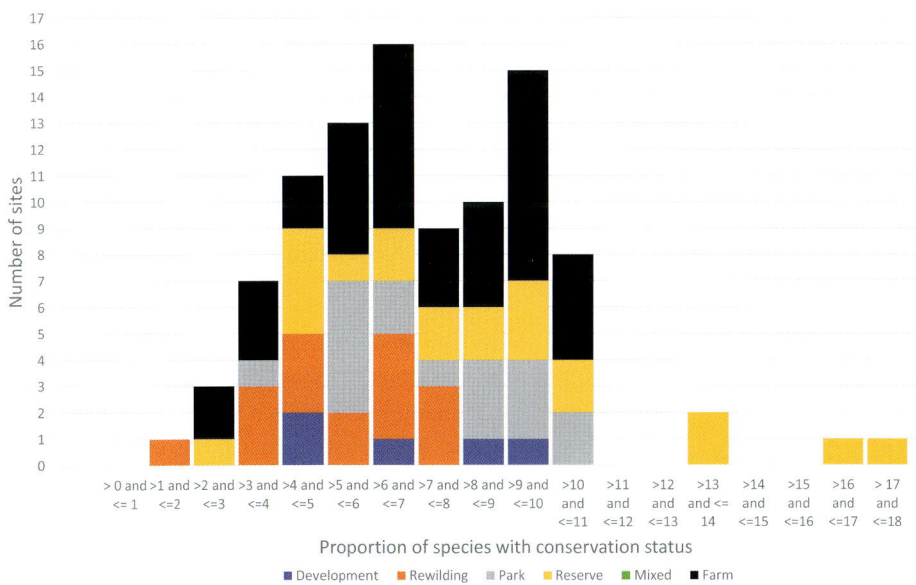

Frequency distribution of the 'proportion of invertebrates with conservation status' from my most recent surveys, displayed in 1% intervals, and further colour-coded by broad site type.

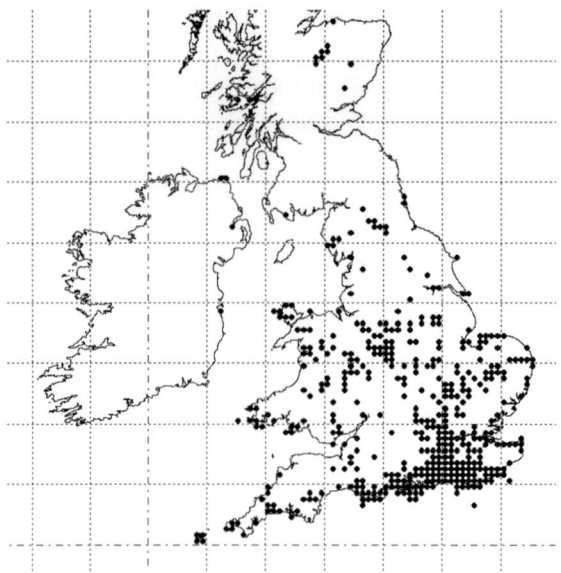

Every hectad in which I have recorded in the UK.

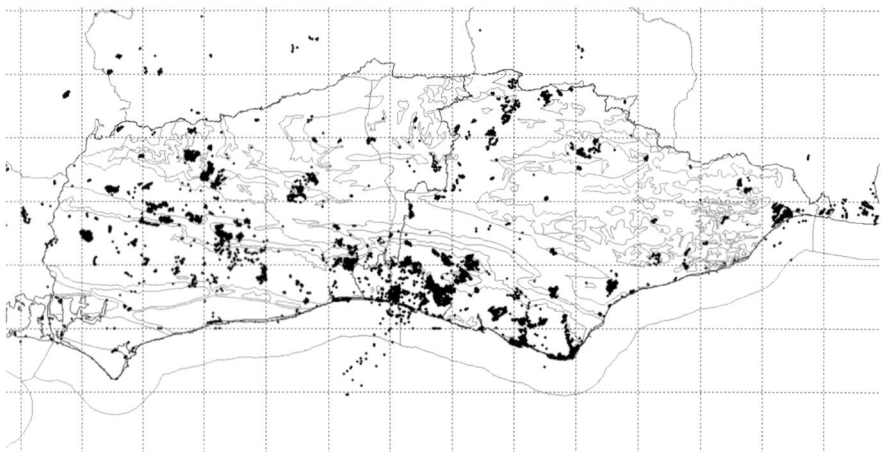

All of my Sussex records plotted. The Hailsham 'record black hole' is clearly visible as the only complete Sussex hectad where I do not have a single record.

to be driven by surveying so many farms on the chalk of the South Downs. This is what I have started to call the 'South Downs effect' – the underlying geology drives the ecology of these sites so much that many farms have a higher than average proportion of rare and scarce species, and they cluster around this second peak. Being on the south coast, where I do most of my work, many species are also at the northern limit of their range, too, again pushing up the proportion compared with the national average. These are all useful insights that I would not have gained if I had not kept these summary statistics in one place.

Generating maps in Recorder 6 (or MapMate) is one of the best things about having a database. I love seeing where I have been over the decades!

Learn how to use pivot tables

If you are a pan-species lister you will need to keep records, and anyone who keeps large numbers of records (whether in Excel, in a database of some kind or on an online recording platform) will need to use pivot tables. These are incredibly easy to use (despite often being described as 'advanced Excel') and take just seconds to work, saving a huge amount of time while also preventing transcription errors. I use them almost daily.

Pivot tables are essential for taking huge lists of records in Excel, a typical output from most databases, and turning them into a matrix that can be read and analysed. Records take the form of rows of data, with each cell being a different part of the record (date, species, location, grid reference, etc.). This is meaningless for most analyses, and not very useful for generating a species list without a lot of manipulation. If you want to display these data in a matrix form (e.g. species in rows, and compartments or sites in columns), which is a vital step if you want to calculate the total number of species per compartment, for example, then you will find that using a pivot table is a very quick and efficient way of doing this. It will only take a matter of minutes, whereas it will take hours if you try to do this the long way, which also has the potential to generate many errors.

Learn how to use GIS

Geographic Information Systems (GIS) are a great way to display spatial data, add information to reports and display records. Most of the commercial systems are costly, with annual subscriptions that can make them unfeasible for most individuals. However, there is also an open-source system that is free, called QGIS. I have used many of the commercial systems over the years and I find QGIS by far the most intuitive and easy to use.

There is a very useful QGIS plugin that is designed for biological recording. The TomBio plugin (short for Tomorrow's Biodiversity, one of Richard Burkmar's creations) will allow you to import records and display them as distribution maps or heat maps, or overlay other mapping layers. I use this in most reports for plotting the more exciting species I have recorded to an eight-figure grid reference, on top of, say, my compartment or habitat map.

You can learn how to do more complex things in QGIS. Digitising say the vegetation communities of a site, is fairly straightforward. It's worth noting that the function 'autocomplete polygon' is slightly different in QGIS – you get to this in 'Snapping Tools' (this tip will save you a lot of time looking for it). If you're struggling to do something in QGIS, a quick Google search will usually lead you to a YouTube tutorial showing you how to do it if you can very concisely describe what you are trying to do.

A useful book on this subject is *QGIS for Ecologists: An Introduction to Mapping for Ecological Surveys* (Pelagic Publishing, 2024) by Stephanie Kim Miles.

Social media and natural history

Social media has changed the face of natural history in the UK over the last couple of decades, almost in parallel with the rise of PSL, and indeed the two are intrinsically linked.

Throughout the individual taxa group accounts in Chapter 6 links are provided not only to useful websites but also to useful Facebook groups. It's worth joining Facebook for the natural history groups alone, even if you don't use the more typical side of the social media platform – we have never been so well connected as we are now.

The PSL Facebook group

Facebook groups have progressed a lot in the last decade, and now pretty much dominate the world of online minority groups. The PSL community was perhaps a little ahead of the curve here, so when so many all-rounders and experts came together in one place it was ideal for posting about species identification across a wide range of groups. At the same time, many of the specific groups were not as active or didn't exist. Over the last decade, these groups have grown enormously – indeed they can sometimes get too big, though this is generally counteracted by having plenty of active admins and clear (but not too restrictive) group rules.

More and more of the experts joined these groups, until eventually most of the active experts, or at least enough of them for the groups to function well, were present in one place. Experts from neighbouring countries will often join, too, both to help and to give a slightly different perspective on things. With plenty of good admins they can do very well. Of course, if you find a rare spider and want to tell everyone about it, the best place is a spider group with over 9,000 members, not a group of pan-species listers with just a few hundred. Although the activity of the PSL Facebook group has waned this is not a bad thing, it has simply moved elsewhere to the largest, most relevant groups. PSL is certainly not for everyone, and there are many people in these groups who are not on the rankings, so this way, many more people get to see your finds and learn from them.

Facebook groups cater really well for niche groups such as specific natural history topics, and with digitising of information it has never been easier to share knowledge, information, data, photos, site knowledge, keys, papers and sheer enthusiasm – as a result it has never been easier to identify species. Along with the fact that so many great natural history books and keys are being published, this means there has never been a better time to be a naturalist (and is why I find the rise of AI apps so bewildering).

Facebook group etiquette

Online communication plays a significant part in many naturalists' lives these days. Obviously, with a range of ages and abilities using these groups, it's important not to make anyone feel denigrated, but rather to lead by example in the way we post and comment.

Always take things as far as you can yourself first, and then if you need some more help, show that in your post. Use photos and text to describe how you have got stuck and why you can't get any further, perhaps showing which species you have narrowed your find down to. It is also vital to state when and where you found it. What you should absolutely avoid doing is just posting a picture and writing "looking for an ID" or worse still, "ID required".

Take this analogy. Imagine you like beer and all the country's top brewers have got together to create a wonderful space for you to come into with infinite amounts of free, top-quality beers and ales. You fancy one of these delicious beverages for yourself, they're totally free of charge after all. You walk into the pub, straight up to the bar and point at the one you want and state, without engaging with the unpaid bartenders in anyway, "beer required" or "looking for beer". It doesn't come across well online either. These groups consist of the sum total of everyone on there's experience and knowledge. We're not an AI app, we're busy people with lots of other commitments, and no one owes anyone an identification so it's important to always be gracious and grateful. So rather than stating "ID required", perhaps try something like "I've run this click beetle that I found on a marsh in West Sussex today through the key several times and I keep getting to a species that I know must be wrong, where have I gone wrong? Help much appreciated" is a good example post from a more experienced naturalist while "I found this huge spider in my house in central Birmingham last week, could someone tell me what it is from this photo please?" as a possible example from a beginner.

Another useful tip is to scroll down the group. Many problem species are often reported by other people at around the same time, so someone else might have already saved you the bother.

Don't use experts to get species identified in bulk or too regularly

Relying on experts too much inhibits your learning. They are there to show you the way and verify, but not to do your work for you. There are many people who enjoy online identification and will do it for you, and though it is tempting to lean on them, they could be using their time to help people who have already gone as far as they can with an identification.

Don't get something identified on one platform and then pass it off as your own ID on another

This really does happen – and people can see right through it. Always be careful to mention anyone else involved in anything that you do.

Check the comments before you comment

It's always surprising how few people check whether someone else has already identified an ID request. It's far better to like someone else's comment or say that you agree with it than just to say the same thing.

If you have given a different determination to what most other people have been commenting, it will be especially important to read the other comments. If you are right, it's a potential learning opportunity for them and the rest of the group by explaining why you believe it's not what everyone else is claiming. If you are wrong, though, then you look a bit of a wally. Avoid writing just the species name, give a little more context. If you are too busy to do anything more than write a species name then perhaps that's a sign that you are over-committing and other people with more time on their hands might be better placed to give more comprehensive feedback.

If you feel like you have to be the first to comment on every post in a group, then maybe it's time to check in with your ego; is this about helping other people, or about showing who knows most?

Don't announce someone else's record before them

If you go out as a group and one person finds something really rare, that is their find to announce on social media, and it's really poor practice for someone else in the group to announce this record first, without the finder's permission, whether or not they mention the actual finder.

Avoid 'post hijacking'

Some groups will already have a ban on this. If someone posts a picture of a species, it's best to avoid posting your own pictures of the same species in the comments, unless you are invited to do so. It comes across as competitive behaviour.

Don't virtue signal

This is especially common with beginners on groups who are not used to the fact that you often need to take specimens in entomology. Trying to make yourself look good by making others feel bad is really unpleasant and should be avoided at all costs.

Chapter 9

Can pan-species listing change your life?

The answer to this question is that it certainly can! It was August 2010 when I first drafted my 'Great List of Everything', as I called it then. Over the following 10 years I pretty much taught myself how to become an entomologist, using the PSL approach as a framework. Although I actually started doing bird and botanical freelance work in my spare time (on top of a full-time job at Sussex Wildlife Trust in 2009), it wasn't until 2014 that I carried out my first freelance invertebrate survey. By 2019 I had become a full-time freelancer, and now I am writing a book about a hobby I helped to create.

These days I am booked up several years in advance for freelance surveys. One area of work that has really taken off recently has been the increasing number of biodiversity surveys on farms. Changes in farming subsidies and a recognition that farmers need to do more to help biodiversity have resulted in growing numbers of farmers wanting broad-brush biodiversity surveys to be carried out on their farms. They want to know as much as possible about the existing biodiversity on their land, while also getting value for money. In other words, they want a standardised survey conducted by an ecologist who knows as many bird, plant and invertebrate species as possible. There has never been a better time to be a pan-species lister.

Now, as an expert in biodiversity surveys, I am documenting methods so that other people can use them, and some of these are being adopted as the national standard. I am better off financially and my mental health has benefited greatly. Freelance life suits me. I enjoy my own company, and I am difficult to manage if my manager is not a passionate naturalist and an idealist, much of which I am only now learning is partly a result of being neurodivergent.

My database, which at the time of writing holds well over 350,000 records, and the fact that I am one of the most experienced and active field arachnologists in the UK, are both entirely a result of PSL. And there have been some memorable PSL meetings, many of them involving just two or three people.

Don't just take my word for it, though. Here I introduce some of the more active members of the current contingent of pan-species listers. I asked 25 people to write a short piece about themselves in a pan-species listing and natural history capacity, highlighting how PSL has affected their life, or how they might have influenced PSL. All list data were captured from the new PSL website in October 2025.

They are presented in alphabetical order, with the exception of the first lister, Mark Telfer, who started the whole PSL movement and has probably made the greatest contribution to its development.

Mark Telfer looking blissfully unaware of the approaching storm behind him. (Mark Telfer)

Mark Telfer, 56 (fourth place with 9,145 species)

I first met Mark when we both worked in the ecology department at the RSPB headquarters in 2003, and the first time I went out in the field with him was in August of that year – the hottest day on record at the time. Our target was an enigmatic and extremely rare beetle, *Emus hirtus*. We got out of the car and were standing on either side of a cow pat when Mark suddenly spotted the beetle between us! It was great to witness his passion for nature, something I could really relate to.

Mark was the person who started the whole PSL movement and, most importantly, gave it its name. His contribution towards PSL has been huge. By making tangible an idea that several different naturalists all came up with at roughly the same time, and providing a place for that idea to take root and grow, Mark breathed life into a movement that is now flourishing. He has been one of the main driving

The rove beetle *Emus hirtus*. (John Walters)

forces ever since, helping with all aspects of PSL along the way, from drafting the first set of 'rules' to defining the taxonomic groups on the old Biological Records Centre (BRC) website, which we still use today on the new PSL site.

Mark is one of the calmest, kindest and most thoughtful naturalists I know, always ready to invest his time and patience in others whom he recognises are passionate and keen to learn. His field skills as an entomologist (and general naturalist) across a vast number of taxonomic groups are quite frankly astonishing. He has clocked up nearly 60 British firsts, and even in the short time we were in Ventnor Botanic Garden together in early 2023 he found a weevil new to Britain. My only regret is that I have not spent more time with him over the decades.

Danny Cooper, 46 (68th place with 3,225 species)

Danny's interest in nature began during his childhood in Surrey, when he visited Chessington Zoo daily. He started recording the species he observed in 1990 (aged 11). His bird list grew as he began to attend evening classes in birdwatching, and in 1999 he joined Surbiton and District Bird Watching Society. His first twitch was a Bobolink *Dolichonyx oryzivorus* in 2002, with Seth Gibson. Soon after this, Danny and Seth half-jokingly suggested that they needed a 'List of Lists' to keep track of all the species they had seen. A week later they had created their lists – they were pan-species listing before it even had a name! While travelling around they saw more mammals, butterflies and dragonflies, and also started to record moths.

Danny studied Countryside Management at Merrist Wood College, Surrey, where he learned about different survey methods and identification of groups he had not tackled before. He then worked for the Isles of Scilly Wildlife Trust, running the Information Centre, where he surveyed marine life as part of his job. After returning to the mainland he studied for a BSc in Applied Zoology. Now based in Cornwall, he continues to record all of the wildlife he observes.

Duerden Cormack, 26 (41st place with 4,189 species)

Duerden has been passionate about wildlife since childhood. At the age of 17 he was awarded a Field Studies Council Young Darwin Scholarship – a residential trip to Preston Montford, Shropshire. Here he learned natural history and survey skills, and he also met the young pan-species lister Harry Witts, who encouraged him to keep a list of all taxa and sign up to the PSL website.

Seven years later he is monitoring biodiversity across a range of taxa for the RSPB at Hope Farm, their nature-friendly arable research farm in Cambridgeshire. When I met with him at Hope Farm in 2023 he told me that his wider taxonomic interest through the PSL movement, his experience of the trickier groups, and the fact that he puts his PSL statistics on his CV were big factors in getting him the job.

Jonty Denton, 59 (first place with 13,090 species)

Jonty is passionate about seeing new species and understanding what they do, in the hope that he can then provide a more balanced opinion on how the sites that he surveys should be managed. One of his earliest memories is of seeing Smooth

Newt *Lissotriton vulgaris* in a horse trough, which he credits with setting him on the path to a PhD on the Natterjack Toad *Epidalea calamita*, which he studied on heaths, saltmarshes and dunes. Pond life is still his favourite niche, but he enjoys square-bashing Surrey (6,985 species to date) and Hampshire (7,740 species to date). He hates killing things, so mapping is his justification. Jonty has been fortunate enough to enjoy a paid hobby and to be sent to most of the recording meccas to swish the nets. He points out that PSL is an important new way into natural history recording, filling the gap left by the demise of the likes of I-Spy, Ladybird and Observer's books and tea card collecting, which were often important to previous generations of naturalists as they were growing up. As it becomes ever clearer that many species are under threat, he urges everyone to get out there and get looking.

Brian Eversham, 64 (currently not on the new PSL website)

Brian started birding at the age of 11 on Thorne Moors, Britain's largest lowland raised mire. Winter birding being sparse, he became interested in lichens by the age of 14, and his interest was encouraged by excellent natural history societies in Goole and Doncaster. After being given a copy of Clapham, Tutin and Warburg's *Flora* for his sixteenth birthday he quickly became addicted to keys. In his first week at Durham University, where he studied zoology, Brian re-found *Arion flagellus* at its original British locality, 200 metres from his room. Within a year he had produced local keys to slugs. He also bought a microscope and set about learning his carabids, ants, Auchenorrhyncha and Heteroptera. Summer vacations included wide-ranging invertebrate surveys, with the addition of many species to the Yorkshire list.

Brian's first job was with the then Nature Conservancy Council's Invertebrate Site Register in London. In 1983 he joined the Biological Records Centre at Monks Wood, Cambridgeshire, first with responsibility for invertebrate recording, and later as research coordinator. There he overlapped with Mark Telfer, and they began surveying Breckland for carabids and other invertebrates, finding *Amara fusca* among many other species.

In 1997, Brian took up a position at the local Wildlife Trust as conservation director, and he is now their Chief Executive. There he has run over 200 wildlife training workshops, and written local keys for a wide range of groups, including ants, beetles, Orthoptera, slugs, land snails and pond snails, *Cladonia* lichens, willows, ferns and umbellifers. In his spare time he has dabbled in microfungi, testate amoebae and other rather unpopular groups.

Brian's pan-species recording slowed down massively with the appearance of the last volume of Sell and Murrell's *Flora* in 2018, which includes 62 controversial new elm *Ulmus* microspecies. He now has a herbarium of about 6,000 elm specimens, tens of thousands of photographs, and new keys for identifying British elms.

Oliver Froom, 47, Thomas Froom, 7 and George Froom, 5 (with 460, 168 and 61 species, respectively)

Oli had a keen interest in nature from a very early age, and this has been revived in more recent years. He finds that being in nature anchors him with a sense of calm and perspective during the inevitable stressful times in life.

The 'Froom family found' Desert Wheatear *Oenanthe deserti* (with a Purse-web Spider *Atypus affinis*) at Beachy Head. (Garry Watton)

In recent years, Oli has been sharing his interest with his young son Thomas, who already has an impressive knowledge of plant and bird names (he knew Yellow Horned-poppy *Glaucium flavum* at the age of two!). During the summer of 2024 I suggested that Thomas might like to take up pan-species listing, and this then became reality at the end of a bioblitz that I carried out in their back garden. I showed Oli how to register and use the PSL website, ostensibly with the aim of setting up Thomas on the site. Now they are both keenly engaged in listing, and Oli's younger son George (aged five) has also recently been recruited, after he unexpectedly pointed out Old Man's Beard on a walk to the play park!

More recently, the family's motivation to look for new species led to their discovery of a Desert Wheatear *Oenanthe deserti* near Beachy Head. Submitting his sons' names as co-finders of this splendid rare bird was a proud moment for Oli!

He is conscious of the risk that he could be appropriating Thomas's formative experiences of listing; Oli doesn't want to tarnish the excitement of discovery for him, or to create artificial memories that he will struggle to recall in the future. However, Thomas has alleviated this concern by being proactively engaged in the process and quite obviously motivated by it. Oli has also tried to support him with integrity, letting him find and identify species himself whenever possible, and by documenting sightings by taking photos of him next to the species.

In view of his strong early start to pan-species listing, I have remarked that by the time Thomas is an adult, he could conceivably sit at the top of the PSL rankings! As his parent, Oli is just happy that he is enjoying it, and he hopes that by giving him this clear structure with the attendant extra level of focus and purpose, he can help him to derive even more happiness and satisfaction from what is a marvellous hobby.

Steve Gale, 66 (46th place with 3,888 species)

Steve is a birder who sometimes dabbled in butterflies and orchids before he became fascinated by moths. The need to recognise these insects' food plants led to a long period of travelling Britain in search of flowering species. In an idle moment he worked out how many species he had identified across all the orders. He then started to maintain this list, adding some of the more easily identifiable wildlife species he encountered, and he became a founding member of the first PSL league table. Although he has never wandered far from his core interest in birds, plants and Lepidoptera, his occasional periods of fascination with other orders have found him on moss and fungi forays in Sussex, knee-deep in Cornish rock pools, and deciding whether or not to invest in expensive insect keys. For him, PSL opened the door to new orders which he would otherwise have neglected, and introduced him to a band of like-minded people who celebrate our natural world through important fieldwork and the sharing of their knowledge and results.

Seth Gibson, 53 (14th place with 7,274 species)

Seth was born into a family with zero interest in nature other than a grandad who used to feed the birds in the garden. His earliest wildlife memory is of watching the 'grey birds with crazy bulging eyes'. Age ten, he bought his first bird book and identified them as Woodpigeons *Columba palumbus*. Birds, and twitching, played a prominent role in his life until he was well into his twenties. He then met a lass who wanted to identify flowers. Seth bought her a wildflower guide for her birthday, let her use it about three times, and then adopted it as his own and never really looked back. Some 25 years later, plants are still a big interest.

It was during his mid-20s that he decided to work out how many species he'd actually seen, complete with first sites and dates. Birds and plants were easy, inverts less so. Obviously, he'd seen 'Bluebottle', 'Greenbottle' and 'Housefly' for Diptera (oh how naïve he was back then). And so, just like that, Seth's 'Organism List' was born – an early version of pan-species listing.

A chance encounter with Ian Menzies on his beloved Epsom Common really kicked off his interest in invertebrates, followed by meeting Jim Porter and Graham Collins. These three talented field naturalists truly opened his young eyes to the world of natural history. At some point during this period, he met Tony Davis – a very dear friend of Seth, with the ongoing misfortune of being his partner in crime for many a PSL (mis)adventure. Through the building success of the PSL movement, Seth has met loads of other pan-species listers, including myself and, of course, the legendary Mark Telfer himself.

PSL has been a lifeline for Seth, introducing him to many folk that he calls friends but, more than that, PSL has very literally kept him alive. For some while, Seth has suffered with less-than-ideal mental health, which culminated in him driving into a bridge at high speed (obviously he messed that up!) and then, some years after that, he found himself stood on the parapet of a bridge late one night, staring at the rocks and river down below. Things really were bad for him that night. He has no idea how long he stood there, he recalls his leg muscles started quivering, so presumably quite some time. Then he heard a Tawny Owl *Strix aluco* calling, as it

Although it may look as if PSL became so successful that it even had stands at motorway service stations, in fact this was a chance encounter with a completely unconnected product (possibly Pumpkin Spiced Latte) that Seth Gibson suddenly emerged from behind, on the way to twitch the Hermit Thrush.

probably had been for quite a while. The call broke through his cloud, he realised he wanted to see this owl. That he badly wanted to see lots of owls. He wasn't ready to end himself, a simple owl brought him down from that bridge. He looked up and saw a bat flitting around a streetlight and, by the time he was home again, he was no longer ready to kill himself. Without his interest in nature, he firmly believes that he would no longer be here. Nature is his ever-present balm, pan-species listers are his people. It's these simple facts that enable him to continue living.

James Harding-Morris, 36 (57th place with 3,496 species)

James has always been into nature. As a child he netted tadpoles and sticklebacks, reared hundreds of caterpillars and obsessively read books about animals. He got his first moth trap at the age of 14, and spent the next few years running it as long and as often as he could without attracting the wrath of his parents' neighbours. Throughout this time he kept a list of everything he'd seen in the garden (three species of amphibian, 14 butterfly species, nearly 400 different moths and dozens of other invertebrates).

While studying English literature at university, James spent his free time splashing after Lobster *Homarus gammarus* in rock pools and hunting for the luminescent glimmer of Glow-worm *Lampyris noctiluca* among orchids on the Yorkshire limestone. It was only when he became a teacher that he discovered PSL. He taught by day and spent his evenings compiling a master list of everything he'd ever seen, or searching for obscure but widespread species (rust fungi, galls and aphids) suggested to him by the PSL community.

After a few years of teaching, he successfully applied for a 12-month post with the RSPB, managing one of their educational outreach projects – Big Schools' Birdwatch. He ended up staying with the RSPB for a decade. During that time, he worked on some of the organisation's biggest engagement projects, including Big Garden Birdwatch. He also branched out into conservation projects such as Back from the Brink – a project that provided opportunities to see species such as Prostrate Perennial Knawel *Scleranthus perennis* subsp. *prostratus* and Barberry Carpet *Pareulype berberata*, and to be present on the day that Chequered Skipper *Carterocephalus palaemon* was reintroduced to England.

Today James works for the Botanical Society of Britain & Ireland (BSBI). It's a fantastic job that involves supporting the network of expert vice-county plant recorders and trying to nurture the next generation of botanists, while also providing him with opportunities to go out and see plenty of plants. It's quite possible that, without a community of people writing long lists of species, none of this would have happened.

Finley Hutchinson, 21 (29th place with 5,318 species)

Finley has always had a passion for nature, but became interested in birding first, before taking a more in-depth interest in invertebrates and other taxa at the age of 15. Lockdown in his GCSE year meant that, with no exams, he had more time to fully develop that interest, helped by the acquisition of his first moth trap and microscope. A pan-species list for his urban garden soon resulted, and rapidly grew to over 1,000 species.

He is now studying for a degree at the University of Exeter's Cornwall campus, where there is no shortage of discoveries to be made, including species new to Britain in double figures, and at least a few species new to science! Jumping between groups of interest and always preferring a challenge, Finley finds that there are no taxa that aren't worth at least an attempt at identification. As a result, his pan-species list includes a lot of obscure species and lacks many obvious ones! Now, with over 5,300 species chalked up in five years (plus a number of lifelong friends met along the way), Finley has just completed a year-long placement in the Scottish Highlands, which opened up a whole range of new species to find and record.

Steve Lane, 62 (fifth place with 8,708 species)

Steve started recording birds and plants as early as 1979, but invertebrates (particularly Coleoptera and Hemiptera) became his passion from 1986 onwards, and he has produced a number of publications over the years. He joined the PSL group in 2013, the same year that he moved to species-rich Norfolk, at which point his list really took off.

He thinks that his main driver for becoming a pan-species lister was the satisfaction of knowing exactly how many species he had seen and having them all listed in one place. He loves lists, writing them every day for all manner of things.

He found that one of the best things about sorting out his original list on joining the PSL group was going through all of his old notebooks and re-living some of the experiences recorded there, which were as vivid as if they'd happened yesterday. His pan-species list is almost a chronological catalogue of his nature-obsessed life, with its dates, locations and species.

Steve thinks that although you don't have to be obsessive to be a pan-species lister, it certainly helps! He feels that PSL has made him a better professional invertebrate surveyor – he now goes the extra mile to find species that he hasn't seen before, and he has become more diligent and thorough in his survey methods.

Sally Luker, 53 (28th place with 5,381 species)

Sally is one of only a handful of female pan-species listers, and describes herself as 'a full-fledged natural history nerd'. When she was growing up, Michael Chinery's *Natural History of the Garden* was her bible, though in her late teens and early twenties she took a break from rearing caterpillars and getting excited about hoverfly larvae to indulge what she describes as her 'embarrassingly pretentious music/art student' phase. It was only when she was being offered formal accountancy training that her childhood memories and fascination with invertebrates returned, and she realised that she wanted to be an entomologist. After that she quickly took to creating lists of birds seen while on holiday, and photographing and attempting to identify everything that she spotted.

Her first biological record was of a Stoat *Mustela erminea* spotted running across a car park, and soon she was generating records by the bucketload. Sally now sits on the board of the local environmental records centre and on the committee of her county biological recorders' forum. A chance meeting with Seth Gibson at a bioblitz resulted in her joining the world of PSL in its original incarnation, and she fondly refers to her fellow listers as 'my tribe'. After returning to university and completing a PhD that happily involved recording all plants and their associated invertebrates in gardens, as well as rearing plant galls, Sally now shares her knowledge with members of the public via the Cornwall-based natural history education business that she and her partner have set up. She also gets to talk about bumblebees in her role with the Bumblebee Conservation Trust.

Ben Mapp, 23 (105th place with 2,214 species)

PSL has improved Ben's overall lifestyle as well as his knowledge and interest in the natural world, turning birding and casual recording into a lifelong hobby. He has always loved collections and listing, and this aligns perfectly with the never-ending nature of wildlife recording. His interest in nature began with birdwatching with his parents at the age of six. Attending the Kids Birdwatching Club at RSPB Rainham Marshes for 10 years not only greatly improved his birding skills but also sparked his interest in plants and invertebrates, which were almost always pointed out on the walks. PSL gave Ben a platform for broadening his horizons beyond birding and improving his understanding and knowledge of British biodiversity.

Ben finds that the unpredictability of PSL keeps him motivated to visit new sites and return to familiar haunts, as he never knows what he might find. The new PSL website has also made compiling lists easier (he was able to create holiday-week lists and compile lists for his house and local country park for the first time), and has motivated him to pay closer attention to all species, no matter how common, rare or unfamiliar they are. In addition, joining the PSL community has allowed him to meet other knowledgeable naturalists who share a similar passion, thus enriching his knowledge and understanding.

PSL has given Ben a sense of personal involvement and fulfilment. It has taught him about the biological recording process and how to make records, and he derives great satisfaction when a record is accepted and he sees a little dot appear on the *NBN Atlas* or another national database.

James McCulloch, 22 (list at 4,104 species as of August 2022 – not on the new site yet due to university commitments)

As a schoolboy, James was passionate about PSL – his list reached around 4,000 and topped the rankings for springtails. Without the encouragement of PSL to look at absolutely everything he might never have discovered his passion for springtails (a rather obscure group), which has culminated in him writing a book on the springtails of Surrey which is waiting to be published, as well as the discovery of a new species for the UK – the as yet undescribed *Katianna* sp. 8 – in his very own garden. His progress has started to slow recently, as his academic commitments as a university undergraduate are not compatible with spending time in the field and the hours of specimen preparation and identification that follow!

Yet PSL has continued to shape his path. The broad-brush approach that is fostered by PSL led him to write papers on springtails, beetles and moths at a remarkably young age. It also gave him the skills to work as a research assistant at Wytham Woods, where he collected and identified specimens for genome sequencing as part of the Darwin Tree of Life Project. It was these precocious qualifications which apparently made him stand out when he was interviewed for a PhD post at the Sanger Institute, a scientific establishment dedicated to genomics research. So James can attribute his career progression in this field almost exclusively to the knowledge and skills that he gained from PSL.

Andy Musgrove, 54 (sixth place with 8,451 species)

Andy decided to write a list of birds that he had seen at the age of 11, and has never looked back. Birds remained a focus throughout his school and university days, and then professionally at the British Trust for Ornithology (BTO), but after John Martin introduced him to moths in the 1990s these became an obsession. After a big pan-species push to see 2,000 species in the year 2000, his listing then slowed for much of the next decade, although a highlight was the creation (with Mike Prince) of BUBO Listing in 2007, as well as its subsequent conversion into the new PSL website in 2023.

Andy recently left his job at the BTO, where he had spent 25 years directing the UK's bird monitoring schemes, to become a freelance entomologist. This development was sparked by his growing interest in sawflies. He has also written the status review for this group for Natural England. Andy maintains that through PSL you can get 'close to the forefront of the taxonomic scientific effort relatively quickly with certain groups' (such as sawflies). In addition, as one of the minds behind BUBO and the new PSL website, Andy himself has had a big impact on PSL.

Louis Parkerson, 21 (51st place with 3,626 species)

Louis has been a birder from a very young age, and soon after he learned to write he started to keep lists of species he had found. In 2019 a birding friend introduced him to butterflies, Odonata and day-flying moths, and in the early months of the Covid-19 pandemic, equipped with a newly acquired moth trap and sweep net, he discovered PSL. After a year or two of focusing mainly on insects he started to obsessively record plants, which now comprise around 40% of his pan-species list. He is currently studying for a degree in Conservation Biology and Ecology at the University of Exeter's Cornwall campus, where a very different suite of species to his home county of Norfolk (and the help of many first-rate naturalists) has rapidly increased his pan-species list. Louis has just completed his placement year in the Scottish Highlands, and is now aiming to reach the 5,000 species mark.

Sarah Patton, 60 (19th place with 6,283 species)

Sarah remembers the first PSL weekend as a wonderful time in the field with some extremely talented and enthusiastic people and a lot of new species (many from groups that she wouldn't have tackled by herself). Starting PSL not only made her look forward to tackling new groups and visiting new places, but also prompted her to look back at over three decades of recording, locating and sorting old records and finding gaps in lists to be filled. She had started with birds back in the 1980s, and while working with keen naturalists at the Arundel Wetland Trust she soon developed a love for the chalk downland in the area. A talented self-taught field ecologist then introduced her to the worlds of botany and entomology.

Although Sarah had worked in the conservation sector for a number of years as a countryside ranger and nature reserve warden, while also pursuing natural history as a hobby, recording was the order of the day, rather the concept of PSL, which was a somewhat revolutionary concept at that time. Sarah has always relished the slightly unconventional, and enjoys telling inquisitive members of the public that she is looking for parasitic flies when she is spotted with a net. She has also gained a reputation for enthusiastically examining animal corpses for insect life. A moth that was a first for Britain, Patton's Tiger *Hyphoraia testudinaria*, was named after her, and she has recorded many unusual fungi and rare insects. Her responsibilities as a carer have limited her recording opportunities in recent years, but she has been using the new PSL website to process a backlog of records.

Stephen Plummer, 64 (third place, with 9,420 species)

Stephen has been drawn to wildlife since he was a very young child, and has pursued this interest as an enthusiastic amateur over the years. During that time he has worked regularly with schoolchildren and with older people to foster their interest in wildlife, and he has been very involved in local natural history societies. He also enjoys meeting with other naturalists from wider groups, such as the Wildflower Society and the British Mycological Society.

When he first heard about it, Stephen found the concept of PSL really appealing because it motivated him to fulfil a long cherished dream – to identify and understand as many of the animal and plant species that we share with this amazing planet as possible.

At that time he was enjoying moth trapping and had been studying hoverflies for a number of years, but PSL took all of this to another level. Previously he had lacked confidence in his ability to identify all kinds of creatures, but being part of a like-minded community gave him the impetus that he needed to overcome those self-imposed limitations.

When he was trying to express what PSL has helped him to achieve, he said that he was reminded of the Bible story in Genesis where Adam names all of the animals. He was a pastor in the Baptist Church for many years, and although he doesn't take that biblical account literally, he does believe that the story points to something deep in our humanity – something that speaks of knowing, understanding and being stewards of the world in which we live. So for him PSL has helped to lead to an even deeper sense of fulfilment.

Kev Rylands, 51 (13th place with 7,438 species)

Kevin Rylands' interest in nature developed during family walks in Hampshire and Surrey and holidays on a friend's farm on the Isle of Wight. He drew up his first bird list at the age of 13, and his knowledge of other taxa slowly increased with the support of others, especially at Rowhill and Yateley Commons. Prior to the establishment of PSL his recording was limited to groups that have vernacular names (such as vascular plants, birds and butterflies), but improved field guides and online resources helped to break down this barrier. Unlike many pan-species listers he has limited himself to field identification rather than specimens, with digital photography being an invaluable tool for this. The majority of his recording is undertaken at Dawlish Warren in south Devon, to maintain a biodiversity audit for this exceptional site, with over 2,000 species recorded annually.

Mark Skevington, 57 (43rd place with 4,095 species)

'Skev' was a reformed twitcher and active moth recorder when he started blogging in 2007, and he soon found a community of like-minded naturalists taking a wider interest in what they found while out and about. Through birding and moth trapping in species-rich habitats, and with the encouragement of like-minded bloggers, his interests soon diversified. Keeping lists was part and parcel of everything he had been doing, so when he heard about PSL it made immediate sense to him. He submitted his inaugural list (1,983 species) in November 2011, and has enjoyed a number of PSL field meetings, including the very first one (at Parham Park in 2012). Skev is always more confident if his subject has legs or vertebrae, but the premise that all species on the list have the same value encourages him to get the most out of even the most desolate-looking habitat, and he has added new-to-county species from many different orders. Juggling this with his family and work commitments has always been a constraint, and PSL is a lower priority now that he is County Moth Recorder for Leicestershire and Rutland, but he is more than happy to keep learning, while identifying and recording on a more casual basis.

Brad Scott, 60 (71st place with 3,120 species)

Brad has had a general interest in nature for years, though he rarely bothered to actually identify anything. He occasionally took photos of plants while on hikes, but it was only when he moved to Sussex and was regularly out in the countryside that he started to pay more attention to what was on his doorstep. He bought some books, and made a start by trying to identify a few plants and insects, aided by the very helpful community on iSpot.

It wasn't long before he realised that he needed some more expert guidance, and that although one can identify plants and animals from books, it really helps to have someone point the way and provide some structure to one's learning. The first Sussex Wildlife Trust course that Brad attended was my 'Grasses, Sedges and Rushes' weekend. It inspired him to further explore the rich and varied habitats in his local area, and he then attended my 'Introduction to Mosses and Liverworts' course, which prompted him to find out more at the local field meetings of the British Bryological Society and tentatively start recording.

Not content with just looking at plants, Brad started a pan-species list. This has provided him with a focus for attempting to understand the diversity of all habitats. He realises that he would only have the time to comprehensively locate and identify a tiny proportion of the UK's flora and fauna, but a pan-species list provides a really useful scaffolding for getting a feel for at least the common species across most taxonomic groups, and has also enticed him into more arcane areas, such as springtails. Appreciating that exploring the natural world is best shared with other people, he has also been involved in setting up a local natural history group, and has become one of the County Recorders for bryophytes in Sussex.

Julian Small, 51 (eighth place with 7,982 species)

Julian started pan-species listing in 2010, several years before he had heard of it! That year marked the start of a new moth-trapping project at the nature reserve where he was based, and he was keen to identify as many of the flies, beetles, caddisflies and other non-moth species that were trapped as possible. After finding out about PSL he became involved in identifying other groups, too, especially parasitic wasps. Julian is the first person to reach 1,000 species of Hymenoptera on the PSL website. There are thousands of species of wasps, and he has discovered that many of them are breathtakingly beautiful and interesting creatures. Julian says is not primarily a naturalist – his job involves managing nature reserves and their biodiversity but I would argue that being a naturalist is important to be able to do this kind of job well. He thinks of PSL as the study of biodiversity, and feels that it has deepened his understanding of what nature reserves are all about.

Bill Urwin, 73 (75th place with 2,500 species, but he hasn't entered his whole list yet)

Bill started birding at the age of 16. After training as a teacher and moving to Norfolk he spent the next ten years there, taking on the role of Honorary Warden at Norfolk Wildlife Trust's Upton Fen reserve and competing in the Country Life 24-hour birdwatches, where he was never on the losing side.

His interests widened to take in plants, dragonflies, hoverflies and woodlice, and in 1984 he moved to Somerset, where at different times he chaired the county moth, bat and invertebrate groups. He has found that being part of the PSL community and learning from its multitude of taxa experts has made his recording at Natural England's Shapwick Heath and as part of the Recorders of the Avalon Marshes (RoAM) group much more comprehensive, as they compile lists for the reserve managers to use when planning their conservation activities.

Simon Van Toller, 61 (20th place with 6,139 species)

Ten years ago, having had a lifelong interest in wildlife, Simon purchased a digital camera and set himself the task of photographing and identifying as many British species as he could. This gave him tremendous pleasure and was good for his mental well-being. However, it turned out to be a much bigger task than he had expected, and he also found it quite a lonely endeavour. As a newcomer he found himself shunned by established specialists and amateurs alike, and because he didn't know many naturalists personally he was finding it difficult to break into this world. He was particularly frustrated by the feeling that he was trying to do too much and should stay in his lane and focus on one taxonomic group.

Then it was suggested that he should sign up to the PSL website. Simon now believes that this is the home of true naturalists, as not only is his broad-brush approach to wildlife welcome – it is positively encouraged. This is a place where he feels that he fits in.

Harry Witts, 25 (39th place with 4,268 species)

Harry is best described as half birder and half general naturalist, as he often gets caught up in birds at peak migration times in spring and autumn, but for the rest of the summer he focuses on insects and everything else. Since he became serious about natural history he has gone through a number of phases in terms of his interest in different taxonomic groups – first butterflies, then birds, then moths and then bees. After that PSL broadened his interest to include all other insects and other taxa. At university he became interested in insect migration, and wrote his Master's thesis on how migrant insects of all orders interact with coastlines. He has been quite itinerant in his recording, moving between Leeds, Hull and London. He relishes how, on even a short walk, PSL can give you a cross-section of the organisms that inhabit an area, and he values the way that knowledge easily built around the framework of PSL gives you a broad ecological appreciation of how all of the species that comprise a habitat interact and fit into their environment.

Chapter 10

Pan-species listing and collaborative competition

There have been various pan-species listing (PSL) challenges over the last 15 years. Some have been opened up to all, many are personal and several have just been between a couple of friends. Although all of these challenges are great fun, they also generate a huge number of records, again using the notion of 'gamification' to inspire people to go to often great lengths to find and record species. Far from being frivolous and competitive, they are collaborative, enjoyable and worthwhile. I shall describe several different PSL challenges in this chapter.

1,000 for 1KSQ

This was not a challenge that I took part in myself, but it was a joy to watch so many people engage with it. It all started just before the old PSL website was built, and was a great way for people to engage with wildlife and recording, usually (but not always) in their home square kilometre. The aim of the challenge is fairly self-explanatory – to try to record 1,000 species in a square kilometre in a year, and the year when it really took off was 2013. One of the rules was that all records had to be submitted somewhere.

One of the best things about this challenge is its low carbon footprint. It also makes you feel more connected with the area immediately around your home, and it's a very good way to generate records of species you probably take for granted and that few people record. It's the gift that keeps on giving!

The results for the 15 people who successfully completed the challenge are listed in Table 9 (a further 22 people didn't quite make it or dropped out before the end of the year).

A blog about the nuts and bolts of biodiversity - finding LOTS of species.

But how many can you find in just a single 1km square?

The banner from Andy's blog. (Mark Lawlor)

Table 9. Rankings of the 15 pan-species listers who completed the 1,000 for 1KSQ challenge in 2013.

Rank	Species	Lister	County
1	1,407	Andy Musgrove	Norfolk
2	1,305	Graham Calow	Leicestershire
3	1,300	Tristan Reid	Cumbria
4	1,170	Rob Yaxley	Norfolk
5	1,113	Tim Strudwick	Norfolk
6	1,138	Mark Skevington	Leicestershire
7	1,109	Keith Robson	County Durham
8	1,102	John Martin	Avon
9	1,100	Mike Mullis	Sussex
10	1,081	Martin Gray	Lincolnshire
11	1,060	Mark Lawlor	Guernsey
12	1,032	Richard Comont	Oxfordshire
13	1,027	Sven Weir	Essex
14	1,013	Paul Bowyer	Somerset
15	1,002	Rupert Higgins	Bristol

1,000 lifers in a year

Many pan-species listers decide to try this personal challenge soon after starting, when lifers are still coming thick and fast. I struggle to add many more than 300–500 species a year now. I still average very consistently around 1.23 species a day, every day, even after 13 years. However, this statistic varies for other pan-species listers, from around 0.1 species a day at the slower end up to a maximum of around 1.5 species a day for those who are really going for it! Of course, for short intense bursts this figure can be much higher, in the region of up to 3.5 species a day, but over many years it is likely to be less than 1.5 new species a day. The average value for 14 long-term pan-species listers who provided data on this was 0.97 species a day.

I nearly managed 1,000 lifers back in 2010, but started to burn out towards the end of the year and had to prioritise my mental health. I had somewhere around 900 new species by the end of the year, though I felt too exhausted by it all to calculate the exact figure at the end.

My chances of getting 1,000 new species in a year are very low now. I could only do it if I planned it thoroughly, treated it like a full-time job and spent half the year in Scotland. I would probably also have to focus on some of my most under-recorded groups, which would be an impossibly large time commitment.

1,000 species in 24 hours

This challenge is a favourite of mine because we actually achieved it! The idea was to form teams of two, with one person doing the bulk of the identification work and the other person meticulously recording and, most importantly, recording in real time (you had to know which species number you were on as you recorded the species, so that you would know when you had reached the 1,000 species mark). All of the listing pairs started at midnight on 10 June 2017 and recorded continuously for a 24-hour period on the same day. Specimens could be taken but they had to be identified within the 24-hour period. There were three other teams crazy enough to give this a go.

The rules were that we could set up as much equipment as we liked before midnight, but we couldn't activate it until the stroke of midnight. Dave Green and I started setting up moth traps at Ebernoe Common at about 11.00 pm, and as midnight struck we turned the lights on. The first few minutes were the most intense. You hit 100 species in a matter of minutes, and at this stage you are significantly impeded by the speed at which you can write things down – you can find and identify species much more quickly than you can record them.

Dave and I had already done a couple of test runs to see if the recording system I had set up actually worked. It involved a detailed species list that I had already created from all of my data and existing records at Sussex Biodiversity Record Centre for the Sussex Wildlife Trust sites we were planning to visit. We had a route that took in all of the different habitats within a small area, in order to minimise driving time and maximise recording time. We started at Ebernoe Common (ancient woodland and species-rich neutral grassland), and then moved to Amberley Wildbrooks (rich and varied wetland), Graffham Common (heathland, bog and conifer plantation), Levin Down (scrubby chalk grassland via arable land) and back to Burton Pond (bog and acid grassland). The system was designed so that we were not double recording and so that we knew (as accurately as possible) what the species count was at any time. We did this by producing the species list in alphabetical order, either by scientific name for species that do not have widely used common names (the vast majority of this list) or by common name. For moths we used a combination of these two approaches, with the scientific names of micro-moths sorted in alphabetical order,

Dave and I at least 12 hours into continuous recording at Graffham Common. (Alice Parfitt)

all mixed up with the common names of the macro-moths. It sounds messy, but it worked.

I printed everything off on waterproof paper, and Dave colour-coded an impressive index system down the side for the different orders and groups. I bound it all together in a clipboard with plenty of gaffer tape to prevent it from falling apart. Presence was denoted by writing down the species number each time we encountered something. For example, if the 327th species was Spotted Flycatcher *Muscicapa striata*, we wrote down 327 in the cell next to Spotted Flycatcher. I hung a bird counter round my neck, and clicked it every time we identified something new, as Dave wrote it down. If I made a mistake and we already had that species (something that happened more often as the day wore on) it was impossible for us to double count as the cell was already filled in. The final results were very close to the score on the bird counter.

I shall never forget botanising by torchlight at 2.00 am on Ebernoe Common, ticking off Bird's-nest Orchid *Neottia nidus-avis* and Snakeskin Grisette *Amanita ceciliae*, with the roding of Woodcock *Scolopax rusticola* overhead in the first few minutes of the challenge, and a sudden deafening blast from a Nightingale *Luscinia megarhynchos* at point-blank range. It was magical. We were up to about 500 species before it got light and before we left Ebernoe. At this stage we were full of energy. Several hours later Dave was plagued with hay fever and my knee had swollen up, but we were close to the 1,000 mark. Then, at around 7.30 pm, we hit 1,000 species with a male *Drilus flavescens*, which was swept from rank, pollen-rich grassland at Levin Down.

We carried on light trapping at Burton Pond, where I hand caught a lifer, Dentated Pug *Anticollix sparsata*, a scarce moth that feeds on Yellow Loosestrife *Lysimachia vulgaris*. However, we really slowed down after that, standing around the moth trap like giant moths ourselves, not quite sure how we got there or what we should do next. It was extremely tough, yet so rewarding.

The final count was 1,035 species (Table 10). We walked 14.2 miles, drove 63.9 miles and recorded on average one species every 73 seconds throughout the day. We raised over £1,000, which we donated to the management of the reserves we recorded on, and we submitted all of the records to the Sussex Biodiversity Record Centre. In total, 460 of the 1,035 species were invertebrates.

This included 114 new site records and three species new to the reserve network. I know this because the year before this challenge I pan-species listed the whole of Sussex Wildlife Trust's reserve network – all 32 reserves. The Trust is still maintaining this list with my help (see page 339).

Table 10. The species totals for the four teams that took part in the 1,000 species in 24 hours challenge in 2017.

Rank	Listers	Species
1	Graeme Lyons and Dave Green	1,035
2	Stephen Plummer and Iain Outlaw	816
3	Sally Luker and Richard Comont	657
4	Seth Gibson	500+

The bird counter was a fairly accurate way to keep track of what species number we were on in real time (I was just nine species over the actual final count).

Dave and I enjoyed fine weather in West Sussex, but Sally Luker and Richard Comont struggled with heavy rain in Cornwall. Yet despite working in what seemed like a different season to us, they still managed an impressive 657 species – and they had taken a dog with them, too! I was barely able to feed myself on the day, let alone be responsible for another organism.

As far as I know, Stephen Plummer and Iain Outlaw were the only other pair crazy enough to attempt this challenge, on the Isle of Wight. They really enjoyed it, beginning alongside moth traps in the middle of Parkhurst Forest. It was always going to be a special time after the first species they noted was a Nightjar *Caprimulgus europaeus*, churring from the top of a nearby tree. They made the most of the wonderful Isle of Wight countryside, exploring many of the special habitats that the island has to offer, including the coastal downs, marshland, estuaries, beaches and rock pools. They ended up in second place with 816 species, and their PSL exploits even got them a photo and a mention in the local paper.

And, of course, Seth had a go on his own (breaking the rules), just within his own home hectad (yet more rule breaking) and only on foot (that wasn't in the

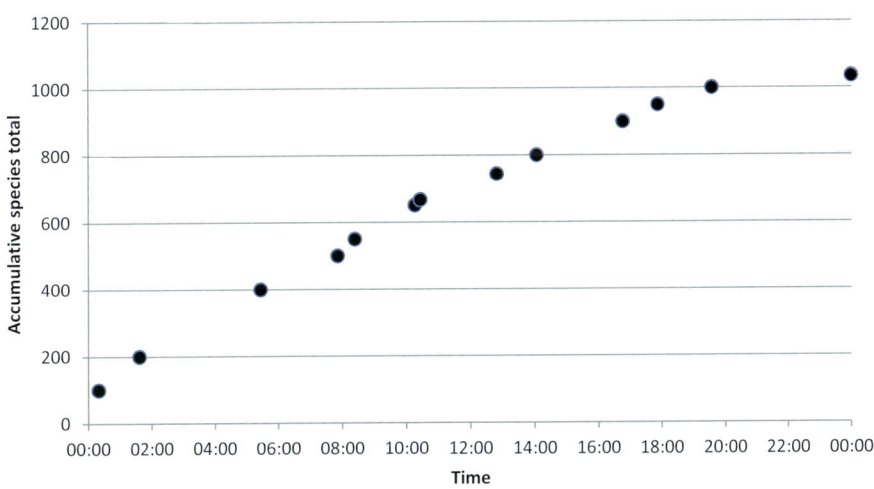

The rate of recording remained steady throughout the day, until we hit 1,000 species!

rules either, so basically it was a completely different challenge). Nevertheless, he just scraped 500 species, which was no mean feat when it had rained for 18 hours that day on Skye!

Over the next few years I often wondered about trying for 1,500 species in a day. Was that even possible? That would be one new species every 57.6 seconds – in other words, at least one new species every minute of every hour. You may not be surprised to hear that we went for it.

1,500 species in 24 hours

Around seven years after we managed 1,000 species in 24 hours we increased the challenge by 50%, to 1,500 species. Dave and I planned a slightly revised route with some new sites. We started where we finished the last time – at Sussex Wildlife Trust's Burton Pond reserve. However, this time we were the only team going for the challenge. We had mostly fine weather on a glorious late spring day. So, did we make it?

No, We really didn't. In fact we got to 1,000 species considerably later than we did in 2017 (at 10 pm, compared with 7.30 pm the last time), and we really struggled to beat our total, ending the day on 1,072 species, only 40 more species than last time (Table 11).

So what went wrong? First, we started in a place that got very cold by about 2.00 am, leading to a thick dew that stopped play until about an hour after sunrise. We then spent too long at a few sites that were not very productive – we simply

Dave Green and I at sunrise, around 6 hours into the challenge already.

Table 11. Number of species recorded in each taxonomic group during the 1,000 species challenge and the 1,500 species challenge.

Taxonomic group	1,000 species challenge (10 June 2017)	1,500 species challenge (25 May 2023)	1,500 species challenge plus microscope specimens (25 May 2023)
Vascular plants	424	403	403
Beetles	116	146	179
Arachnids	54	118	131
Moths	113	87	91
Birds	67	77	77
Bugs	51	60	67
Flies	28	35	43
Molluscs	26	25	25
Bees, ants and wasps	20	23	39
Butterflies	10	15	15
Bryophytes	39	15	15
Crickets, grasshoppers and allies	14	15	15
Other insects	6	12	17
Mammals	7	9	9
Crustaceans	7	8	8
Fungi	15	6	6
Myriapods	5	5	5
Springtails	1	4	4
Dragonflies and damselflies	9	4	4
Lichens	15	4	4
Fish	1	1	1
Reptiles	2	1	1
Amphibians	1	0	0
Total number of species	1,035	1,072	1,158

couldn't find as many species as we did the last time. We really had to work hard right up to midnight even to beat 2017's total, which was very tiring.

However, it wasn't a complete failure. Although the challenge was about the number of species we could find and identify in 24 hours (1,072 species), when we added all of the specimens that we identified *after* the 24 hours we reached 1,158 species (trying to key specimens out at the microscope after being on the go for 24 hours was not as much fun as I thought it would be). On the day we recorded 555 species of invertebrate, though I really wanted to get to 1,000 invertebrates in a day. However, I now realise how tough that challenge is, especially on cold nights when there are very few moths on the wing.

If Darth Maul were a beetle, he would probably look like *Pilemostoma fastuosa*.

The standout site was a client's farm that I had surveyed the previous year – Hoyle Farm on the Greensand in West Sussex. It was incredibly productive, and we even found around 50 species there that I didn't record during the survey last year, including one of my favourites – *Pilemostoma fastuosa*. Best of all, though, we raised £2,925 for Sussex Wildlife Trust!

Spider year-listing challenge

You could argue that this wasn't really a PSL challenge, as it involved only one invertebrate order, but it was born out of a challenge between several pan-species listers in 2019, and is very much in the gamification camp, with huge biological recording benefits. I started looking at spiders in detail around 2011–12. It all started when I bought 'Big Roberts', the three-volume book that shows all the UK spiders and their genitalia. Spiders soon became one of my main interest groups, so in 2019 when I saw a message on Twitter from Matt Prince asking if anyone was going to join in with a spider year-listing challenge, I had to go for it. The challenge was to see who could record the highest number of spider species in 2019. I started five weeks after Matt, and despite catching up at one point in the summer, by the autumn I had almost given up as he was so far ahead. Then around October I started to head out spidering in any spare time that I had. I was still about 25 species behind but, more importantly, I was making records of spiders that had not been seen in Sussex for decades.

In December of that year, I decided to give myself a two-week arachnological sabbatical after I left my job at Sussex Wildlife Trust, using my home town of Rugeley as a base for the north end of Staffordshire (perhaps the most southerly location for many montane spiders).

Oh boy, that was one of the best two weeks of my life. I searched for spiders in some beautiful places on my own all day long, and then spent each evening at the microscope. I ended the year on 379 species, and even ended up getting a lifer on New Year's Eve (*Gongyidiellum murcidum*). I also made a friend out of it, whom I soon recruited to PSL – Simon Van Toller. I beat Matt by 11 species and I really don't think he was expecting it. Neither was I, to be fair!

I now couldn't help wondering whether it was possible to see 400 species of spider in a year. An initial attempt in 2020 had to be aborted due to the pandemic, but by 2021, Tylan Berry (now one of the best arachnologists in the country) and I were ready to give it another go. This time I approached it with the same intensity as at the end of 2019, but for a whole year. I found that it was getting easier and easier, as I was becoming better at finding and sampling for spiders, more efficient at recording them and more confident about their identification (especially field identification). I believe I have now seen more spiders than anyone else in the UK, due in part to these three years. There are currently 50,655 spider records in my database, yet 36,931 of these have been made since 1 January 2021, so over 70% of all the spiders I have ever recorded have been in the last four years or so. Although Ty and I dominated the rankings, everyone else who took part also benefited greatly from the approach. I ended the year on 477 species of spider, which is over two-thirds of the species present in the UK.

During the challenge I made some amazing discoveries. For example, one day I visited Amberley Chalk Pit to look for two rare spiders, *Eratigena picta* and *Iberina candida*, and failed to get either of these targets. However, when I arrived home and went through the specimens I noticed a tiny golden money spider (less than 1.5 mm long), which was clearly *Centromerus albidus*, not seen in Britain since 1969! This species is designated as Critically Endangered (and was Presumed Extinct). There are still some people who cannot get past the competitive element, labelling the endeavour as "egotistical and pointless", but I say to such people: just look at the outcomes of such challenges and see the bigger picture here. The knowledge I have acquired through such challenges has permanently embedded, greatly benefiting

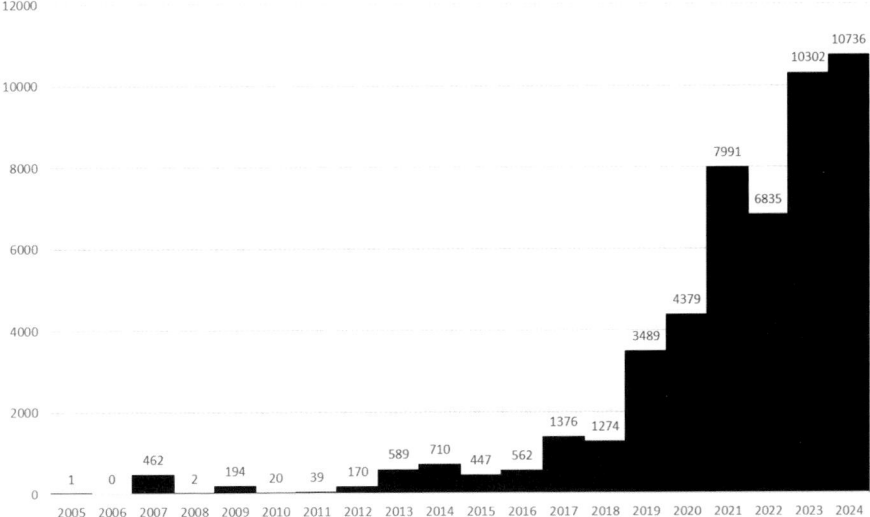

The first year of spider year listing tripled the number of spider records I made compared with previous years, but this figure has nearly tripled again since.

my work. It's also had many beneficial effects on spider recording in the UK. For instance, it is literally going to change the conservation status of some species, and it has generated tens of thousands of records, numerous county firsts and some species that have not been recorded for decades.

Here's something really interesting. Without actually trying to list spiders in 2023 (although I did target about 20 spiders I'd not seen before), I reached 401 species! This is more than the total I reached in 2019 that involved nearly every free day dedicated to the challenge in the second half of the year. This really shows how much I have been able to improve my knowledge of this fascinating group in just four short years. Gamification works.

Pan-species year listing

On the new PSL website it's now very easy to create a 'pan-species year list.' However, a word of caution is needed – PSL is really about a lifetime of effort, and regular pan-species year listing could well become a distraction from this. I year listed birds for around 20 years, and this involved a lot of travelling to the same places each year – places that a large number of other people were also heading to, often to look at the same birds. I think these records are not nearly as useful for wildlife recording as under-recorded groups on under-recorded sites. I have always liked the idea that PSL could help to divert some of the time and energy of many birders away from annual year listing and towards more under-recorded groups. I appreciate that this might seem to contradict what I have just said about the benefits of year listing spiders, but there are two big differences here. First, spiders are a really under-recorded group, and secondly, I only year listed them for two or three years to receive the benefits.

PSL is a lifelong pursuit to amass the largest list possible over the course of your life. If 'pan-species year listing' becomes too central a focus, and you start doing it year after year, it could end up defeating the object. You might spend so much time chasing the year-listing targets from the website that you stop recording new species, which is what PSL is all about. It is also not practicable to tick off each species on the website each year, as this would eat into time that might be better spent finding new species, or generally recording new areas.

Blind pan-species year listing

What I would suggest then is give it a go, but don't make it an annual thing. Another way to year list, is to do it without using the website as you go, that is generate your list, only when the year is out. In 2023, I made 55,110 records of at least 4,419 species but I wasn't year-listing. Most of this was from paid surveys, volunteer surveys and lots of travelling around the UK interviewing people for the book and a few PSL-style holidays. I didn't even check the total number of species until I had finished all of 2023's specimens in mid-March 2024 (I call this 'blind' year-listing). There are so many common species that I didn't see, yet in 2023 I saw half of everything I had ever seen in my life. Of these 4,419 species, 3,267 were invertebrates (including 1,013 beetles and 401 spiders). My records from 2023 alone would have put me in 30th place on the rankings. This figure of 4,419 species seen in a year is likely the record to date, so I challenge the PSL community to beat me, is there someone out here who can get to 4,500 or even 5,000 species in a year!? Maybe I will make 2026 the year I give PSL year listing a proper once-in-a-lifetime go; is 6,000 species in a year possible?

City Nature Challenge

Some projects are adopting the PSL approach without realising it, and City Nature Challenge, which is an international project, is one of them. This community-based challenge takes place over three days from late April to early May each year, and uses iNaturalist as its base.

For the City Nature Challenge I have to submit my records without photos (as I didn't take any), so that they are then not classed as 'research grade'. This way I get to submit them as usual from my database straight to my local record centre, without fear of duplication.

This is an interesting challenge, and it's well worth taking part in it. I would encourage you to restrict the use of iNaturalist to a minimum, as iRecord has been set up as a one-stop shop for anyone to enter their records, from where they will be forwarded to the many recording schemes, records centres and conservation organisations in the UK. I challenge the UK natural history community to do better, so that we can have a recording platform as slick and appealing as iNaturalist, perhaps with more of a community vibe. It clearly is much more suitable for these kinds of challenges than iRecord, which probably partly explains why it has taken off so rapidly in recent years. In addition, some counties haven't adopted iRecord, or have come to it late, which means that the verification process doesn't work well

The Sussex Tiger *Nephrotoma sullingtoniensis*, one of my highlights of 2023.

everywhere. In Sussex we were one of the early adopters, so it has always worked fairly seamlessly here.

One unexpected advantage of the City Nature Challenge in 2023 was that the 'city' of Brighton and Hove recording area now extended as far north-west as Pulborough, and as far south-east as Seaford. This meant that the city now included the RSPB-managed Wiggonholt Common, an area of heathland just south of Pulborough. I headed there one mild afternoon to do some recording, and was quite surprised by the large numbers of *Nephrotoma* craneflies that were flying up out of the Heather *Calluna vulgaris*. It wasn't until I got home that I realised these were something rather special. They were the Sussex Tiger *Nephrotoma sullingtoniensis*, a very rare cranefly which had only ever been known from the nearby Sullington Warren, and even then was represented by only three records. People have searched for it in late June for many years to no avail, probably because its flight time was actually much earlier in the year. This shows why it's important to record at a time when other people are perhaps not typically recording. April is my favourite month for biological recording, especially for spiders but also for the high rarity content in samples across a range of taxa. You also never know what you might find in early spring, even if the general species richness is lower than in the following months.

I have helped the city of Brighton and Hove to record the highest number of species in any city in the UK over the last three years, by making sure that my farm bioblitzes fell during the City Nature Challenge. I made 992 field records (rising to 1,044 records when the identifications were added) in a single day on 1 May 2022. In fact, over the whole weekend in 2023, I made around 2,801 records of 1,014 species, which included 570 invertebrates, 303 plants and 77 birds.

Tony Davis & Seth Gibson's annual challenges

Perhaps the most challenge-heavy duo of all time are Tony Davis and Seth Gibson. Their love/hate relationship is a joy to behold. In 2022 they did quite a simple challenge: which one of them could see the most species of vascular plant in a year. When the Walrus *Odobenus rosmarus* turned up relatively close to Tony's home in late December 2022, I called Tony to let him know – only to find Seth (the lucky git) just happened to have stayed at Tony's house that very night. By the time I arrived on that wet and icy December morning, they were already there. And what were they going to do after? Try and find some Agrimony *Agrimonia eupatoria* as Seth needed it for his vascular-plant year list (it's not so common on Skye where Seth lives). Now, there might not have been much of a conservation gain to that specific record, but there will have when you consider all of the other great finds they will have made over 2023 with this kind of challenge.

I've seen them do all sorts – the one I liked most was flies & beetles. Pulling out a couple of orders like this, and really drilling down into just those, is a great idea. It's great how some challenges are just between a couple of people, especially when they share their progress on blogs and bring people along with them. The new website is perfectly set up for such ad-hoc challenges.

Chapter 11

Pan-species approaches to surveying and monitoring

How to pan-species survey everything, everywhere, all at once!

Going freelance at the start of a global pandemic might seem like a cataclysmic mistake, but an opportunity arose from this situation – the private landowners and local authorities kept me going when much of the NGO work was postponed or simply disappeared. In June 2020 an existing client asked me to do a 'biodiversity survey' on her large farm – over 600 ha, divided into two discrete areas across two distinctly different landscapes, downland (chalk) and Weald (clay).

This is how I designed the method for surveying biodiversity on farms. Large farms (around 200–300 ha) get the gold standard – six visits spread out about a month apart from April to September. Each site is split into six, roughly similar-sized compartments, divided on the basis of habitat type where possible. An awareness of existing management and proposed management is vital when setting up such monitoring, and it's important to get these compartments right from the start, as they are hard to change after the first visit.

During each visit exactly one hour is spent recording within each of these large compartments (40–50 ha each for a 250–300 ha farm). The key is to keep moving, so as to be as close as possible to the next compartment when the hour is up. The compartments are tackled in a different order on each visit. This ensures that no one compartment gets preferential treatment at the best time of day, or when your energy levels are at their highest. Also, by tackling the individual compartments in a different order, you will start and finish at different places, which will push you to find new areas on these large sites. This method works well at the landscape level, which is a popular concept with conservationists, who often struggle to collect biological data systematically at this scale. It is also modular and can be scaled both upwards and downwards.

Starting the list from scratch in each new compartment means that you will record the common species six times (or more) over the course of a day. Although it might seem as if you will not pick up as much as you would if you were targeting rarities or focusing solely on the biggest species list, that is not the case. One of the most important principles of this approach is that *it keeps you moving* and makes you cover a lot of ground. In 2024 I recorded 893 invertebrate species during one such survey. In fact it was the highest invertebrate total I had ever recorded at the time, even including those occasions when I only surveyed for invertebrates. In total there were 74 rare or scarce species – one of the highest figures for any invertebrate survey

I've done, including all SSSIs, where I am just surveying for invertebrates. Overall, I recorded 1,237 species from that site. The fact that it's a working dairy farm and not a nature reserve makes this all the more astonishing.

These surveys do effectively become invertebrate surveys with birds, plants and mammals tagged on. It is inevitably a compromise – for example, lichens, most fungi and bryophytes are not covered in as much detail as I would like – but in general I cover as much as I can, and more than most. The sudden interest in biodiversity on farms is highly compatible with the 'broad-brush' approach that PSL can offer. Farmers are unlikely to be in a position to pay a botanist, an entomologist and a birder to survey their land, especially when their brief is simply to find out what they have, where it is and how they can manage it better than they are doing at present. The benefit of having one person conduct the survey with a broad-brush approach is that the core principle of PSL, that every species counts as the same, will also be applied to the management recommendations.

I'm working my way through the entire Downland Estate (c.5,000 ha) for Brighton and Hove City Council using this methodology, I have surveyed six farms so far, and am just over a third of the way through the Estate. One of the big advantages of adopting a standardised approach is that I can make comparisons between sites and amalgamate all of the data to produce a landscape-scale assessment, too (see Table 11 on page 321).

Even with six visits you won't manage to get everywhere, and this work is physically and mentally exhausting. On each visit I regularly walk over 10 miles while carrying about 15 kg of equipment, making up to 1,000 records. However, this approach really does capture a flavour of what is happening on these big farms and estates, and by visiting throughout the year you are in a position to see what is going on in terms of management, and then provide targeted recommendations based on these findings and observations. Much of the management happens late in the year, so although the September visits return the lowest biodiversity totals of the year, they often give you the most insight into what the farmers are up to. 'Oh, that's why those fields have nothing in them!' is the kind of thing you will hear me saying a lot in September!

So many farmers want to do the right thing for wildlife, but the industrialisation of farming over the last 50 years or so has led to a mindset that often tries to industrialise any wildlife-friendly management, too. The three pieces of advice I most frequently give are never to do everything all at once, to try to be less tidy, and to 'make sure you remove those arisings!' Yet we shouldn't expect farmers to know all this – that is why they are employing people like me. It is just so heartening that so many people are interested in wildlife on their farms. This is a very exciting time for biodiversity, and there is a need for farmers to be more widely recognised for the work they do here. The sites that I survey are often wildlife havens, a far cry from the barren deserts that the public often perceives UK farms to be.

For smaller farms, six visits are harder to justify so I make just three visits, using exactly the same method. With fewer visits the timing is more crucial – you will need to carefully space them about six weeks apart to maximise the biodiversity you will record throughout the year. However, having fewer visits is a compromise, and you will record less than with the six-visit survey.

A very unexpected find of around 300 plants of Great Pignut *Bunium bulbocastanum* at Paythorne Farm.

It's possible to use the modular nature of the six- or three-visit survey to cover several small farms of roughly similar size (*c.*40–50 ha) if they are clustered close enough together to avoid the need to drive long distances between them.

On the farm survey I mentioned at the beginning of this chapter I found a thriving colony of Great Pignut *Bunium bulbocastanum*, known in Sussex only very recently from a dozen plants a few hundred metres away on National Trust land, whereas on the farm there were around 300 plants. I also found a colony of Cistus Forester *Adscita geryon*, a tiny struggling population of Adonis Blue *Polyommatus bellargus* and important arable plants such as Night-flowering Catchfly *Silene noctiflora* and Narrow-fruited Cornsalad *Valerianella dentata*, as well as a few species I had not seen before, such as the weevils *Leiosoma oblongulum* and *Tychius lineatulus*. This is the stuff that I need to sustain my energy levels, it's as important as my lunch! I do start to get tired when I am not finding very much, and I am sure this is related to a lack of dopamine. It is amazing how much dopamine can sustain me; it can push me through some very physically demanding situations that I barely notice as long as I am finding lots of wildlife.

My summary data from the three-visit and six-visit farm surveys are listed in Table 12 and Table 13. I am very grateful to all of the farmers and their representatives for giving me permission to reproduce these figures.

Table 12. Data from three-visit biodiversity farm surveys in Sussex (the grey-tinted sites denote sites that were surveyed as groups or clusters).

Site	Undisclosed site 1A	Undisclosed site 1B	Undisclosed site 1C	Iford arable	Iford chalk	Iford brooks	Cockhaise	Woodsland	Clapham Farm	Haymans Farm	Undisclosed site 2	Alciston Court	Hoyle Farm	Hermitage Farm	Hackhurst Farm
Area (ha)	180	370	370	464	109	297	151	128	178	60	265	180	100	53	39
Year	2022	2022	2022	2021	2021	2021	2020	2020	2020/21	2022	2022	2022	2023	2023	2023
County	East	East	East	East	East	East	East	East	East	West	East	East	West	East	East
Geology	Sand/clay	Sand/clay	Sand/clay	Chalk	Chalk	Clay	Clay	Clay	Chalk/clay	Clay	Clay	Chalk/clay	Sand/Clay	Clay	Clay
Records	1,679	2,038	2,046	1990	1995	1709	1,697	1,871	1,635	1,719	2,049	1,956	2,169	2,342	2297
Total species	663	879	936	719	657	650	746	768	782	799	883	726	904	933	925
Mean number of species per compartment	264.0 ± 9.60	318.8 ± 9.2	330.0 ± 12.6	257.8 ± 15.8	267.5 ± 9.0	245.8 ± 15.5	282.3 ± 13.2	296.7 ± 24.3	249.8 ± 9.5	277.3 ± 6.70	333.3 ± 12.9	260.7 ± 9.7	293.5 ± 13.6	313.5 ± 15.8	294.7 ± 19.3
Invertebrates	431	558	588	472	433	396	473	464	499	538	531	441	621	633	603
Invertebrates with status	22	25	26	27	43	19	14	15	40	43	20	23	56	43	29
Proportion with status	5.1%	4.5%	4.4%	5.7%	9.9%	4.8%	3.0%	3.2%	8.0%	8.0%	4.0%	5.2%	9.0%	6.8%	4.9%
Spiders	110	96	115	90	84	71	78	83	77	113	113	78	111	118	109
Beetles	93	168	182	152	118	135	151	135	158	151	149	130	181	189	184
Hymenoptera	23	28	29	44	42	17	30	34	30	28	33	34	44	36	36
True flies	39	55	52	33	24	25	39	45	21	34	47	39	49	36	37
True bugs	57	73	86	66	58	45	78	69	81	82	77	67	108	98	86
Moths	36	40	41	26	37	20	31	27	26	54	36	31	45	58	62
Butterflies	13	23	16	16	22	15	15	19	21	20	16	21	22	15	18
Dragonflies	9	14	15	2	2	13	7	12	7	8	12	1	2	8	8
SQI (Pantheon)*	114	115	116	122	132	113	108	108	124	129	110	112	135	120	114
Mammals	4	9	10	5	5	3	6	5	3	4	8	4	2	7	5
Birds	42	61	60	54	46	65	58	50	49	48	56	48	42	46	41
Birds with status	18	27	25	28	23	33	21	16	25	22	22	20	18	20	14
Plants	158	235	257	182	164	174	200	241	223	185	271	228	214	199	159
Plants with status	1	0	0	2	5	2	0	0	7	0	1	2	5	0	0

*Pantheon is the Biological Record Centre's online database for analysing large lists of invertebrates, the SQI is a score of how rare the assemblage is.

Table 13. Data from my six-visit biodiversity farm surveys in Sussex (the four grey-tinted columns represent a landscape-scale study of the Brighton and Hove Downland Estate; the dark grey column represents the amalgamation of data from these four farms).

Site	Truleigh Farm	Perching Manor	Shaw Farm	Marshalls Farm	Ovingdean Court	Farm cluster A	Farm cluster B	Court Farm East and Pickers Hill	Downland Estate assessed so far
Area (ha)	350	275	40	285	220	283	197	246	949
Year(s)	2020–21	2020–21	2022	2024	2022	2022	2023	2023	2022–2023
County	West	West	East	West	East	East	East	East	East and West
Geology	Chalk	Clay	Clay	Clay	Chalk	Chalk/clay	Chalk/clay	Chalk	Mixed
Records	2,348	2,698	2,378	568	2,815	2,287	4,324	3,995	13,422
Total species	822	974	877	1,237	862	855	1,076	962	1,618
Mean number of species per compartment	353.3 ± 16.0	402.0 ± 12.0	364.5 ± 10.3	517.5 ± 5.8	382.7 ± 17.8	367.2 ± 12.3	437.8 ± 11.6	397.7 ± 18.0	396.5 ± 9.0
Total invertebrates	547	633	616	893	562	547	786	657	1,160
Invertebrates with status	53	33	27	74	50	35	76	78	128
Proportion with status	9.7%	5.2%	4.4%	8.3%	8.9%	6.4%	9.7%	11.9%	11.0%
Spiders	77	101	104	145	84	89	114	101	156
Beetles	185	188	194	284	187	192	258	219	395
Hymenoptera	42	51	31	60	46	37	79	54	102
True flies	46	50	61	81	46	32	74	58	105
True bugs	86	108	94	146	83	94	108	104	165
Moths	45	44	37	81	43	43	48	48	111
Butterflies	22	23	21	17	24	19	30	23	31
Dragonflies	3	7	12	7	2	1	6	4	6
SQI (Pantheon)*	131	116	115	126	128	119	129	134	133
Mammals	3	6	6	7	2	4	6	4	7
Birds	50	64	61	58	52	59	55	53	78
Birds of Conservation Concern	19	25	31	30	30	31	27	28	42
Vascular plants	219	257	182	252	237	229	205	234	337
Plants with status	9	1	1	2	7	5	3	5	9
Chalk-grassland indicators	Not recorded	Not recorded	n/a	n/a	42	43	36	49	55
Arable plant index	56	28	n/a	13	59	33	13	49	74

Recording multiple taxa in practice

Suppose you have one hour to cover a 40 ha compartment on a downland farm (which could be one huge field, or a number of smaller ones), the clock is ticking and you want to get as many species as you can for your client, in the most standardised way possible. First, you need to have all the necessary equipment with you. I always carry the following:

- *A light but sturdy sweep net and frame* (from Watkins & Doncaster) *with a white Velcro butterfly net* (from NHBS) *attached*. They do tend to rip, but this system works well for me. If I see a day-flying moth, butterfly, dragonfly, fly or bee that I need to individually catch to see up close, a standard sweep net with a calico bag isn't going to work here, but a clear butterfly net can also act as a sweep net, which is why I get through at least six of them a year. My approach is 'one net fits all'. Spin the net round through 180 degrees and you have a beating stick. This means there is one less thing to carry and all my kit is out of my bag and in my hands, all of the time.

- *Suction sampler*. This is fairly heavy, but I have learned to carry it under my left arm, tucked in under my armpit. You might want to find a different way to carry it that works for you (some clip it to their belts or shoulder straps), but the key thing for me is that I have my right hand free, holding my net ready for one-armed sweeping, while my left arm is free to carry my tray.

- *Beating tray*. My sorting tray (it's called a 'confectioner's tray') is larger and heavier than many people would want, but it works for me. It serves both as a beating tray for beating foliage and as a tray for sorting through samples collected from the butterfly net and the suction sampler. Its large size is particularly useful on the few days when I go out in the field with other people, as it's nearly the size of a dinner table! Flip it over and you can use it for sorting through wet samples, such as sieved *Sphagnum* or aquatic invertebrate samples.

- *Binoculars and hand lens*. Pentax Papilio 8.5× magnification binoculars and a hand lens are always tied around my neck.

- I only ever take a pond net and a sieve when I need them. I usually take a sieve out in winter and/or when there is moss to be sieved.

The key thing about this set-up is that I am mobile – I don't have to stop frequently to deploy or pack away equipment, as it's always in my hands. I can cycle from sweep netting to beating to suction sampling as I am moving. This is vital for covering as much distance as possible over large areas of land. One-armed sweeping is not for everyone though. If I have to spring after something, I simply drop my tray and suction sampler and launch after it. Writing things down quickly on the move in this way takes some creative use of my limbs but it is achievable. It works. It really shouldn't, yet it works.

What does an hour of such intensive surveying actually look like? As the clock starts I usually begin with all of my field kit on the ground and I walk around in a big loop, recording all of the plants, birds and large invertebrates that I can see

In action with my suction sampler under my left arm and my tray in my left hand. (Dave Green)

– in other words, the easy stuff. After 5 to 10 minutes, or as soon as these records start to dry up, I head to any distinctive feature in the landscape (maybe a chalky bank, a dew pond or a large patch of Gorse *Ulex europaeus* scrub. In arable blocks, there's usually only one route available, the margin of the field). The key point is to keep sampling as you move towards or along the feature, switching between the different vegetation layers that can be sampled and the relevant survey techniques. If something isn't generating records, switch to something that is. A longer sward would involve me switching to a period of sweep netting and then back to suction sampling, whereas very short grass with no sweepable sward is better sampled just with the suction sampler.

When you reach the feature, work it before heading to the next feature (perhaps a fallen dead tree next to an oak with low and beatable branches). If I find something unusual, it gets a more detailed, eight-figure grid reference. Everything else is just recorded to a site centroid in the middle of the compartment or field. You have to be reactive, adapt to the seasons and use your experience. You most definitely do not walk a transect with this approach. You almost have to think like an invertebrate! With practice, this can become a surprisingly standardised method for collecting vast amounts of biodiversity data within a short time, and it's also hugely rewarding and enjoyable.

Of course, during my invertebrate surveys (when I am not covering all other taxa) the areas that I cover tend to be much smaller and I am recording a lot less, which makes everything a little easier. However, I still always use timed counts across carefully selected compartments.

I put everything I can't identify in the field straight into alcohol in a push-cap tube on my left hip on my utility belt for ease of access, which also means there will less processing to do on the same evening. Alcohol is not as good as ethyl acetate for bees (I have to fluff them up with a little paint brush I call my 'fluffer'), but it's better for spiders, and just a lot easier if you have a busy field season. I identify birds mainly by ear (I wonder how many silent soaring raptors have drifted over my head while I am face down in my tray!). Effectively, this method is an invertebrate survey with other species tagged on to it.

Regular notebook stops to write things down are vital, by three or four species, I start to forget them. My record total number of species to date for a survey like this was during a farm bioblitz day that coincided with the City Nature Challenge – I made 1,042 records of 408 species during a six-hour recording period. This method comes with a significant time commitment to enter just the field data into Excel, so that I can then import them into my database (over four hours for that number of records).

Of course, all of these data need to be analysed and interpreted so that they yield something meaningful. This is often hard with baseline surveys, as you have nothing to compare the data to, but that is where the internal analysis comes into play. All of that seemingly pointless repetition, starting the list from scratch in each compartment, comes into its own here. These internal comparisons are great not just for identifying biodiversity hotspots, but also for highlighting where the poorer areas are.

I never miss an opportunity to convert biodiversity records into standardised numbers and indices. Some of these are universal and I will use them in every survey, such as the total number of species, total number of invertebrates, proportion of invertebrates with conservation status, ubiquitous species (those found in all compartments) and unique species (those found in just one compartment). Other metrics are rather more site-, habitat- or taxa-specific, such as total number of ancient woodland indicator plants, Plantlife's arable plant index, the total number of dragonflies or the Saproxylic Quality Index.

In future years, if you use exactly the same method, you can make direct comparisons not only between the years but also between the compartments. A further form of analysis involves the bulking of all years by compartment. In such an accumulative analysis, with each round of surveying (each coat of paint on the masterpiece) the assemblage of species recorded edges a little closer to (but never reaches) the actual, absolute species assemblage present. What I love about the absolute species assemblage for any site is that we can never ever know it: an absolute list exists only within the fabric of the Universe. All we can do is a make a crude attempt at that list which gets a little more accurate for every additional day we spend on that survey.

This method is not for everyone. Rarity hunters might struggle with the repetition, counting all the common species at least six times, and possibly as many as 36 times over a six-visit survey. Yet with such huge sites, which are almost always unknown quantities before the first visit, this method is really great for keeping you moving. It unquestionably generates huge numbers of records across the landscape, and these have a meaningful, spatial element. Without a method designed to keep you moving – in other words, if you spend six hours going wherever you want on a large site – you are likely to spend a disproportionate amount of time near the starting point or at the first hotspot you find, and far too little time in the areas you would otherwise reach towards the end of the day. There would probably also be some places that you wouldn't get to at all.

Field identification

For the last six years or so, on each day I have spent in the field I have been noting down the total number of invertebrates that I have field-identified that day. For each species that is new for the day, I write the number next to it in a circle. This takes only a moment to do, and is a really worthwhile exercise for the following reasons:

1. I don't need to kill so many invertebrates to identify them (they are very happy about this).
2. This reduces the amount of time I need to spend at the microscope in winter (I am very happy about this).
3. I become a more effective naturalist, by understanding what I am seeing in the field as I see it, which benefits my understanding of autecology. I can start to make observations of what species are present in the field, where they are exactly and how they are using the places where they live. Identifying all specimens at home now feels like a very detached way to record invertebrates.
4. It helps to push the boundaries of what is considered field-identifiable material. This is particularly relevant to spider recording but affects all areas of entomology.
5. It allows me to assess how good the day's surveying has been, by comparing the day's total with all of the other totals I have recorded that year (anything over 200 always feels like a good day, while anything over 250 is exceptional).
6. It keeps my energy levels up and is a lot of fun! PSL is all about the 'gamification' of natural history, and this really is an extension of that.

This approach gives me an incentive to take partially memorised keys from the desk into the field. However, it's important to note that I can only do this because I have spent years keying these species out at the microscope first. If you think you can skip this part completely and jump ahead to only field-identifying species, you have missed the point. No entomologist can do a thorough survey without taking a lot of specimens. Even when trying to minimise my daily specimen collection, I still come home with hundreds each day, equating to over 10,000 a year.

Deciding whether it's worth putting in the effort to field-identify a species is ultimately a judgement call. If it takes you 10 minutes to field-identify something, this is not a good use of your time, whereas if it only takes 10 seconds then perhaps it is. Clearly a good hand lens is vital for this approach, but more important still is knowing where that one ID feature that you need to see is located on the species, and being able to see it quickly. You will need to know the angle at which to view that feature, and therefore you will need to know the best way to hold or store the individual. All of these things come with experience, but the more you work at it, the better you will get.

One such survey, which I carried out in 2023, resulted in the largest number of field identifications I have ever made in a single day of surveying. My previous record was at Hurston Mill in 2022, where I recorded 271 invertebrate species in the field. These high totals are almost always achieved in May and June, so I wasn't expecting to beat this record in August! After a six-week drought in late spring and early summer, many sites sprang into life between July and September. In the first of six one-hour recording blocks (bear in mind I was also surveying plants, birds and mammals) I reached 100 field-identified invertebrates, including several Nationally Rare species. I still had five hours of recording time left to find another 171 species to beat my previous record. The game was on!

After those five hours of recording I had hit 298 species. Hoyle Farm is a small to medium-sized farm in West Sussex and has some of the best acid grassland I have seen in the county, as well as being home to many fantastic invertebrates. I think finding a new location for the Nationally Rare spider *Philodromus margaritatus* was the highlight of the day. The site has a varied underlying geology, and all of the associated habitats that this geology supports are in a relatively small, almost circular layout. Everything came together – a warm sunny day after rain, a well-managed site with a variety of habitats, and because this was my third visit I already knew where many of the hotspots were. With the 69 microscope identifications added to the 298 field identifications, the total for the day was 367 species of invertebrate

Philodromus margaritatus, unexpectedly beaten from Ash at Hoyle Farm.

and around 789 records altogether. Moreover, 32 of the 367 invertebrate species were found to have some form of conservation status, this high proportion being a very useful indicator of the site's invertebrate quality. The site was so special that we went out to visit it during the 1,500 Species Challenge in 2023, where it was the standout site of the entire 24-hour period.

I can write down species very quickly in a hardback notebook, placed in my combats' right thigh pocket. Yet I am not sure I would have ever felt like I could regularly survey in this way, if it wasn't for days like the 1,000 Species Challenge or other 'one-man bioblitzes'. Would I be doing this without the pan-species listing approach? I can confidently and categorically say I would not. I often get asked "but surely, you're the only person who can do this?" No. Just look at the more experienced end of the pan-species listing community, there will be some who are driven and experienced enough to follow this method. To me, the whole point of PSL is to bring about this kind of change.

I am regularly told that I need a trainee by my clients. So here I am doing something even better, sharing my methods in the hope that others might adopt them and help farmers and biodiversity in the process.

Chapter 12

Pan-species listing of sites

Pan-species listing (PSL) has many indirect benefits for conservation, such as increased numbers of records of under-recorded groups, and the production of many more skilled naturalists to generate such records, but there are some more obvious and direct benefits, too. In this chapter I shall look at how a PSL approach can be applied not to personal rankings but to a site or collection of sites. This approach might appeal more to those who have problems with the perceived competitiveness of PSL. Of course, site lists were around long before the advent of PSL, but very few site managers have a full pan-species list for their sites even now.

Creating a pan-species list for a site or location

A small number of people do sign up to the new PSL website just to manage one or more site lists, rather than to keep their own personal pan-species list. Yet, at the time of writing, you can enter data for a personal site such as your garden very easily, but it's not that easy to make a collaborative site list. The only way to do this is to set up an account with multiple users, but it's not possible to make site lists public for anyone to update. Nor are there 'site rankings' at present, but rest assured – one day it will be possible to look down the rankings of, for example, pan-species listers' gardens.

There are currently only 281 location lists on the new website, and just 24 of the 410 people (5.9%) on the main rankings also have garden lists. I am surprised by this, as garden PSL is immensely satisfying. My own list has only 348 species, but it is a tiny garden, and as with all these things the goalposts shift, making the mundane exciting. Memorable garden wildlife encounters include the time I hand caught a Privet Hawk-moth *Sphinx ligustri* one summer evening, or when a male Banded Demoiselle *Calopteryx splendens* flew through (I am miles from any running water), or when a Dunlin *Calidris alpina* called overhead on a foggy morning. Meanwhile Plumed Fan-foot *Pechipogo plumigeralis* and Golden Twin-spot *Chrysodeixis chalcites* have been lifers in the last two years.

On the old PSL website there was a place to write some notes about your site, such as the habitats present, their size, and whether there are any designations or any relevant history. Yet sites and locations never quite took off in the way that the personal rankings did. I entered some of Sussex Wildlife Trust's reserves on to the old website, soon after I pan-species listed the entire reserve network, but it's tricky to keep it updated when there are so many sites.

In early 2022, the top site on the old Biological Records Centre (BRC) website was Wicken Fen in Cambridgeshire, with 8,674 species, followed by Esher Commons in Surrey, with 7,945 species, but neither of these had been updated since 2015. After

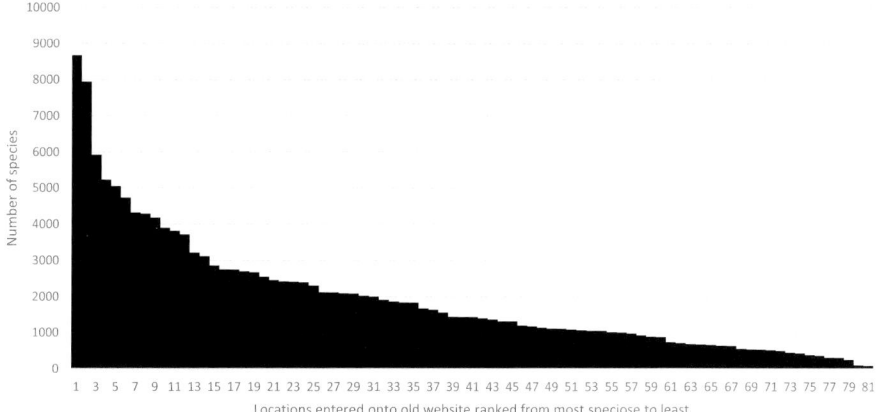

Spread of the 81 sites on the old PSL website's location rankings.

this, there was quite a big fall in numbers, with RSPB Minsmere in Suffolk having 5,928 species. It will be a while before the new PSL website can build up species lists on this scale. At the time of writing, the top location on the new website is Dawlish Warren, at 5,574 species and, although not yet added to the new website, staff at Wicken Fen told me that their list reached the 10,000 mark in 2025.

Pan-species listing an entire reserve network

Around 2016, when I was the senior ecologist at Sussex Wildlife Trust, we seemed to be becoming too concerned with wildlife we didn't have, while at the same time not knowing exactly what wildlife we did have. Therefore, reserve by reserve, I began to build a spreadsheet that pulled all of the available data together in a user-friendly way.

I had diligently kept records from my time at the Trust, all of which up to this point I had shared with Sussex Biodiversity Record Centre. This, along with years of data collected by many other specialists, volunteers and amateur naturalists meant that the data were all there. They just needed to be compiled in a logical way.

There were 32 reserves, some of which were small and lacking designations, while others were large with many designations. Some were managed by multiple organisations, such as Amberley Wildbrooks. For an exercise like this you do have to draw a line somewhere, and defining the reserve boundaries accurately from the start was key to this. The data I received from Sussex Biodiversity Record Centre included a buffer (in order to capture records that were clearly on the reserves but recorded at a very low resolution, such as the hectad level), which for small sites can result in a huge number of records being generated that are not actually on the site and need to be filtered out. Therefore these small sites (and especially the sites with complex ownership) required a surprising amount of effort to process.

I presented the data in one huge worksheet in Excel (using pivot tables), ordered by the broad taxonomic groups generated by Recorder 6. I listed the sites as 32 columns and designed the spreadsheet to generate totals for each group, totals for each site and overall totals for the entire network. It was just shy of 10,000 species

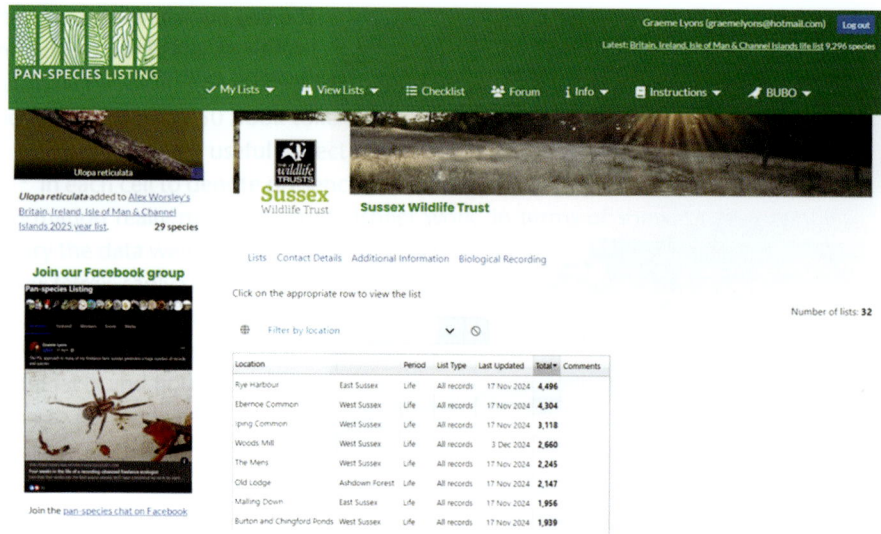

Screengrab of the top eight most diverse and well-recorded Sussex Wildlife Trust sites.

Clearly, comprehensive site species lists pre-date the world of PSL, but I now find a full pan-species list for a site (or sites) much easier to create, analyse, update and navigate as I became comfortable managing my own personal pan-species list, especially the additional levels of taxonomic experience necessary to build and maintain such a list. I even offer the creation of annotated site lists as a service for clients, as these lists are such a useful tool. Once set up, it's fairly easy for people to manage these lists themselves from that point on, but they only remain useful if they are updated regularly.

You might not have the time or expertise to manage a list of 32 sites, but I strongly believe that anyone managing land for nature conservation purposes should create and maintain an annotated pan-species list for their sites. It's a great way to learn about taxonomy and the species you have present, even if you don't know how to recognise them. There is also a tangible benefit when talking to specialist stakeholders on nature reserves – it goes down very well if you can at least speak the same language as them. Not only will it help you to better manage the site, but also it will help to connect you to the biodiversity of your sites, and with biological recording seemingly being seen by many conservation institutions and site managers as a low priority for their site-based staff, this is invaluable. After the initial time commitment such lists are very easy to manage if new additions are tackled as and when they are discovered.

We have recently added the 32 Sussex Wildlife Trust site lists to the new PSL website under one account, Sussex Wildlife Trust, which I shall continue to manage along with the Trust's ecologist. You can now see at a glance whether you've found a species that is new to any of these reserves. Hopefully, over time, this side of the website will continue to grow.

Chapter 13

Public engagement and PSL field meetings

TinyRecorder

In early 2020, just before the pandemic struck, I decided to promote engagement with wildlife in a new way, in the form of a little Lego figure I created called TinyRecorder (actually some three years after a friend of mine Barry Yates had done something similar at Rye Harbour with TinyBirder – I just turned it up to 11 and added *a lot* of humour and banter). What started as a bit of a joke on social media ended up reaching 28,000 followers in just over a year. It was great fun to do, and

TinyRecorder meets a Cockchafer *Melolontha melolontha* (top left). Tiny gets eaten by a Strawberry Anemone *Actinia fragacea* (top right). Tiny and his friend Fungi walk past some Fly Agaric *Amanita muscaria* (bottom left). Tiny goes out at night to look for *Alopecosa fabrilis* (bottom right).

so well received, allowing me to communicate some really complex topics. The approach was always in line with the PSL ethos, and covered a huge range of different taxonomic groups, but it was carried by the humour, which was unsurprisingly cheeky, irreverent and self-aware. TinyRecorder's fungal adventures were particularly well liked. Sadly, in recent years, I've not had time to keep this up, maybe now this book is published though he'll make a comeback…

The Lyons Share

Since 2010 I have tirelessly promoted the concept of PSL on my blog, the Lyons Share, and have encouraged many to get involved through my blogging escapades. Before the creation of the old PSL website I regularly used to post updates of my entire list, so looking back through the 'Pan-species listing' tab is fascinating. I now carry out single surveys where the total number of bugs, spiders and other invertebrate groups are greater than my PSL list after several years. It's great to have this information stored somewhere that would otherwise have been lost, as it's helpful to see how far I have come. I also used to provide regular updates on 'The state of pan-species listing' at the end of the year, to see how the PSL community and movement were evolving.

Global Birdfair

In July 2024 I attended my first birdfair in over 30 years, to run a PSL stall with Mike Prince and Andy Musgrove. It was very enjoyable, and the highlight for me was running six one-hour mini bioblitzes in the surrounding countryside, twice a day (at 10 am and 2 pm). We took a computer attached to a big screen so that we could

Me (centre) showing Zak Spaull (left) some invertebrates and Robin Sandham (right) the new PSL website on our stand at Global Birdfair 2024. (Mike Prince)

tick off species as we found them on a Global Birdfair 2024 site list, in near real time. This was a great idea of Mike's, as it really helped to show the functionality of the PSL website live in the field.

Many people took part in each mini bioblitz, and I met some really promising young naturalists, some of whom were already pan-species listing (and some who weren't then but are now).

We again attended in 2025, recording some 841 species collectively over the weekend, beating our 641 species total in 2024!

How to run a bioblitz

I really dislike the term 'bioblitz' with its negative connotations which suggest that the land will be scorched and barren after the ecologists have been through and collected everything, leaving behind only a species list, nailed to a dead tree. However, we seem to be stuck with it. When I first encountered the term, it was always about amassing together as many specialists as possible across a wide range of taxa, usually to hit a site hard in a short period, typically one that had had little or no recording on it. Sometimes these events were for a specific reason, too. Often, but not always, there was also a degree of public engagement, but this was secondary to the recording element.

Over the last decade or so I have witnessed some real 'mission creep' around the term bioblitz. It seems that many organisations, which are often pushing more and more of their output in the direction of engagement, seem to be less concerned about the records that they make at such events and more interested in showing wildlife to members of the public. These bioblitzes are increasingly being carried out on well-recorded sites, with plenty of public access by people who are not experts. Of course encouraging new people to engage with wildlife is immensely important and worthwhile, but it's not a bioblitz unless it also generates records. It's also not necessarily the best use of specialists' time, and possibly not always the best way to engage with the public either.

If you are planning any recording event or bioblitz, first advertise it well with local specialists, and get the local PSL community on board. On the day, bring along the existing species list (if there is one) in a user-friendly format that one or more people can edit or update as they go along. This is much more difficult with a large group, especially as they split up, but it is really encouraging to be able to shout out to people when a species new to a site has been found. Having one person collate everyone else's records in a standardised format after the event is the best approach, including the inevitable chasing up of records. It's vital to leave plenty of time for this collation, as most specialists will be giving their time for free and identifying material from such records while also having their own work commitments. Allow months rather than weeks or days for this – it will be worth the wait as you'll get a better species list.

It's also a good idea to give feedback to the recording community in the form of a report that shows what the collective effort was, any highlights, the total number of records, the total number of species, and so on. However, my number one recommendation is to stop hijacking bioblitzes with an agenda that is primarily about public engagement at the expense of making any actual biological records. Alternatively, you could just call those events something else, and make it clear to specialists, the public and yourselves what the actual outcomes are. Even a primarily public engagement-focused event that is about teaching people to make biological records should also generate such records.

If you have a site and want to attract experienced biological recorders because you don't wish to pay surveyors, be careful. Unstructured recording over a single day or weekend, no matter how many records it produces, is no substitute for a considered professional survey carried out over a number of visits throughout the year. When approaching experienced recorders, especially active surveyors, acknowledge that you are getting something for free from them – you might be providing access to a site, but they are providing a wealth of hard-won experience, and quite possibly another day of their time spent at the microscope.

Try to keep people moving during the bioblitz, this will involve having at least a vague route planned (and timed), though this will be like herding cats. The group will probably get split up and you may not see most of your bioblitzers after you leave the car park.

PSL field meetings

The first organised PSL meeting was in May 2012 – a weekend of biological recording in West Sussex, arranged by me. After that, and in the open, democratic way that is characteristic of the PSL movement, we felt that it would be best to have new people planning the field meetings in their area on each occasion. The meetings don't happen every year, and some of them are invitation-only gatherings of small groups of like-minded friends. A few of the larger, more official meetings are described below.

2012: Parham Park and Heyshott Escarpment, West Sussex: Lead, Graeme Lyons

Even before the old website was set up there was a thriving PSL community, and the first PSL meeting was held on 25–26 May 2012 in West Sussex, and organised by me. For many of us this was the first time we had met in person, and the social aspect was quite a draw for some. On the first day we met at Parham Park, a really species-rich area for saproxylic beetles, with plenty of old, open-grown oaks, while there was a soundtrack of Field Cricket singing away on the nearby acid grassland.

I selected this site because it sits in the hectad TQ01. This had once been assessed as the most biodiverse hectad in the country, so it seemed to be the perfect landscape for the inaugural pan-species listers' field meeting.

A highlight included the larva of Variable Chafer, found in a recently fallen oak. We set the moth traps, headed to the pub for a meal and came back to find traps full of invertebrates! The list of species for the day was impressive, and I have many fond memories of that field meeting. For me the most exciting moment was when I stumbled across a large weevil that looked rather like a lump of tar. It was the most extraordinary looking creature (and was immediately christened 'Mr Lumpy' by Penny Green). Mark Telfer soon identified it as *Syagrius intrudens*. The exciting thing about this species is that it's a non-native endemic. It's clearly non-native due

We found some great saproxylic species in this tree. From left to right: Dave Green, Sarah Patton, Penny Green, Neil Fletcher, Jonathan Newman, the late Simon Davey, Clive Washington and Martin Harvey. (Mark Skevington)

The weevil *Syagrius intrudens*, also known as 'Mr Lumpy'.

Mark Telfer and Mark Skevington wait excitedly to see whether or not, after three hours of intense coaching, a feral Sussex man can be trained to use a smartphone to identify a weevil.

to belonging to a genus only known from Australia, yet it has not been found in Australia! It feeds on Bracken and it was common where we found it. Despite working extensively in similar habitat in Sussex during the intervening 13 years, I have not seen this species since then. It was a lifer for every person on the field trip!

My other vivid memory was of Mark Telfer using his 'Autokatcher' – a large net that is mounted on the roof rack of a car. Mark deployed it at dusk, driving up and down the drive to the house at Parham Park (surrounded mainly by acid grassland but with plenty of trees, too). We were not disappointed, and Mark even managed to catch a beetle new to Sussex.

After going through the traps we set about searching the tree trunks at night, finding exciting species like the saproxylic beetle *Opilo mollis*.

On the second day we walked up to Heyshott Down, a north-facing area of chalk grassland with important populations of Duke of Burgundy *Hamearis lucina*, orchids and bryophytes. This was an opportunity to see all sorts of interesting species, including *Ozyptila claveata* (little did we know that this was a first for Sussex) and Fly Orchid. Nicola Bacciu found a rare spider, *Araniella alpica*, that has not been recorded in the UK since then. This irks me somewhat, as I have surveyed Heyshott Down twice subsequently and been unable to find it – it's the only orb weaver in the British Isles that I have not seen.

Mark Telfer using his Autokatcher to gather up the last remaining pan-species listers on site (Sarah Patton, Jonathan Newman, Martin Harvey and Neil Fletcher).

From left to right: Penny Green, Neil Fletcher, Dave Green, Mark Skevington, Graeme Lyons, Steve Gale, Nicola Bacciu, Matt Prince, Jonathan Newman, Sarah Patton, Clive Washington and Mark Telfer. This photo was taken by a passing walker.

A group of pan-species listers argue over the identification of a mammal. From left to right: Bill Urwin, two strangers on quadrupeds, Mark Skevington, Nicola Bacciu, Mark Telfer, Sarah Patton, Jonathan Newman and Neil Fletcher. (Sami Webster)

2013: Surrey: Lead, Sarah Patton

Sadly, I missed this meeting. A trip to Surrey Heaths was arranged and there was plenty to be seen, including *Steatoda albomaculata* (a spider I still need) along with *Cryptocephalus bipunctatus* (a rather nice beetle). Bill Urwin famously pooted a horse fly off a horse and surprised it so much, it nearly flung the rider off! I guess this happened seconds after this photo was taken.

2014: Woodwalton Fen, Cambridgeshire: Lead, Jonathan Newman

This was the third PSL field meeting, and it was held at Woodwalton Fen on 1–3 August 2014. One of the main targets for the group was the impressively rare, rainbow-coloured Tansy Beetle *Chrysolina graminis* (*Chrysolina* is one of my favourite beetle genera, and they were not hard to find here). A well-behaved Musk Beetle *Aromia moschata* at eye level was another great moment for many of us. According to my blog I added 27 new species to my list that weekend, and wrote this at the time:

> I wonder where the next event will be and who will host it? These are great opportunities to meet like-minded naturalists and learn from those around you, but attendance was surprisingly low, with only 12 people maximum, I hope next year there will be even more people! A massive thank you to Jonathan Newman for organising the event. Having arranged the first one back in 2012, I know that these things take up a lot more of your time than

The multi-coloured Tansy Beetle *Chrysolina graminis* is a thing of beauty.

Crikey, we all carry so much stuff, don't we? From left to right: Nicola Bacciu, Keith Lugg, the late Richard Shotbolt, Jonathan Newman, Graeme Lyons, Seth Gibson, Matt Prince, Marilyn Abdulla, Robert Smith and Graham French. (Mark Skevington).

you would think. It was really well organised and I was pleased to relax and not feel like I was at work (even though I relentlessly surveyed all weekend and got a lot of people a lot of new species). That's what this is all about, though: helping those further down the rankings, even if it means everyone is closing in on me. All I want from PSL is to be a better naturalist, help make lots of other competent naturalists and have a bloody good laugh along the way! So far, so good I say!!!

2015: Portland: Lead, Seth Gibson

I didn't attend this meeting, but I did receive this great testimony from Bill Urwin about Keith Lugg's fishing skills:

> We all went to the tackle shop and bought those little £10 fishing rods and some bait. We then went up the East Cliffs and stood there as a group, casting out to improve our fish list. Keith Lugg was an absolute fish magnet – every time he cast out it was only a few seconds before his cry of 'Fish On!' rang out over the clifftops. I think we added to several people's fish lists on that trip.

One of the main targets was the tenebrionid *Omophlus pubescens*, which was abundant on Thrift *Armeria maritima* in the Ferrybridge area.

2016: Holkham, Norfolk: Lead, Bill Urwin

I was not at this meeting, although I had really wished I was there until I found *Calosoma sycophanta* in a field in Sussex. Have I mentioned that already? (see page 17).

The meeting was held on 17–19 June, and was mainly centred around Holkham and also Burnham Overy – sand dunes were certainly a big focus. The major highlight was that everyone got to see the summer-plumage Great Knot *Calidris tenuirostris*!

Omophlus pubescens on Thrift. (Bill Urwin)

Greater Streaked Shieldbug *Odonotoscelis fuliginosa*.

There are some great sand dune specialists local to that part of the world, such as the small malachite beetle *Clanoptilus barnevillei*, and this was the second PSL field meeting to turn up the Nationally Rare spider *Steatoda albomaculata*.

2018: Cornwall: Lead, Sally Luker

Sally and the team managed to record an impressive 1,037 species over three days on 8–10 June 2018. The most speciose group was vascular plants (387 species), followed by moths (145 species) and bryophytes (121 species).

A long gap followed this meeting, due to the pandemic.

2023: Sandwich Bay, Kent: Leads, Kev and Debs Ryland

This was the first organised PSL field meeting in five years and I was determined not to miss it. The event was held on 23–25 June 2023 at Sandwich Bay Bird Observatory (where our host Steffan Walton was hugely helpful), and it proved to be a huge success for the 12 people who attended (one of whom was not even a pan-species lister). Attendance has always been rather low on these events, as so many of us are crazily busy in the summer so don't be disheartened if you organise one, and only 12 people attend, this is fairly typical.

We did a lot of moth trapping around the Observatory, where most of the group were staying, but the main excitement was along the coastal strip. The suction sampler was essential for many of the small coastal species there, including some nice coastal spiders and a long overdue lifer, Greater Streaked Shieldbug *Odonotoscelis fuliginosa*. Bright Wave *Idaea ochrata* and Restharrow *Aplasta ononaria* moths were everywhere.

Just when you think there are only going to be eight of you, a contingent of four young whipper-snappers in their 20s and 30s turned up; Duerden McCormack, Harry Witts, Robert Jacques and James Harding-Morris, it's great seeing how much the movement is taking off with the younger generations. I headed off with a local, Steve Reynaert, to the north end of the beach, not realising just how long a walk it was going to be and that no one else followed us, it's hard to stay as a cohesive group at these events, especially on such huge sites. I really wanted to see Dune Tiger Beetle *Cicindela maritima* again (my last sighting was in 2005 at Titchwell RSPB). We did find it, but stumbling upon one of my most sought after spiders, the little ant-mimic jumping spider, *Synageles venator*, was a real highlight.

I recorded four new bees from that stretch of the coast despite collecting very little material, including an all-black *Bombus ruderatus*. On the way back I bumped into the four younger listers, this was a great opportunity to show them some of the rarer species we had found by suction the previous day. Someone grabbed what looked like a carabid, and on closer inspection it was the coastal tenebrionid, *Crypticus quisquilius*. The sun had got to everyone by this point. Robert Jacques started laughing and said, "what did you just say – Crispy Squidwilliams?" We never take ourselves that seriously.

Kev totalled up all the species we recorded as a group over the weekend and it came to 935 species.

2025: Spurn, Yorkshire: Lead, Duerden Cormack

Duerden Cormack organised a fantastic field meeting at Spurn Bird Observatory in Yorkshire from 30 May to 1 June 2025. As well as getting all attendees lots of lifers, he did a really great thing in harnessing our combined expertise to help record and inform future decisions about the land that the observatory manages. It's great on

From left to right we have, Matthew Secombe, Sam Buckton, Simon Van Toller, Duerden Cormack, James Harding-Morris (who could barely contain himself having just ticked the weevil *Conarthrus praeustus*), John Poland and Alex Payne.

Armadillidium album.

such trips how we all swap our experience and get each other lifers; there isn't a shred of competitiveness to it – there are no losers in PSL! I did very well on galls, aphids and springtails thanks to Seth Buckton, James Harding-Morris and Duerden Cormack respectively.

At the time of writing, I have not even had chance to empty my notebook, let alone start on the specimens. It was a resounding success, though, with a typical maximum of 12 people in attendance. This time I was one of the oldest listers present; I think the average age was probably in the 30s. The fact there is a pub just a few hundred metres away makes any stay at this observatory great fun!

The wildlife was awesome too. Highlights for me included this scarce coastal woodlouse *Armadillidium album* found under the log just visible in the middle of the photo opposite. I got a new spider on the last morning (*Baryphyma maritimum*), but the most memorable encounter was jumping down a small cliff to get a look at a Starry Smooth-hound just caught by a fisherman, who looked utterly bewildered as we tumbled down the cliff towards him with our cameras raised. Only for it to happen again ten minutes later when the same guy caught a Small-spotted Catshark *Scyliorhinus canicula*!

As I write this, I realise that we have never had a field trip outside of England. We really need to start thinking about a field trip to Scotland. If anyone could make this happen, I would be forever grateful to them!

Chapter 14

Representation and demographics in pan-species listing

Why are there so few women in pan-species listing?

The question of why there are so few women in the rankings really deserves more than just a few pages, but I hope that by analysing the data in the rankings and asking for the views of a number of women in pan-species listing (PSL) and the wider natural history community I can attempt to cast some light on this situation.

Currently, just one woman (Yolanda Evans) is listed as a top lister for any of the 38 categories (see Table 4 on pages 38–39). On the new PSL website there are no women in the top ten, one in the top 20 (5%), six in the top 100 (6%) and 17 in the top 200 (8.5%). At the time of writing, then, women represent around 9% of the total (37 out of 410). On the old website there was a slightly higher proportion of women, peaking at 12%, but again this is a long way from parity.

Clearly this situation needs to be improved. To examine its possible causes I have consulted widely among the small sample of female pan-species listers, and also among a number of my female naturalist friends who are not pan-species listers. I asked them why they thought there were so few women in PSL, especially at the top of the rankings, and seemingly so few women in jobbing entomology, as there is a connection here with PSL.

Their responses revealed a number of different reasons:

- concerns about personal safety when out alone
- reluctance to engage in such a male-dominated hobby
- fear of being patronised when seeking help
- not wanting to go deep into a subject, as women are not allowed to make mistakes in the same way that men are
- specifically, when it comes to entomology, a greater reluctance to kill and take specimens
- lack of time due to work and childcare commitments
- the gender pay gap resulting in less disposable income for hobbies
- the competitiveness of PSL not being appealing
- being steered away from such hobbies in childhood

- being less inclined than men to engage in obsessive listing and collecting behaviours.

The following discussion of each of these reasons is based on opinions expressed by the women whom I consulted. Of course, many of the issues here are societal and PSL alone isn't going to change them, yet we can all do our bit – I see talking about this openly as a small part of that.

Concerns about personal safety when out alone

It makes me really sad, as the countryside is personally the place where I feel safest, that many women are unable to share that sense of freedom and safety. Clearly there is a feeling among quite a few of the women I consulted that PSL, which so often involves working alone in the field, could potentially be a threat to personal safety. After hearing their views I set up the WhatsApp group 'PSL Sussex', and have encouraged others to set up similar groups in their own regions (there's now a London group and talk of a West Midlands one too). The aim is to provide a place for groups of pan-species listers to get together more frequently and arrange field excursions. This would be a good way for all of us to stop spending so much time in the field alone, myself included.

Reluctance to engage in such a male-dominated hobby, fear of being patronised when seeking help and a reluctance to go deep into subjects

The higher proportion of women further down the rankings suggests that the upcoming generation of pan-species listers is less likely to view the current male domination of PSL as a problem. However, we all need to be aware of patronising attitudes – our own and those of others – and if we witness this behaviour towards women (or any 'othered' groups) we need to call it out, and check in with ourselves if we find we are doing it too. It is very important that as a group we are approachable, that we positively encourage others, whoever they are, to feel they can ask for help, and that we provide support without showing any kind of bias.

Lack of time due to work and childcare commitments, and the gender pay gap resulting in less disposable income for hobbies

This situation also seems to be slowly changing, as reflected by the fact that there are increasing numbers of younger female listers lower down the rankings, but it could take a generation or two for them to populate the top end of the rankings. The gender pay gap seems to be gradually closing, but the cost-of-living crisis continues to have a very negative impact on disposable income.

The competitiveness of PSL not being appealing

Concerns about the competitive element in PSL may have been heightened by the fact that the new website is run by the same software as BUBO (famous for being a bird-listing site). However, I hope that this book has shown that PSL is about far more than just listing. We have discussed the possibility of making the main rankings less prominent on the homepage, and perhaps showing other aspects of the site there instead.

The most important thing we can do when meeting other people is to show that we are a collaborative community first and foremost, and avoid focusing on competitiveness and rankings. This means being as welcoming and accepting as possible, and leading by example. Sharing your knowledge and enthusiasm is a great way to help others and encourage them to become involved in PSL.

Being steered away from such hobbies in childhood (this may be particularly relevant to special interests developed by neurodivergent children), **and being less inclined than men to engage in obsessive listing and collecting behaviours** (there is no evidence for this either way, but I will discuss in detail below)

While researching autism and neurodivergence since my diagnoses, I spoke to a female neurodivergent naturalist and therapist friend of mine about what might be going on here, and I think discussing this through the prism of autism provides insight on this point for all.

As a young child, I readily memorised the names of dinosaurs. I would line my Star Wars figures and their weapons up in the order that I was bought them, then I finally moved onto birds, plants and moths. I might not have been given a lot of help from my family *but it was never discouraged*. When I look back at all this, these were the strong early signs of autistic special interests. Yet I only obsessed over Star Wars because I was a boy born in 1978 with a parent who exposed me to such things. Clearly if I had been born 50 years earlier, it would not have been Star Wars; if I had been born female, it would quite probably not have been Star Wars either. The things we develop an interest in have to be shown to us by our primary caregivers, and they have a huge sway over what exactly those things are (or aren't) – even if those decisions themselves are subconscious.

One of the reasons that autism is thought to be so hard to detect in women is that it often manifests differently to how it does in men. As well as young women masking more, making autism harder to detect, the special interests developed by autistic boys often involve unusual (for their age) pursuits that typically involve lists (or collecting). In the case of girls, these are often replaced by equally intense (but much more typical for their age) concerns, for instance a strong enthusiasm for a particular TV show, a popular icon or reading.

Perhaps this is a reason that so few women have so far found their way to the upper reaches of the PSL rankings? At the top of the rankings as *they are now*, they are likely well represented by neurodivergent men exercising their special interests. And is this because we men were allowed to and encouraged by our primary caregivers, while many of our female counterparts were not? And we have been doing so unhindered since we were children – obsessing over lists and collections. Meanwhile, our female counterparts have mostly (but clearly not all) been doing other things, and possibly not coming to PSL until a little later in life.

Are these differences in autistic male and female special interests biological or social? In my opinion, I am sure it is the latter and my therapist friend strongly agreed with this. I think acting like we are biologically different on this front is not helpful, as it would lead to the assumption that whatever changes we made to try and reach parity would fail, so wouldn't be worth bothering with in the first place. There must be plenty of women (neurodivergent or not) that have been steered away from a developing a passion for slugs, snails or beetles etc., because it wasn't seen as ladylike, while it was fostered in their male equivalents. There is literally a nursery rhyme that we have used to teach children what they should and shouldn't be interested in. Female hyper-focused, list-orientated types clearly exist but are not finding their way into the extreme end of natural history in the numbers that men are.

PSL is great for fostering enthusiasm in any budding young naturalist. Maybe, then, it's the parents of autistic (and equally, neurotypical) children that need to be more flexible and let their young girls develop more left-field interests whatever they may be, from an early age and nurture them – and to stop being so judgemental as to what is and isn't suitable for any person. If so, you (and they) won't go wrong with pan-species listing. Neurodivergence is discussed in more detail in Chapter 15.

People from ethnically diverse backgrounds

People from ethnically diverse backgrounds are perhaps more under-represented in the conservation, natural history and PSL worlds than any of the other groups mentioned in this chapter. At the time of writing I know of only two people of colour who are regularly recording invertebrates, one of them being my friend Esmond Brown. I recently had a conversation with Esmond about this. He is on the PSL website but is only really listing spiders, yet he's a good all-round naturalist and entomologist with a broad taxonomic interest, so I consider him to be aligned to the pan-species listing approach, enough to have relevant experience here.

Esmond pointed out that most second- or third-generation immigrants are unlikely to be attracted to a low-paid occupation. Of course, if you have made the huge step of moving your family to another continent you are unlikely to want to settle for anything but the best options for your children. Perhaps if we valued the natural world more, we might start to pay conservationists a more competitive salary.

There are the more predictable unconscious biases that people from ethnic backgrounds can encounter, such as there potentially being less likelihood that their records might get accepted, to being denied access to certain sites, or even having to think more carefully than I would about walking around on their own with unusual equipment like a suction sampler, which could easily lead to a run in with the police.

Yet it is very difficult to draw conclusions from talking to just one person here and I did not want to cast the net wider and talk to naturalists in general, I wanted this to be specifically about pan-species listing and talk to people that I knew had very detailed taxonomic interest across multiple taxa and I sadly don't know of anyone else I can ask for their experience.

Disabilities

Did I get into pan-species listing because my brain already worked like a database, or did I train it to be like this, through years of entomology? When I read this back, it's *really* obvious I am autistic and have ADHD but my diagnoses were a huge surprise to me at the time. I wonder how many more prolific and accomplished naturalists (and especially entomologists) are neurodivergent? I suspect it is a very high proportion. So much so that I recently ran a poll on the PSL Facebook page to ask how many people had either been confirmed as being neurodivergent or strongly suspected they were/or had already self-diagnosed. Just over 10% of the group took part in the poll but, some 57% considered themselves to be neurodivergent. Although this figure is clearly only provisional, it strongly suggests that pan-species listers include a very high proportion of neurodivergent people among their number, probably close to parity and possibly a majority.

I suspect that PSL is particularly compatible with AuDHD, as it allows you to study natural history in immense detail (an autism trait) while simultaneously providing a source of dopamine whenever you see something new, which can interrupt your life anywhere and at any time (greatly benefiting people with ADHD). Although you obviously don't need to be neurodivergent to be a successful naturalist, I am not gonna lie, it has helped me!

I should add that it's important to acknowledge that autism and ADHD are both recognised disabilities in the UK. They could well help you as a pan-species lister, but it is possible they might not, and they are very likely to have major impacts on other aspects of your life.

I can't remember a single family member's birthday and I'm terrible with people's names, I can't find the time to clean my flat or make my own food much of the time. I have almost certainly given myself carpal tunnel syndrome in recent years from very repetitive actions like sweep netting and data entry. I am terrified of airports, not because of the planes but the bureaucracy with which they are associated; filling out new forms causes me disproportionate anxiety. Certain smells can illicit uncontrollable reactions, but worst of all, it has negatively impacted many of my relationships and friendships. So, believe me, it is not all rosy – as any neurodivergent person can tell you – even though it might come with a set of abilities that can really benefit the pan-species lister.

People with impaired mobility will almost certainly find access to all areas of the countryside difficult, but I struggled to find anyone from within the PSL community to discuss this with. However, I did speak to Joe Myers, not a pan-species lister but a naturalist whom I met at a British Arachnological Society field meeting at Chippenham Fen. This is what Joe said.

> Getting out and about into nature isn't easy; the better areas tend to be the harder ones to navigate over, such as those with boggy or rocky ground. If I remember correctly, we met at Chippenham Fen, a naturally wet and muddy site, meaning it was a workout to get anywhere. Whilst it would be nice if there were hard and flat paths everywhere, that would degrade the quality of the habitat – so it's a trade-off. Another issue is when people have intervened and the task isn't done with disabled people in mind; stiles in particular tend to be nothing more than a couple of planks between a narrow gap, so only people able to walk can pass them. Then there are times when I wonder if any thought went into the design process, and things are inaccessible when a small tweak could have made them accessible, such as a hide with steps when they could have built it set into the ground to make it level access. Facing barriers gives me anxiety about trying new places – is it worth the effort only to find out I can't get around? These days I rarely venture to anywhere unfamiliar unless I've researched it or I'm going with others who may be able to help.

Clearly there are many reasons why people with significantly impaired mobility might be or feel excluded, and Joe's comments and suggestions need to be addressed so that the PSL and natural history communities can become more accessible and inclusive.

People from more disadvantaged backgrounds

In my experience, the majority of pan-species listers come from middle-class backgrounds, but there are also a few of us from deprived backgrounds. The best advice I can give to anyone coming into PSL who feels that their social background could be a barrier is this: you can do anything you want to do, so don't let anyone tell you otherwise, including (perhaps especially) yourself.

The anti-intellectual mindset so pervasive in my town was very hard to push back against. Most of the kids on my street were part of a loose 'gang' that hung around the entrance to my school of an evening. The progression went something like drinking, smoking, casual drug use, through to harder drug use, to robbing cars, occasionally burglary and then, in some cases, prison. I had to be very strong-minded to resist this; I watched a close childhood friend, with whom I used to go birding, get sucked into that life. I often wonder how different it could have been for him if he'd stuck with the birding. As far as I know, no one else from my class even went to university.

PSL is about supporting each other, and one of the ways to remove barriers for those from disadvantaged backgrounds is by the sharing of resources and equipment, or by making second-hand books or equipment available at an affordable price, or even for free.

Another great help is to provide lifts to sites for those that do not have a car. This might seem like a small thing but honestly, Ewart Gardner changed my life with his generosity in this way. I saw so much of Staffordshire and further afield that I never would have seen without him, it literally opened my eyes to so much more than I would ever have been exposed to.

Sadly, due to the ongoing cost-of-living crisis in the UK, social mobility has declined significantly in recent years, and the rising burden of debt among university graduates is another huge pressure on young people today (I doubt I would have been able to afford to go to university if I was 18 today). For these reasons it is likely that those from more comfortable backgrounds will always be at an advantage, so we in the PSL community need to redouble our efforts to be more inclusive of others.

Young people

Reassuringly, there is no shortage of young people involved in PSL. On the old website there was an option for providing your age on your profile, as I thought it would be worth capturing these data to get a picture of the age demographic. The mean age of the sample of 162 people who provided this information was 41 years, while the mode (the most frequent age) was actually 19! The data are displayed in the graph overleaf. Encouraging as this discovery may seem, I should point out that providing one's age was optional, and if older people were less inclined to provide this information that could have been a driver of this skewed age distribution.

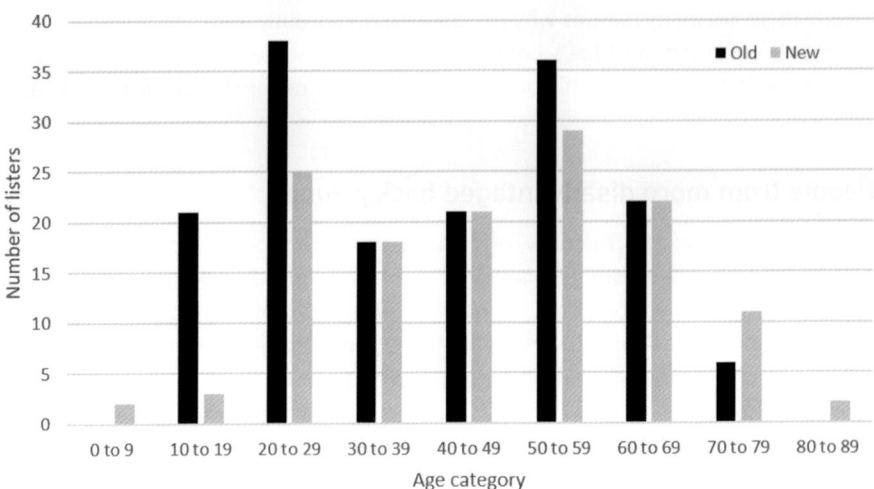

Frequency distribution of the ages of pan-species listers on the rankings from the old website (February 2024) and the new website (October 2025).

The new website is unlikely to have a field for displaying your age (as identity theft and data protection are both greater concerns these days), but analysis of the anonymised raw data from the new site (dates of birth with no names attached were provided by the team) has revealed the mean age to be a little higher, at 47 years with the 50s being the most populated age group. Despite this, as with the old site, the mode is much younger, with 29 being the most frequent age.

Although we don't have all the answers to the imbalances in our PSL community, we are very open to new ideas, so please do get in touch if you have any suggestions that you think might help in any of the areas that have been discussed in this chapter.

Chapter 15

Neurodivergence, natural history and pan-species listing

A caveat

I must state here I am very new to all this. Writing about my experiences of neurodivergence, just over a year after being diagnosed, is an attempt to help other such naturalists to feel seen. I apologise if you are neurodivergent and don't feel represented here. Indeed there is a saying that 'if you've met one autistic person, you've met one autistic person'. You can also say the same of ADHD.

If you are not neurodivergent you might perhaps feel that this is all completely irrelevant to you, so there is no point in discussing it. I would gently suggest that, given how prevalent neurodivergence is in the pan-species listing (PSL) community, you probably already have some neurodivergent friends, and learning a little about this subject might help you to understand them better. Furthermore, like me you might actually discover that you are neurodivergent too, by recognising some of what I discuss here.

Neurodivergence explained

The health and well-being benefits of being in nature are now well known. For me, it is very much the place I feel most alive and at peace. Yet it's not enough for me to just be *in nature*. In fact the idea of being in a wonderful landscape rich in unique wildlife and not being able to relate to it by sampling, recording or directly interacting with it fills me with anxiety, if not dread. As a child, and before I had ever been abroad, I regularly had nightmares about being in a foreign country where I was surrounded by strange and magnificent birds but had no field guide to identify them.

For me, and probably for many pan-species listers, superficial interaction with the countryside is rarely going to cut it. It is the very act of searching for something, recording intensely or the thrill of the twitch that feeds the more extreme naturalist – the hunt for something new which delivers that dopamine hit. As I mentioned earlier, I have only relatively recently been diagnosed with autism and attention deficit hyperactivity disorder (ADHD) – a combination known as AuDHD, which is a distinct neurotype. According to the World Health Organization, around 1% of the population are autistic and around 3–5% have ADHD. Around 50–70% of people with autism have ADHD, whereas most people with ADHD are not autistic (though a significant number of them are). This suggests that people with AuDHD make up less than 1% of the population, but I suspect there are probably a lot more of us in the world of PSL than we realise.

In this chapter I shall use the term 'neurodivergent' to refer collectively to autism (ASD), ADHD, AuDHD, obsessive–compulsive disorder (OCD), dyslexia and anything else that falls under the term.

I found a recent discussion with Sarah Whild, pan-species lister and author of *The Biological Recording Handbook*, very enlightening. Sarah is neurodivergent herself, and when she was a lecturer at the University of Birmingham she discovered that her course, on biological recording, had a considerably higher proportion of neurodivergent students than any other course.

What exactly is neurodivergence? My own experience and understanding of it is that our brains are wired differently from those of neurotypical people. I think differently (I can't think in straight lines), process information differently, see connections that others don't, and can focus like a laser if I'm interested in something. However, I can't engage at all with things I'm not interested in, I can talk for ages and interrupt too often, I can appear blunt and rude, and I tend to crash after too much social interaction. Anxiety is common among those of us who are neurodivergent, and we commonly experience a deep sense of 'not fitting in'.

If someone came around to visit the family home, I was the kid who would talk at them about moths for three hours. It's called 'info dumping' and can be a wonderful thing if you are lucky enough to be the recipient of it, yet you might not feel that way if you have other things to do. This behaviour lives on in my blog, The Lyons Share. Neurodivergence is different for everyone, though, and I feel mostly lucky.

From the time I was given my first dinosaur book I have been memorising and classifying names and images. As an adult, I have a near photographic memory for things that interest me. As an entomologist, I have memorised the conservation status of most British invertebrates and their food plants, and have also memorised most of the palps and epigynes of British spiders. In addition, I try to memorise as many keys as I can. I am unusually good at recognising faces (known as a super-recogniser – something that is typically not an autistic trait and one of the reasons diagnosis took so long) and have also memorised many other facts, figures and dates over the years. All of these things act as a source of dopamine when I get a match to one of the many databases in my head and I will (subconsciously) go to incredible lengths to get that dopamine. I'm like a dopamine vampire, my hunger is sated after getting a new or unusual species, recognising someone in the street or in a film, but after a while it fades and I have to go out on the hunt again!*

So what exactly is dopamine?

Dopamine is a neurotransmitter known as the 'feel-good' hormone, which is released by the brain when we do and achieve things, it's a natural reward system. Neurodivergent people typically have low levels of dopamine, or more accurately an impairment in the way dopamine is processed, which results in many such people finding novel ways to get that dopamine. How about a system of collating

---

* Dons cape, laughs maniacally and flies out of the window... to the nearest nature reserve.

experiences of natural history that will be with you throughout your whole life and at all times? Pan-species listing is just that, one huge dopamine-filled Easter egg hunt! If you have an addictive personality, and many neurodivergent people do, then this is a much better thing to become addicted to than say, alcohol, drugs or gambling.

All this got me thinking. Did we (a group of naturalists including a significant proportion of mostly undiagnosed neurodivergent people), create a lifelong calling for ourselves and other neurodivergent people without even knowing what we were doing? Might PSL actually be about as tailor-made for neurodivergent people as it is possible to be then, with its constant stream of dopamine rewards and its near infinitely branching rich veins of special interest? As I have researched this unexpected chapter of this book, I have strongly come to believe this.

Anxiety

When I'm not out in nature for prolonged periods I become very anxious, but after one day in the field the anxiety disappears. It is a combination of the physicality of being in nature (literally getting my hands dirty, sieving moss or wading through mud, and getting cold or wet) and being able to get out of my own head that is so vital here. This doesn't work though unless it's both the physical and cerebral aspects of being in nature that surveying, recording or pan-species listing provides. Of course, some may recognise this as a form of 'passive mindfulness' but I am quite late to the party in understanding this.

The anxiety usually occurs when my laser-like focus moves away from what I am working on (when it can be an incredibly useful tool for getting large, complex tasks done efficiently) and turns in on myself. Long periods of time spent inside my own head (for example, when at my desk) and not being physically active are most likely to cause it. Physical signs of anxiety typically include shallowness of breath, numb lips, strange sensations in the hands and a lump in the throat, all of which can be symptoms of other conditions, and so exacerbate the anxiety. Once this cycle starts it's hard to escape from it without getting into the flow state (see page 274), which I find very hard to do at my desk and very easy to do in the field. A period of low productivity often follows, which only increases the feelings of failure and associated anxiety, making me feel even more unwell. I almost always feel better in the morning, though – it is as if there is some kind of daily factory reset, apparently a known ADHD trait. It's like I've lived 17,358 24-hour-long lives, stitched together by the vague idea of who I am. Learning that all of these feelings are strongly connected to neurodivergence has really helped me to control them, and therapy has helped me to listen to my body and then adopt strategies aimed at reducing them or preventing them altogether. More recently, I have learned that cold showers and meditating are incredibly useful tools for regulating over the winter, something especially helpful if you have a relentlessly busy mind like I do. However, anxiety and impostor syndrome have been particularly challenging during the writing of this book.

Getting a neurodivergence assessment

For about five years I had a niggling feeling that I was neurodivergent. However, my own erroneous preconceptions about autism and ADHD, and the results of quick online tests which suggested that I was borderline autistic but did not have ADHD (tests that would turn out to be wildly inaccurate) meant that I didn't take it any further. Still unable to dismiss the niggling feeling and, importantly, feeling that there was a need to inform this book, I reconsidered my decision and underwent an assessment. I also have to thank Chris Packham for helping me to reach this decision. I watched his BBC documentary, 'Inside Our Autistic Minds', and was particularly affected by his description of walking around a woodland and noticing when even a single leaf was out of place. I could really relate to that experience, but with the addition of invertebrates, plants, birds, songs, calls, smells, subtle changes in management, climate, soils and hydrology!

After I received my assessment everything suddenly made sense. Now I understood why I would always hear birds a few seconds before everyone else, why I would spot even the most well-camouflaged moth on a tree, and why certain strong smells make me convulse. Most importantly, though, there was a reason for that feeling of 'not being quite right', and through that realisation I was able to stop being so hard on myself. I wasn't a broken outsider after all – as a neurodivergent person I was just hard-wired differently. For the first time in my life, I was able to accept that there are some things I am not good at, and find ways to action them. A tendency to overwork, push myself too far and maintain a level of fierce independence is almost certainly as much connected to childhood neglect as it is to neurodivergence, which often manifest in the same way (again leading to a late diagnosis). I include this here to show how complex mental health is. Nothing happens in a vacuum.

I am lucky, I have some skills for which I am grateful, yet it's uncomfortable for me to think of them as gifts from my neurodivergence, as I have worked really hard to hone them over the years.

Neurodivergence and shame

Where AuDHD has been a particular challenge for me, is with social interactions. I can happily give a talk to 250 people about something I know well because everyone else is quiet, while going out for dinner with five other people fills me with dread. I crave intense, one-on-one conversations, which I know many people find exhausting. I often feel like I have only a short amount of time to get some complex information across to people, but that moment when you realise you've been talking about spider's genitalia to a complete stranger for an hour again, and they're backing away from you to the nearest exit is really soul-crushing. As one of my clients once described it, being neurodivergent is like having a brain like a Ferrari, but with the brakes of a Nissan Micra.

I have always said that I don't think in straight lines, long before my diagnoses. I am a nightmare for starting a conversation about one thing, then moving onto something *seemingly* totally different, while actually they will be connected in some way. Throughout my life I have been accused of going off topic or being disruptive when much of the time it's just that other people haven't connected the dots in the way I have. When someone asks how I am, I tell them. This is invariably not what they are expecting, especially if I have had some recent bad experiences. I over-share because I have that honesty and naivety that many neurodivergent people have.

And with this regular criticism comes a lifetime of shame: that you are disruptive, don't think like everyone else, don't play by the same rules, aren't good enough, or are simply wrong in some way. We are different but we are not wrong, and neither are we broken and in need of fixing – we just have a different operating system.

From my point of view, I feel like the one who *isn't* broken in a world that doesn't make a lot of sense, and it's because of this, if you are neurodivergent, you've probably been surrounding yourself with other neurodivergent people your whole life, and this is certainly true of many of my natural history and PSL friends. We tend to make more sense to one another and subconsciously seek each other out.

Meltdowns and triggers

My neurodivergence journey is so new that I have only recently realised that I've been experiencing meltdowns my whole life, although they've been happening much less often recently.

The triggers for a meltdown often involve failure on my part, such as navigation errors in the car or dipping on a twitch. Once I am in an irrational state like this, I lash out at the people closest to me in a very childlike way. I can see very clearly how upsetting this behaviour can be to the person on the receiving end, even though things said in the midst of a meltdown are not rational and do not reflect what the neurodivergent person really feels or thinks.

I have found that the only way I can stop a meltdown is to get out of the situation that triggered it and try to press the reset button. This usually involves being in nature and doing some intense recording, all in an attempt to reach the flow state so that I can forget who I am briefly. Having things to look forward to is really important for people with ADHD, otherwise they can find themselves in a tough place. Dipping on twitches is a particularly strong trigger for me, so I tend to limit my twitches and instead focus on blind recording, more 'guaranteed' lifers or targets where there is a plan B.

I have come to understand why I am so triggered by critics of PSL who incorrectly claim we are 'just a bunch of twitchers', as if that is something to be ashamed of. When an autistic person is shamed for their information dumping or list keeping this can be particularly painful if, as is the case for me, *the list is the wildlife* – it's one of the ways I have chosen to interact with and express my love for the complex biodiversity of these islands, and no one else gets to decide if this means I care about it any less than they do. For example, a survey is so much more than just the species list, yet it's utterly meaningless without it. The situation is no different for pan-species listers and their lists.

Embracing complexity and nuance

A colleague once said to me 'You are about species and I am about processes', to which I angrily responded 'No. You are about processes and I am about species *and* processes'. Long before I ventured into the world of neurodivergence I adopted the motto 'Embrace complexity and nuance', and it has served me well. Yet others are quick to pigeonhole me as someone who can't see the bigger picture because I am so focused on the detail. In fact this couldn't be further from the truth, the two are not mutually exclusive.

I prefer to think in terms of guilds of species, rather than species. This is a very useful way to make sense of a seemingly overwhelming list of species. I have always thought in terms of the collective rather than the individual. I love analysing the data I collect, as this is the final part of the job. The survey is not complete when the last visit is finished, or when the last specimen is identified, but only when you've made sense of all that noise by completing the final report or, even better, presenting that report to a client or other audience.

I think the 'noise' puts a lot of people off. It must seem like an impossible mountain of complexity and information, but for entomologists like myself it's anything but – we embrace the noise! Interpreting large data sets and long species lists and transforming them into tangible, meaningful actions laid out in simple terms is definitely one of my strengths.

I believe some of this comes directly from the AuDHD neurotype: the intense laser focus, detailed ID work and species lists are benefited by the autism. While the large-scale analysis, discerning meaning out of the noise, fearless creative thinking and a need to challenge everything all of the time comes from the ADHD. I'm somewhere in the middle, both these things spinning me around and pulling me in often opposite directions, while I try and get them to work together in my best interests – which only ever seems to happen for me when natural history is involved.

Neurodivergence and the workplace

One thing I have come to realise since leaving my last job is how well the freelance life suits me, and how I really should have taken this step earlier, while also highlighting the things I really struggled with in a medium-sized organisation. The endless meetings I found particularly difficult, especially when there wasn't really a key wildlife focus to them. I often struggled with the way people interacted with me during meetings. I would often feel rushed, or criticised for going off topic, going into too much detail or talking too much. Yet my argument has always been that unless you understand the detail you can't possibly understand the processes. This is what ecology is really about, and I noticed that detail, especially about species, was starting to seem superfluous to some, whereas more nebulous concepts and processes without defined outcomes were trending.

In meetings I would always be scanning the horizon for issues directly related to what we were talking about – with one eye always set on the long term, and how what is happening now might look in years or decades to come.

Neurodivergent people often take unwarranted criticism badly, but are likely to respond well to considered, constructive criticism. Being misrepresented or misunderstood is often really triggering for us. With hindsight, I don't feel that I always received the support or understanding I needed from the institutions I worked for, despite being fortunate to have two great bosses who themselves were really supportive.

They obviously didn't know about the neurodivergence, but they both nurtured my talent, and allowed me to work remotely when I needed to. Fresh challenges and a long leash to investigate new areas of natural history and conservation were really important, and it was great to be allowed time to teach myself about the wildlife on the reserves I worked on.

Most neurodivergent people are biologically hard-wired to respond very badly if their integrity is called into question. Therefore trust is of huge importance to us, and of course in a job where you are on your own in a field situation for much of the time it's essential to be trusted. I just needed someone to point me in the right direction, wind me up and let me go!

Nurturing young neurodivergent naturalists

As a child I was very focused on natural history, yet I was always made to feel guilty for being selfish, obsessed and single-minded, which apparently dominated every holiday and weekend. In hindsight, I was a really good kid – talented, hungry for new information and experiences, albeit rather restless and relentless. All redeemable qualities really, and useful skills for a naturalist – but I was never allowed to feel that. So if I could give one word of advice to any parents of a neurodivergent child with a passion for natural history, it would be to do everything you can to encourage that interest, in whatever form it takes, and above all to share that journey with them, so that they don't feel like an outsider. Something as simple as just listening, engaging and facilitating them to get to more distant sites could make a huge difference.

Coming out as neurodiverse

Would I ever write it on my CV? After a year of digesting my diagnoses, I would, especially now I have spoken about it publicly. If I personally were to see that someone did write their neurodivergence on their CV while applying for a role in a natural-history related post (or any other, for that matter), there is no way that I would see this as a negative – quite the opposite. At its very least, I would see that that person is open, honest and brave. At its best, it could come with an impressive set of skills, even if some of them might seem a little different. Your organisation is likely to flourish with these kinds of people on board; in fact, it probably already is and you (and maybe even they), don't even know it.

The same goes for the world of pan-species listing. Neurodivergent people make up a much higher percentage of the PSL population than we do at the national level. Clearly, there are also many pan-species listers who are not neurodivergent and we are likely a stronger group because of this mix.

Getting assessed, self-assessment or just leaving it

If you feel like you don't quite fit in, and recognise some of yourself in what I have talked about above, you may well be neurodivergent too. If it's never negatively impacted your life, and you don't carry a great burden of shame or guilt with you all of the time, there may be little value in investigating it further.

But if you do carry that shame or guilt, know that it's not yours to carry. Sloughing this is off is an invaluable part of being assessed, and I would argue that self-assessment, although valid, may not provide the powerful and sudden change that emerges from this period of enlightenment.

I felt that my life was not sufficiently negatively affected to warrant pursuing assessment via the NHS, as my pursuit of a diagnosis was initially about informing this book and I really didn't have the typical two or more years to wait for an assessment. I went private, which was costly but very worthwhile and this whole chapter of this book (and my life) would not be happening if I had not gone down this rabbit hole! I now realise how important it was for me, and how much I did need that diagnosis to be happier; in retrospect, it would have been totally acceptable to do this via the NHS. I am very happy to have opened this box, but it was a bit of a challenge immediately after diagnosis.

One thing I would definitely advise is to get diagnosed early. The world is a different place from what it was some 20 years ago. Not only is the available support much-improved, but it will potentially spare you the difficulty of what I went through immediately after the diagnoses – that is, the entire reframing of my life (all 45 years of it) through this new lens. It's a difficult process and I really hadn't given myself much time to do it either. Much of this process seems to happen in the background, taking up vital bandwidth, but it also requires you to do some active work too.

Essential text

Joe Harkness (2025) *Neurodivergent, by Nature: Why Biodiversity Needs Neurodiversity*. London, Bloomsbury.

For a deeper dive into many of the topics mentioned above, Joe Harkness's recently published book uses different case studies to discuss why so many neurodivergent people end up working in conservation and how this ultimately benefits biodiversity. I find it remarkable that we were writing such similar things at roughly the same time but in such different ways.

Chapter 16

Threats to pan-species listing

AI identification apps and pan-species listing

In recent years, AI identification apps (I prefer the term 'imitation engines', given what they actually do) for smartphones have become very popular, and are clearly a way of enabling people who might never have engaged with natural history before to do so. As iNaturalist has built-in AI, as well a global remit, many users worldwide rely heavily on it, especially in countries where there are no existing field guides, or when people are going on holiday and it wouldn't be practicable to purchase guides.

Although I am clearly in a minority here, I am less comfortable with their casual use in natural history in the British Isles, especially in pan-species listing (PSL). I ran a poll on the PSL Facebook page and was quite surprised to find that around 75% of people were in favour of them, with only around 25% having reservations. The most popular option in the poll was 'If beginners use them to save time at the start, that's fine as long as it leads towards keying out and more traditional learning afterwards', with around 40% of people agreeing with this. The second most popular option was 'They are a tool like other tools and are a useful labour-saving device', with 30% agreement.

There are two main reasons for my concerns about the use of such apps. First, their regular use goes against the ethos of PSL, and secondly, they could have significant long-term effects on our individual and collective learning.

One of the strengths of PSL is that as a community of naturalists we have amassed a huge amount of personal knowledge, most of which we have obtained and contributed to ourselves by personally identifying the majority of species that we add to our lists. Pointing a smartphone camera at something and using an app to identify it is a very long way from picture matching or keying something out, or even asking for help. I can see the benefits of very occasional use of such apps, but only when all of the other available options have been exhausted, so regular use goes against what we are trying to do. From a fairness point of view, I really don't mind whether people add AI-identified species to their PSL lists or not. I think such use of these apps will always be self-limiting, as any pan-species lister who is focusing only on things that can be identified by apps will, after an initial flurry of identifications, find that they soon run out of species that can be identified this way.

The issue for me is not about fairness, rules or principles, but about one of the main benefits of PSL – the creation and facilitation of competent, all-round naturalists. Some might argue, so what if there are lots of useful records being generated but PSL is not just about creating large numbers of records of under-recorded groups. We also need large numbers of competent all-round naturalists and ecologists to make sense of all that information. I don't believe that freeing up time by not having to learn the salient identification features of different species will leave you space to learn about the ecology of those species – the learning of one is closely linked to the learning of the other.

AI identification is not the same as picture matching

At the age of around five I learned the joy of picture matching a Goldfinch *Carduelis carduelis* from a bird book that I found at home. This very first experience of looking through the possible species and whittling them down to one by a process of elimination was vital to me for everything that came afterwards. If apps had been available then, and I had used one to identify that finch, I don't know if I would be here writing this book. Perhaps I would just have become hooked on apps and maintained a much more superficial engagement with nature. What I can say for certain is that one of the greatest joys of natural history is figuring things out for myself with the available literature, and with occasional help from the wider community when necessary. I've never needed more than this. I would encourage pan-species listers and naturalists alike to go down the same route, using apps only sparingly or when there is no alternative. For me, repeat viewing of field guides has always been a very effective way to learn species, so that when I finally see them in the field I can identify them straight away. Just one session of picture-matching will involve looking between your specimen and the images in the book many times. Before you have figured out what it is, you'll have figured out what it isn't and over many sessions, these really stack up, resulting in long-lasting changes to your brain.

AI identification is not the same as using keys

Regular use of keys will cement in your mind exactly why something is a particular species, and why it isn't. It will also expose you to the (often scarcer) closely related confusion species and their salient features, long before you might ever encounter them, so that when you do finally see them you will be able to recognise them quickly. Finally, you will learn in great detail about the different parts of the species you are looking at. I fear that regular use of apps will mean that people miss out on all of these things in the future. At present, image recognition of such species, which are often very small, or need to be identified by underside characteristics, or features that are hard to photograph (such as a spider's palp), is rather beyond the capability of many ID apps. However, as the software improves, regular use of such apps could become problematic for all the reasons mentioned earlier.

AI identification is not the same as asking someone for help

When you ask for help with an identification there is a two-way of flow of information – the verifier learns from you at the same time as you learn from them, and in a social setting, lots of other people get to share the knowledge too. Apps tend to encourage people to make identifications in isolation, as a result of which they are likely to become less engaged with the wider recording community, rather than more so.

AI identification channels us down a photographic route

I have concerns that when an image becomes the main focus of an identification, the value of the associated data becomes almost meaningless. Information such as location, date, habitat, food plant, grid reference all become superfluous and will probably not even be considered, let alone collected. This kind of information would normally be used to help to clinch an identification, improve the value of records and to help make us more knowledgeable ecologists.

AI identification contributes little or nothing to surveys

One of the major benefits of PSL is that, as a result of years of study across multiple taxa, experienced pan-species listers are able to identify hundreds of species in the field within a short space of time. For me this has culminated in the 'one-man bioblitz' surveys of farms that I described in Chapter 11. Regular use of apps would not have helped me to reach the point of being able to do these surveys, nor would they add anything to the findings if I used them during the actual surveys. If I take any photos for identification purposes in the field, these are almost always of early lifecycle stages, micro-moths or of galls and identifying these back at my desk is often a highlight of my day. Occasionally at field meetings, someone jumps in here using an app and takes this enjoyment away from me (and ultimately the associated learning for me, them and the rest of the group, as well as the associated dopamine reward). As far as I am concerned, an impatient need for instant gratification (as opposed to waiting a few hours until you get home to do some research and picture matching) is not a valid reason for using these apps and suggests that many people use them far more often than they admit. Such 'metacognitive laziness' could lead to cognitive decline in the long run if used excessively.

Graeme Lyons is marked safe from AI

It is for these reasons that I have decided to ban all use of apps on any day at work, course, guided walk or field trip I am leading. I give people the choice: 'You can have my experience, or the app's.' They suck all of the enjoyment out of natural history for me, and I know I am not alone in feeling this. From clients using them to 'test' my field identifications, to people reaching for them faster than a wild west gunslinger when I make the rhetorical statement, 'What the heck is that?!' (I'm not asking for help here, it's an expression of excitement!). They add nothing, while they take away what few opportunities I have to learn new things.

I strongly encourage other naturalists to do the same. We might not be able to stop their use but we can at least make it clear how negative we think they can be. PSL isn't about instant gratification, it's about hard-won knowledge. If you don't have the time to invest in learning species identification, maybe you're in the wrong place.

Long-term effects of regular use of apps on our collective learning

Regular use of apps (including asking for confirmation of identification) could prevent us from developing a very important skill – *learning to know when we are right*.

With each new identification that I make myself, that sense of knowing when I am right (and when I am wrong) grows stronger. Regular use of an app will not allow this ability to develop to the same extent – it's a bit like learning to ride a bike but never reaching the point where you can remove the stabilisers, something that might be a hindrance if you want to work as a cycle courier, or win the Tour de France.

It's definitely better for naturalists to have a touch of impostor syndrome than to overestimate their knowledge and skills (known as the Dunning-Kruger effect). Good naturalists always question themselves – it's important always to remember that you might be wrong, and that you can change your mind. I still suffer from impostor syndrome even after 35 years of living and breathing natural history and have come to consider it a friend that keeps me on my toes; just don't listen to it so much that it stops you from doing things.

There is evidence that acquisition of knowledge changes the structure of the hippocampus in our brains, akin to 'the knowledge' of taxi drivers in London – this kind of learning can actually increase our IQ. We need to ensure that we only incorporate AI apps into PSL after careful consideration, so that we don't adversely affect our collective learning. Seeking help when you are stuck, attending courses and spending time with naturalists who know more than you (and with those who know less than you) are all a key part of learning.

Perhaps my greatest concern is that the regular use of apps may cause some people to start unlearning what they have already learned at a critical juncture, when they should be acquiring new knowledge (rather like losing your knowledge of the UK road network once you start using satnav systems). This has the potential to cause major repercussions for natural history in years to come, by which time it may be too late to do anything about it.

The kinds of problems that could beset a naturalist armed with a smartphone and an ID app include running out of batteries, loss of signal, the potential to miss new species, being restricted by the photographic approach to identification (as there would be a tendency to avoid anything very small or distant), and being slowed down while waiting for new photos to be uploaded and processed.

I was pretty horrified to find that many environmental consultants use these apps in their day-to-day work for identifying common plants, including some staff at Natural England using them on their SSSI assessments. However, AI apps only really work on species that you can picture match – the 'easier' end of natural history. Using these in a professional sense is categorically not the same as employing a competent botanist and significant mistakes will be made, nor is it equivalent to teaching yourself with a book. Most serious botanists working in the field will have learned how to identify the majority of common plants by eye (only needing to key out the ones that they rarely see, difficult groups, or are seeing for the first time), as well as learning their associated ecology, conservation status and habitat management advice, and where to look for them. This will be near impossible to learn on the job with an app and it's vital that employers recognise this is a highly skilled job that can't be replaced by people with apps. Someone with such a wealth of knowledge and skills will always be faster, more efficient and more accurate when recording in the field than someone who is heavily or entirely dependent on the apps.

Use of apps for identification when there is no alternative

This is the situation in which apps are most likely to be useful – for example, to identify the many naturalised plant species that are simply not in the keys. However, as more books, keys and resources are updated to include such species, the need for apps to fill this gap will decrease.

I can see that people who cannot afford a library of books might see this as an alternative in the short term but, in the long term, this is likely to lead to many of the issues I mention above. Although it might seem like it looking at my list, I have never tried to tackle all of the groups at once. I have waited until I have either bought the books, found the keys or acquired the time to take on new groups comprehensively, and I strongly recommend this approach. Someone pointed out on the Facebook group that you would need to spend £450 to cover all beetles by buying Duff's four-volume set as a reason as to why apps are such value for money, but AI is not an alternative to these keys. You could only get a proportion of UK beetles to species using image recognition, and even that would come with a huge margin of error. There are lots of free online resources for identifying beetles mentioned earlier in this book and no one is going to become a professional entomologist by using AI alone.

Conclusion

AI apps are clearly of benefit if they result in far more people engaging with the natural world, even at a relatively superficial level (e.g. through using birdsong apps). However, many of the pan-species listers and naturalists I know grew up without these apps, and most of us still don't use them today, or only use them occasionally, in the kinds of situations outlined earlier.

I can, however, see the benefits of having far more people engaged with the natural world, even if that is mostly in a superficial sense – the Merlin birdsong app certainly helps do that. But will this be worth the price we could pay in the future regarding our collective learning? Will the many more people recording at the entry level offset the number of people that might be prevented from taking it to a higher level, for example?

To me, just as I got to the point of being able to write a book about how to train your brain to operate like a walking biological image-recognition database, a device comes along and tries to take that opportunity away from you, by making you believe there is an equivalence between the two things. It's really important that anyone reading this recognises that they're totally different things; one is a method that has provided to me and the natural world, all of the benefits I have written about in this book, the other a tool I have never once felt I needed to use.

Although I do believe regular use of apps pushes us beyond the 'your list, your rules' red line, with its wide blurred edges, I do not think that occasional use does, but even that is not for me. In the coming decades it is likely that we will increasingly outsource knowledge and learning to the machines that we create, our brains shrinking in the process... and that makes me feel really terrified for the future. One thing is certain – love them or loathe them, AI apps are here to stay.

eDNA

An interesting question came up recently in the PSL Facebook group. Should we include species identified by eDNA? After some lengthy discussion the general conclusion was that we were against this idea. The key reason for me is that these species would not be identified by us, and PSL is all about doing identifications ourselves.

I then had an interesting conversation with Sam Thomas, who pointed out that this area of identification is going to expand rapidly in the coming years, and also

A fundamental difference exists between the primary purpose of iNaturalist and that of the various biological recording schemes. iNaturalist is first and foremost a vehicle intended to help people forge closer connections with nature. It encourages people to slow down, take a closer look at non-human organisms, photograph them and then share their observations with the online world. It is a way to get more people thinking about and hopefully caring for the state of nature.

Generating valuable scientific data, on the other hand, is of secondary importance to iNaturalist. Ken-ichi Ueda, the founder of iNaturalist, emphasised this point when he wrote that 'the data iNaturalist produces is a byproduct. If it's messy but people are outside looking at stuff, then the system is fulfilling its purpose'.

That having been said, I know a number of dedicated iNaturalist users, myself included, who chose to primarily use it to create biodiversity data that might be of value to scientists, conservationists, researchers and fellow naturalists. I know for a fact that my data have been used for scientific purposes, because over the years I have been contacted by multiple researchers as well as an ecologist from the Pennsylvania Natural Heritage Program. Thanks to my iNaturalist observations, the creeks and forested areas around my home have been designated as the Kelly Station Road Natural Heritage Area.

I have also found value in iNaturalist as a convenient way to maintain a PSL list. While it is true that iNaturalist is based on users providing a digital voucher such as a photo or an audio recording, there is a workaround – users of iNaturalist have the option of submitting what is termed a 'casual observation'; these observations do not require a digital voucher. For example, if I see a Pileated Woodpecker *Dryocopus pileatus* fly overhead but I do not get a photo, I can still add it to my list as a casual observation. Casual observations are generally excluded from scientific data sets, but the organisms will appear on your personal list if you do the proper search of the iNaturalist database.

It is clear that a version of PSL is developing (or could develop) in countries around the world within iNaturalist, including some of the gentle competition that works so well to spur people on. Where PSL differs, however, is that we have a bespoke system of maintaining a list, whereas iNaturalist is a way of engaging people with wildlife and making records, with any listing facility being secondary to this.

Another feature of iNaturalist is its flexibility. It's very easy to see rankings for a state, country or even a species. So although I have my concerns about its use in the British Isles, where we already have a more appropriate system (iRecord), I can see that its wider use around the world is a very good thing for the recording of all taxonomic groups. If there are already both countrywide and state species rankings across all taxa in the USA, we could perhaps say that PSL is already happening there under a different name.

PAN-SPECIES LISTING IN OTHER COUNTRIES 379

Pileated Woodpecker *Dryocopus pileatus*. (John Boback)

Transparent Burnet *Zygaena purpuralis* on Bloody Crane's-bill *Geranium sanguineum* on the Burren.

North-western Europe

Of all these countries, Ireland, Germany, Belgium, the Netherlands and Luxembourg are probably most likely to tick enough boxes to make PSL a realistic possibility. If anyone in north-western Europe is reading this, do get in touch! We have a template ready to help you to set up something similar to our version of PSL.

Ireland

As PSL is set up for Ireland on the website, I reached out to Liam Lysaght at the National Biodiversity Data Centre of Ireland, who provided the following information:

> One of the great benefits in Ireland is that there is one records centre for the whole country, and one very obvious way to enter records into this – Ireland's Citizen Science Portal. We encourage people to use that in Ireland, not iRecord.

Website

Ireland's Citizen Science Portal: www.records.biodiversityireland.ie
> The records centre covers the entirety of Ireland. Therefore a PSL recording area has been set up on our website that allows for this – just select 'Ireland' from the drop-down menu.

Facebook group

PSL: Ireland
> It's still early days, but we have set up a Facebook group to help the embryonic PSL community to develop in Ireland. There are currently just eight people on the rankings and the top lister is Karl Woods, with 1,141 species, but I hope numbers will grow over the coming months and years.

Chapter 18

Lifetime strategies for pan-species listing

I've already discussed at length how pan-species listing (PSL) is a hobby that can provide enjoyment and a sense of purpose for a lifetime, as well as being good for your physical and mental health. But what are the best strategies for getting a good list, or the best possible list for you?

Start as early as you can

This is the number one strategy for getting a big list. Of course you can start PSL at any age, and if you are in your eighties and have discovered that you have been pan-species listing for years without realising it, do get signed up now and add your species to the new website! I was 32 when I pulled my list together, and the first draft only included 2,748 species, despite all the years I'd been pursuing a passion for natural history. Several of our most prolific pan-species listers (including Duerden Cormack, Finley Hutchinson, Nathan Jackson, James McCulloch, Louis Parkerson and Harry Witts) are only in their early to mid-twenties, but have already seen more species than I had by my early thirties, such are the benefits of starting early. They are all great naturalists already, and likely to be formidable ones in a decade or so.

Win the lottery and quit your job

Unless you inherited a vast financial empire and are able to fund a lifetime of pan-species listing antics, then this might be your only option for quitting your job and going full time PSL. Seriously, though, I do keep waiting for a pan-species lister to emerge with a near infinite expendable income, and they could be unstoppable; but only if they learnt as they went along and did not just simply go around ticking things off. For the time being, however, this is not a thing, and it's probably better for everyone else's morale that it stays that way.

Get a job associated with biodiversity

If you want to spend almost every day outside seeing and identifying species that are new to you, you might achieve this through a job in the conservation side of the ecological sector. These jobs may be considerably fewer in number than the ecological consultancy jobs out there, but unless the consultancy is one of the rare ones that focus on conservation work, I would not generally recommend that kind of work as a way to see large numbers of species. Certainly as a pan-species lister I

What is the largest possible pan-species list that one person could archive?

This is really hard to quantify, but I suspect that some of our existing pan-species listers might be the first to reach 20,000 species. It probably won't be sufficient to work hard across the groups that you are really interested in – you will definitely need to be thinking strategically. The new website is a great resource for seeing which taxa contain the most species. For example, targeting algae is likely to be a very effective way to boost your list, but algae are hard to get into, and extremely time consuming, pulling you away from other more accessible taxa. Certainly no one is going to reach 20,000 by focusing solely on invertebrates, or solely on lower plants and fungi – you will need to become competent at both, something that will be a real challenge for the jobbing entomologists who currently dominate the top rankings.

And finally...

It is my hope that this book has inspired you to start pan-species listing. Now sign up on the website, get out there and do some recording – this could be the beginning of a lifelong love affair with the natural world, and you'll never be bored again. Happy listing!

www.panspecieslisting.com

Guess what? I caught up with my dream species – a Rainbow Sea Slug *Babakina anadoni* at Falmouth! – just in the nick of time before going to print. What amazing wildlife we have in the British Isles.

Index

Page numbers in *italics* indicate illustrations.

abbreviations, recording   265
Abdulla, Marilyn   351
Acanthodrilidae   110
Ackers, R. Graham   88
Acorn Barnacle *Semibalanus balanoides*   138
acorn worms (Hemichordata)   243
Acroceridae (hunchback flies)   187
aculeate hymenoptera   164
aculeate wasps   165–6
Adder *Vipera berus*   226
Adonis Blue *Polyommatus bellargus*   196, 329
Aeolidia filomenae   14
Aesculapian Snake *Zamenis longissimus*   227
Agapanthia villosoviridescens   175
Agile Frog *Rana dalmatina*   228
Agrimony *Agrimonia eupatoria*   326
Agromyzidae   193
AI identification apps and PSL   371–3
    picture matching and   372
alderflies (Megaloptera)   211, 213
algae   57–60
    diatoms (bacillariophyceae)   58
    freshwater algae   57
    seaweeds   60
    silicate-encased algae   58
    stoneworts (charophyta)   59
Alloeotomus germanicus   204
Allolobophora chlorotica   110
Alpine Bartsia *Bartsia alpina*   81, 85
Alpine Cinquefoil *Potentilla crantzii*   85
Alpine Newt *Ichthyosaura alpestris*   228
Alydus calcaratus   341
Amberley Wildbrooks   317
Ampedus elongantulus   286
amphibians   228–9

amphipods   292
Ampullaceana balthica   103
annelids   110–14
Annual Meadow-grass *Poa annua*   10, 265
ant-lions   211
ant-mimic jumping spider *Synageles venator*   354
Anthophora plumipes   263
Antigastra catalaunis   206
ants (Formicidae)   168
Anurida maritima (Finley Hutchinson)   141
anxiety   365
aphids (Aphididae)   161–2
APHOTOMARINE website   54, 114
aquatic bugs   158–9
Aquatic Heteroptera Recording Scheme   159
aquatic hyphomycetes   76
aquatic invertebrates   248–9
arachnids   119–28
    Halacaridae   126
    mites   126
    opiliones (harvestmen)   123–4
    pseudoscorpions (pseudoscorpiones)   125
    scorpions (scorpiones)   124
    spiders (Araneae)   119–23
    ticks (Ixodida)   128
    water mites (hydrachnidia)   128
Araniella alpaca   348
Archaeognatha   144–5
Arched Earthstar *Geastrum fornicatum*   74
Archidoris pseudoargus (Sea Lemon)   14
Argogorytes mystaceus   166
Armadillidium album   355
Armadillidium depressum   134
Armadillidium pulchellum   134
arrow worms (Chaetognatha)   242

Ascomycetes 54, 74
Ash *Fraxinus excelsior* 26
Ash Key Gall *Aceria fraxinivora* 126
Asian Desert Warbler *Curruca nana* 1
Asilidae (robberflies) 187
Aspen *Populus tremula* 244
Athericidae (water snipe flies) 187
Atherton, Ian 77
Atlantic Cod *Gadus morhua* 223
Atlantic Puffin *Fratercula arctica* 3
attention deficit hyperactivity disorder (ADHD) 5
Auchenorrhyncha (hoppers) 159–60
autism and attention deficit hyperactivity disorder (AuDHD) 5–6, 363–4

Bacciu, Nicola 348–51
Banded Demoiselle *Calopteryx splendens* 338
Barberry Carpet *Pareulype berberata* 308
Barkham, Patrick 197
barnacles (Cirripedia) 138–9
Barnard, Peter 211
Barnes, R.S.K. 250
Barrel Jellyfish *Rhizostoma octopus* 92
Basidiomycetes 54, 70–2
Basking Shark *Cetorhinus maximus* 224
bats 235–7
  active recording 236–7
  passive recording 236
Bechstein's Bat *Myotis bechsteinii* 236
Bee, Lawrence 120
bees 164–5
Bees, Wasps and Ants Recording Scheme (BWARS) 172
beet bugs (Piesmatidae) 158
Belshaw, Robert 194
Bentley, Chris 3
Beosus maritimus 158
Beroe cucumis 90–1
Beroe gracilis 90
Berytidae 158
Bibionids (St Mark's flies) 185
biological recording 259–66
  backlog processing from old notebooks 264
  basics 259–60
  casual recording 262
  collecting metadata 264

determiner 260
iNaturalist 260–2
incidental records 262
iRecord 260–2
measure of abundance 260
notes 260
patch listing 262–3
recording codes and abbreviations 265
sampling technique 260
sharing data 263
structured surveys 262
vice-counties 265–6
Biological Recording Handbook, The (Whild) 364
Biological Records Centre (BRC) 22
Birch Catkin Bug *Kleidocerys resedae* 158
birds 230–3
  getting a big bird list 232–3
  patch listing and surveying 232
  sea watching 232
Bird's-nest Orchid *Neottia nidus-avis* 318
bivalves 99–107
Black Goby *Gobius niger* 223
Black Grouse *Lyrurus tetrix* 1
Black Sea-bream *Spondyliosoma cantharus* 224
Blackcap *Sylvia atricapilla* 292
Blackthorn *Prunus spinosa* 35
Blair's Mocha *Cyclophora puppilaria* 283
Blamey, Marjorie 2, 83
Bloody Crane's-bill *Geranium sanguineum* 379
blow flies 186–7
Bloxworth Snout *Hypena obsitalis* 262
Blue-fin Tuna *Thunnus thynnus* 224
Blue Jellyfish *Cyanea lamarckii* 91–2
Blue-rayed Limpet *Patella pellucida* 12, 103
Bluebell Rust *Uromyces muscari* 73
Boat Bug *Enoplops scapha* 15
Boback, John 377
Bohemian Waxwing *Bombycilla garrulus* 4
Bombus ruderatus 354
Bombyliidae (bee flies) 187
Bootlace Worm *Lineus longissimus* 241
Bordered Straw *Heliothis peltigera* 206
Bosanquet, Sam 77
Botanical Society of Britain & Ireland (BSBI) 265
Bottle-nosed Dolphin *Tursiops truncatus* 225

INDEX   389

Bowe, Sarah   87
Brachypera zoilus   16
Brackish Water-crowfoot *Ranunculus baudotii*   250
Braconidae   172
Bradbury, P.   65
Brambling *Fringilla montifringilla*   262
Branchiopoda   140
Breadcrumb Sponge *Halichondria panicea*   87
Bright Wave *Idaea ochrata*   353
Brill *Scophthalmus rhombus*   223
Brimstone *Gonepteryx rhamni*   196
Brine Shrimp *Artemia salina*   140
British Arachnological Society (BAS)   123
British Bugs website   155
British Mycological Society (BMS)   75
British Trust for Ornithology (BTO)   265, 310
brittlestars (Ophiuroidea)   217
Broad-bordered White Underwing *Anarta melanopa*   200
Broad-shouldered Shieldbug *Cydnus aterrimus*   256
Brock, Paul D.   156
Brown-banded Carder Bee *Bombus humilis*   281
Brown Hydra *Hydra oligactis*   98
Brown Rat *Rattus norvegicus* and *Zilla diodia*   263
Brown Trout *Salmo trutta*   220
bryophytes   77–80
    liverworts   79
    mosses   78–9
bryozoans   108–9
'BTO-style' codes   265
BUBO listing   25–7
Buckton, Sam   354–5
Buff-breasted Sandpiper *Calidris subruficollis*   291
Bullhead *Cottus gobio*   220
Burton Pond bog and acid grassland   317
Bush-crickets   149–50
Butterflies   196–8
Butterfly Isles, The (Barkham)   197

caddisflies (Trichoptera)   210–11
Calipobrola speciosa   190
Calliphoridae   186–7
Calosoma sycophanta   17, 352, 383
Calvadosia campanulata   94
Campodea staphylinus   142
Candelabrum cocksii   98
Candy-striped Flatworm *Prostheceraeus vittatus*   116, 247
Carpal Tunnel Syndrome   360
Carthusian Snail *Monacha cartusiana*   99–100
Cassida rubiginosa (Thistle Tortoise Beetle)   179
casual recording   262
Catriona aurantia   246
Cat's-ear *Hypochaeris radicata*   84
cattle   273
Cave Spider *Meta menardi*   253
centipedes (Chilopoda)   130–1
Centromerus albidus   323
cephalopods   107
Ceratapion carduorum   292
Ceratapion gibbirostre   292
Cestoda (tapeworms)   115, 117
chafers   181–2
Chalcids (Chalcidoidea)   171–2
Channel Islands   256–7
Chequered Skipper *Carterocephalus palaemon*   197, 308
Chimney Sweeper *Odezia atrata*   84
Chinery, Michael   309
chitons   99–107
Chrysoperla carnea agg.   211
Chthonius ischnocheles   125
ciliate protozoa   67
Cistus Forester *Adscita geryon*   329
City Nature Challenge (CNC)   261, 266, 325–6
Cladocera   140
Clancy's Rustic *Caradrina kadenii*   283
Clanoptilus barnevillei   353
Clement, Eric J.   85
click beetles (Elateridae)   176–7
cliffs   254
Clifton, Jon   202
clown beetles (Histeridae)   177–8
clubmosses   81
Clytra quadripunctata   179
Cnidarians   92–8
    freshwater hydrozoans   98
    hard corals (Scleractinia)   96–7

hydroids (Hydrozoa) 97–8
jellyfish (Scyphozoa) 92–3
sea anemones and corals (Anthozoa) 95
sea fans (Alcyonacea) 96
soft corals 96
stalked jellyfish (Staurozoa) 94
Cockchafer *Melolontha melolontha* 343
cockroaches 152–3
codes, recording 265
Coleophora pennella 207
Coleoptera (beetles) 173–83
    click beetles (Elateridae) 176–7
    clown beetles (Histeridae) 177–8
    dung beetles and chafers (Scarabaeidae and Geotrupidae) 181–2
    false clown beetles (Sphaeritidae) 177–8
    ground beetles (Carabidae) 180
    ladybirds (Coccinellidae) 174–5
    leaf beetles (Chrysomelidae) 179–80
    longhorns (Cerambycidae) 175–6
    rove beetles (Staphylinidae) 182–3
    soldier beetles (Cantharidae) 176
    water beetles 178–9
    weevils (Curculionidae) 180–1
Coleopterists Society of Britain and Ireland (ColSoc) 183
Collaborative competition and PSL 315–26
    1,000 for 1KSQ 315
    1,000 lifers in a year 316
    1,000 species in 24 hours 316–20
    1,500 species in 24 hours 320–2
    blind pan-species year listing 324–5
    city nature challenge 325–6
    pan-species year listing 324
    spider year-listing challenge 322–4
    Tony Davis & Seth Gibson's annual challenges 326
Colour Identification Guide to Moths of the British Isles 2
Columbus Crab *Planes minutus* 136
Common Bird's-foot-trefoil *Lotus corniculatus* 289
Common Cuttlefish *Sepia officinalis* 225
Common Earwig *Forficula auricularia* 153
Common Goose Barnacle *Lepas anatifera* 136
Common Green Grasshopper *Omocestus viridulus* 292
Common Hawker *Aeshna juncea* 33
Common Hydra *Hydra vulgaris* 98
Common Knapweed *Centaurea nigra* 84
common names on social media 288
Common Prawn *Palaemon serratus* 136
Common Squid *Loligo vulgaris* 107
Commophila aeneana 203
Comont, Richard 318
Compass Jellyfish *Chrysaora hysoscella* 92
Conarthrus praeustus 354
coneheads or telsontails (Protura) 142
Conger Eel *Conger conger* 224
Connemara Clingfish *Lepadogaster candolii* 222
Conopidae (bee-grabbers) 185
conservation status 280–1
contribution 43
    concept of 43–4
Cooper, Danny 303
Cooper, Steve 2
copepods or sea lice (Copepoda) 139
Corallimorpharia 96
corals (Anthozoa) 95
Cormack, Duerden 303, 354–5
Corncrake *Crex crex* 231
crabs, lobsters, shrimps and prawns (Decapoda) 136–7
Crambids 201–2
craneflies 192–3
Crawley, Mick 265
crickets 149
Crossbill *Loxia curvirostra* 283
Crucian Carp *Carassius carassius* 221
crustaceans 133–40
    barnacles (Cirripedia) 138–9
    Branchiopoda 140
    copepods or sea lice (Copepoda) 139
    crabs, lobsters, shrimps and prawns (Decapoda) 136–7
    freshwater crustaceans 134–5
    large marine crustaceans (Malacostraca) 135
    marine amphipods 138
    marine isopods 137–8
    seed shrimps (Ostracoda) 139
    water fleas (Diplostraca) 140
    woodlice 133–4
Crypticus quisquilius 354
Cryptocephalus bipunctatus 350
Cryptocephalus rufipes 255–6

ctenophorans (comb jellies)   90–1
Curculio villosus   245
cuttlefish   107
cyanobacteria (blue-green algae)   23–4, 56
Cycloporus papillosus   116
Cylindroiulus londinensis   130
Cylindrotomidae   192–3

Dace *Leuciscus leuciscus*   221
Dahlia Anemone *Urticina felina*   95
Dalman's Leatherbug *Spathocera dalmanii*   156
damselflies   146–8
Dark Crimson Underwing *Catocala sponsa*   37
Dark Green Fritillary *Speyeria aglaja*   198
Dartford Warbler *Curruca undata*   232
data flow   263–4
Davey, Simon   347
Davis, Tony   33, 326
day-flying moths, recording   206–7
Dead Men's Fingers *Alcyonium digitatum*   96
Death's-head Hawk-moth *Acherontia atropos*   10
Deer Tick/Castor Bean Tick *Ixodes ricinus*   128
Dendrocoelum lacteum   116
Dendroxena quadrimaculata   177
Dentated Pug *Anticollix sparsata*   318
Denton, Jonty   303–4
Dermaptera (earwigs)   153
Dermestes haemorrhoidalis   204
Desert Wheatear *Oenanthe deserti*   305
Devil's-bit Scabious *Succisa pratensis*   84
Devonshire Cup Coral *Caryophyllia smithii*   97
Dewick's Plusia *Macdunnoughia confusa*   206
Diastrophus rubi   244
diatoms (Bacillariophyceae)   58
Dicranum majus (acrocarpous moss)   78
Dictyoptera (cockroaches)   152–3
digital resources   294–7
   data   295–6
   Geographic Information Systems (GIS)   297
   photography   294–5
   pivot tables   297

Dingy Skipper *Erynnis tages*   21
Diplocephalus latifrons   288
Diplocephalus picinus   288
Diplostraca   140
Diptera (true flies)   184–95
   Agromyzidae   193
   bibionids (St Mark's flies)   185
   blow flies   186–7
   Conopidae (bee-grabbers)   185
   craneflies   192–3
   Empididae (round-headed flies)   189
   Eudasyphora cyanella   192
   Muscidae (house flies)   191–2
   picture-winged flies (Tephritidae)   189
   Sarcophagidae (flesh flies)   188
   Scathophagidae (dung flies)   186
   Sciomyzidae (snail-killing flies)   188–9
   soldier flies and allies   187–8
   Syrphidae (hoverflies)   189–91
   Tachinidae   194–5
   wing-wave flies (Ulidiidae)   189
Discodoris rosi   106
Dobson, Frank S.   68
dogs   273
Dolichopodidae (long-legged flies)   191
Domino Sun-jumper *Heliophanus kochii*   122
Doris verrucosa   14
Dover Sole *Solea solea*   223
dragonflies   146–8
Drilus flavescens   318
Duke of Burgundy *Hamearis lucina*   348
Dune Tiger Beetle *Cicindela maritima*   354
dung beetles   181–2
Dunlin *Calidris alpina*   338
Dusky Cockroach *Ectobius lapponicus*   152
Dusky Doris *Onchidoris bilamellata*   12
Dusky Sallow *Eremobia ochroleuca*   205
Dwarf Snail *Punctum pygmaeum*   100
Dwarf Thistle *Cirsium acaule*   84
Dyer's Greenweed *Genista tinctoria*   157
Dysmachus trigonus   187

Early Sand-grass *Mibora minima*   82
earthworms   110–14
Ebernoe Common   317
echinoderms   215–17
   sea cucumbers   216
   sea stars   217
   sea urchins   26

longhorns (Cerambycidae) 175–6
Loxostege sticticalis 206
Lugg, Keith 351–2
Luker, Sally 309, 318, 353
Lumbricidae 110
Lygaeoidea (seed bugs) 158
Lyme disease 272
Lyons, Graeme 349, 351
    Lyons' razor 267

Mackerel *Scomber scombrus* 223
macro-moths 199–201, 204–8
    early stages 204–6
Magpie Inkcap *Coprinopsis picacea* 71
Malacostraca 135
mammals 234–8
    larger mammals 237–8
    small mammals 237
Mantodea 153
Manuel, R. 249
Mapp, Ben 309–10
marine amphipods 138
marine bivalves 106–7
marine fish 221–5
marine gastropods 104
marine isopods 137–8
marine leeches 112
marine molluscs 103
marine oligochaetes 114
marine turbellaria 117
marine worms 110–14
Marram Grass Chelifer *Dactylochelifer latreillii* 125
Marsh Frog *Pelophylax ridibundus* 228
Marsh Ragwort *Jacobaea aquatic* 4
masking 6
Matcham, Howard 57
Mauve Stinger *Pelagia noctiluca* 92
mayflies (Ephemeroptera) 209
McCormack, Duerden 353
McCulloch, James 15, 310, 340
Meadow Bug *Leptopterna dolabrata* 157
Meadow Crane's-bill *Geranium pratense* 84
Mediterranean House Gecko *Hemidactylus turcicus* 227
Megabunus diadema 123
meltdowns 367
memory training and brain hacks 288–93
    active recall techniques 288–90

    amphipod or isopod? 292
    Blackcaps sound like R2-D2 292
    emotional connections and cultural references 291–2
    mnemonics 290–1
    orthoptera sounds 292
mental health 274–5
Merryweather, James 83
Mesosa nebulosa 173
metadata, collecting 264
Metatrichia floriformis 61
Metellina mengei 290
Metellina merianae 253, 290
Metellina segmentata 290
Micrasterias rotata 58
micro-moths 201
    early stages 207–8
Micrommata virescens 120
microscopes 267–8
microspecies and difficult families 86
Microworld (website) 66
Midwife Toad *Alytes obstetricans* 228
Mike's Insect Keys 54
millipedes (Diplopoda) 130
Minnow *Phoxinus phoxinus* 221
Miridae 157–8
mites 119–28
    Aceria squalida 244
    gall mites 127
    soil mites 127
mnemonics 290–1
molluscs 99–107
    cephalopods 107
    freshwater bivalves 102
    freshwater molluscs 102
    freshwater snails 102–3
    marine bivalves 106–7
    marine gastropods 104
    marine molluscs 103
    sea slugs (Nudibranchia) 104–6
    terrestrial slugs 101–2
    terrestrial snails 99–100
Monkey Orchid *Orchis simia* 16, 20
Monogenea 115
Moon Jellyfish *Aurelia aurita* 92
mosses 77–8
moths 199–209
    adult day-flying moths, recording 206–7

by-catch 204
crambids 201–2
dissection 203
light trapping 204
macro-moths 201, 204–6
micro-moths 201, 207–8
plume moths 203–8
pyralids 201–2
tortrix moths (Tortricidae) 202–3
Mottled Umber *Erannis defoliaria* 205
Mountain Ringlet *Erebia epiphron* 197
mountainous areas 258
mud dragons (Kinorhyncha) 242
Muscidae (house flies) 191–2
Musgrove, Andy 14, 25, 310–11, 344
Musk Beetle *Aromia moschata* 350
Myers, Joe 360
myriapods 129–32
centipedes (chilopoda) 130–1
millipedes (diplopoda) 130
pauropods (pauropoda) 131
symphylans (symphyla) 132

Narrow-fruited Cornsalad *Valerianella dentata* 329
National Vegetation Classification (NVC) communities 4
Natterjack Toad *Epidalea calamita* 304
Natural History Book Service (NHBS) 53
Natural History Museum (NHM) 25
NatureSpot 55
NBN Atlas 55
nematodes (Nematoda) 240
Neobisium carcinioides 125
Neptune's Heart Sea Squirt *Phallusia mammillata* 219
Nesticus cellulanus 253
Neurigona quadrifasciata 191
neurodivergence and PSL 363–70
anxiety 365
coming out as neurodiverse 369
complexity and nuance, embracing 367–8
dopamine 364–5
getting assessed 369–70
meltdowns and triggers 367
natural history and 363–70
neurodivergence assessment 366

and workplace 368–9
young neurodivergent naturalists, nurturing 369
neurodivergence and shame 366–7
New Forest Cicada *Cicadetta Montana* 159
Newman, Jonathan 347, 349–51
Night-flowering Catchfly *Silene noctiflora* 329
Nightingale *Luscinia megarhynchos* 232, 318
Nightjar *Caprimulgus europaeus* 318
Niphargus aquilex 134
Noctule Bat *Nyctalus noctula* 236
Norfolk Hawker *Aeshna isoceles* 147
North-western Europe 380
Nostoc commune 56
notebooks in the field 271
Nottingham Catchfly *Silene nutans* 84
nudibranchs 104–6
Nursehound *Scyliorhinus stellaris* 222

Oak Apple *Biorhiza pallida* 245
obsessive-compulsive disorder (OCD) 364
Occam's razor 267
Octopi 107
Odonata 146–8
Oedemera femoralis 16
Ogden, C.G. 65
Okenia nodosa 14
Old World Webworm *Hellula undalis* 283
Omophlus pubescens 352
Ophonus melletii 204
Opiliones (harvestmen) 123–4
Opilo mollis 348
Orange Bird's-foot *Ornithopus pinnatus* 85
Ormyrus nitidulus 171
Orthoptera (crickets, bush-crickets, grasshoppers and groundhoppers) 149–52
sounds 292
orthopteroids 149–53
Dermaptera (earwigs) 153
Dictyoptera (cockroaches) 152–3
Mantodea (mantises) 153
stick-insects 152
Osprey *Pandion haliaetus* 3
Otter *Lutra lutra* 16
Outlaw, Iain 318
Ovenden, Denys 149

Oxeye Daisy *Leucanthemum vulgare* 84
Ozyptila claveata 348

Painted Top Shell *Calliostoma zizyphinum* 136
Pallas's Warbler *Phylloscopus proregulus* 1
Palpita vitrealis 206
pan-species approaches to surveying and monitoring 327–37
  biodiversity farm surveys in Sussex 330–1
  field identification 335–7
  recording multiple taxa in practice 332–5
Pan-species Listing (PSL)
  aliens 42
  alive or dead 41
  benefits 9–21
    each species counts as one 10
    improves species ID skills 10–11
    improves taxonomic knowledge 11
  building and maintaining 44
    on PSL website 45
  and collaborative competition 315–26. *See also* Collaborative competition and PSL
  contribution 43–4
  covert PSL 376
  data back up 50
  developmental stage 42
  downsides to 19–21
    competitiveness 19
    twitching 19–21
  filling in the gaps and deleting species 49
  free or captive 41
  from scratch or pulling all old records together 45–6
  goals of 9–10
  guiding principles 37–43
  hybrids 42
  life changes by, question of 301–14
  pending species not yet on website 42
  representation and demographics in 356–62
  of sites 338–42
    creating 338–9
    listing an entire reserve network 339–42

  taxonomy 41
  teaches handling large data sets 11
  timeline 22–7
    beginning 22
    and BUBO listing 25–7
    old website 22–3
    taxonomy of the old website 23–4
    UK Species Inventory (UKSI) 25
  unexpected finds 15–17
  in unusual habitats and specific situations 244–58
    aquatic invertebrates 248–9
    caves, tunnels and other subterranean habitats 252–3
    channel islands 256–7
    cliffs 254
    galls and leaf mines 244–5
    hot houses, botanical gardens and garden centres 251–2
    London 255–6
    microscopic invertebrates 249
    mountains 258
    quarries and scree 253–4
    rock pooling 245–8
    saline lagoons 250
    urban settings 255
  updating and keeping records 49
  website. *See* PSL website
  See also lifetime strategies for PSL; PSL website; threats to PSL
pan-species tourism 386
Panorpa communis 212
Panorpa germanica 212
Pantaloon Bee *Dasypoda hirtipes* 164
Parasitic Anemone *Calliactis parasitica* 95
Parkerson, Louis 13, 311
Parsons, Mark 201
Pasqueflower *Pulsatilla vulgaris* 81
patch listing 232, 263
Patton, Sarah 311, 347, 349–50
Patton's Tiger *Hyphoraia testudinaria* 311
pauropods (Pauropoda) 131
Payne, Alex 354
Peacock *Aglais io* 198
peanut worms (Sipuncula) 114
Pediciidae 192–3
Pedunculate Oak *Quercus robur* 206
Pelagella castanea 14
Pellenes tripunctatus 121

INDEX 399

Pemberley Books   53
Pennsylvania Natural Heritage Program   378
Perch *Perca fluviatilis*   220
Perennial Knawel *Scleranthus perennis*   308
Perforate St John's-wort *Hypericum perforatum*   207
personal offline database   46
Petersen, Jens H.   71
Phasmatodea (stick-insects)   152
Philodromus margaritatus   336
Philoscia muscorum   133
Phoenix *Eulithis prunata*   26
Pholcus phalangioides   262
Phosphorescent Sea Pen *Pennatula phosphorea*   96
photography   294–5
Phytomyza hellebori   193
picture-winged flies (Tephritidae)   189
Piesmatidae   158
Pignut *Conopodium majus*   84
Pilchard *Sardina pilchardus*   222
Pileated Woodpecker *Dryocopus pileatus*   379
Pilemostoma fastuosa   322
Pine Ladybird *Exochomus quadripustulatus*   290
Pine Marten *Martes martes*   237
Pink Water-speedwell *Veronica catenata*   244
pivot tables   297
planarians   117
plant bugs (Miridae)   157–8
    and allies   157
Planuncus tingitanus   152
platyhelminths   115–17
Platystomos albinus   181
Pleurobrachia pileus   90
Pleurozia purpurea   79
plume moths (Pterophoridae)   203–8
Plumed Fan-foot   283
    Pechipogo plumigeralis   338
Plummer, Stephen   62, 311–12, 318
Plumose Anemone *Metridium dianthum*   95
Poland, John   85, 354
Polleniidae   186–7
Polycera quadrilineata   14, 104
polychaetes   113

Pompilidae (spider-hunting wasps)   166–7
Pondweed Leafhopper *Erotettix cyane*   160
Pool Frog *Pelophylax lessonae*   228
Porcellionides pruinosus   133
Porcupine Marine Natural History Society   89
Porifera (sponges)   87–9
Portuguese Man o' War *Physalia physalis*   92
potassium hydroxide (KOH)   68
powdery mildews   75
Praying Mantis *Mantis religiosa*   20, 153
Prince, Matt   349, 351
Prince, Mike   25, 344
Privet Hawk-moth *Sphinx ligustri*   338
Procumbent Pearlwort *Sagina procumbens*   84
protists other than algae and slime moulds (polyphyletic group)   63–4
pseudoscorpions (Pseudoscorpiones)   119–28
PSL field meetings   346–55
PSL website   28–50
    blockers   35–6
    list   29
    Listing Milestones   32–3
    new website   28–9
    personal database/records in Excel, managing   46–9
    personal offline database   46
    preparatory work   44
    species new to the British Isles   31–2
    statistics box   29–31
    taking part in   28–50
    targets   33–5
        with the birds removed   34
        at the start of October 2024   33
        top 20 pan-species listers   34
    and UKSI   31
Psocodea (bark, book, feather and body lice)   160–1
Psylloidea   162
Ptychopteridae   192–3
public engagement and PSL field meetings   343–55
    global Birdfair   344–5
    Lyons share   344
    running a bioblitz   345–6
    TinyRecorder   343

Purple Emperor *Apatura iris* 26
Purple Marbled *Eublemma ostrina* 206
Purple Moor-grass *Molinia caerulea* 81
Pycnogonids (sea spiders) 118
Pygmy Backswimmer *Plea minutissima* 159
Pyralids 201–2
    Pyrausta ostrinalis 202

Quadrulella symmetrica 63
quarries and scree 253–4
Queen of Spain Fritillary *Issoria lathonia* 197

Radford's Flame Shoulder *Ochropleura leucogaster* 283
Radix auricularia 103
Rainbow Sea Slug *Babakina anadoni* 98, 404
Ranking News 32
Raper, Chris 31
Red Dead-nettle *Lamium purpureum* 263
Red-eared Terrapin *Trachemys scripta* 227
Red Fox *Vulpes vulpes* 289
Red Sheep Tick *Haemaphysalis punctata* 128
Red Squirrel *Sciurus vulgaris* 35
representation and demographics in PSL 356–62
    disabilities 359–61
    people from ethnically diverse backgrounds 359
    people from more disadvantaged backgrounds 361
    women in 356–9. *See also* women
    young people 361–2
reptiles 226–7
Restharrow (moth) *Aplasta ononaria* 353
Reticulated Dragonet *Callionymus reticulatus* 223
Reynaert, Steve 353
Rhabditophora 115
Rhagionidae (snipe flies) 187
Rhagonycha fulva 176
Rhiniidae 186–7
ribbon worms (Nemertea) 241
Richard's Pipit *Anthus richardi* 1
Rivularia bullata 56
Roach *Rutilus rutilus* 220
Roberts, Michael J. 120

Robin *Erithacus rubecula* 35
Robinson trap 204
Rock Goby *Gobius paganellus* 222
rock pooling 245–8
Rock Speedwell *Veronica fruticans* 85
Roesel's Bush-cricket *Roeseliana roeselii* 151, 292
Roper, Charles 22–3
Rose, Francis 83
Rose Chafer *Cetonia aurata* 17
Rosy Feather-star *Antedon bifida* 217
rotifers (Rotifera) 239, 240, 250
Rough Hawkbit *Leontodon hispidus* 84
Round-headed Rampion *Phyteuma orbiculare* 84
rove beetles (Staphylinidae) 182–3
    Emus hirtus 302
Roy, David 22
Royal Entomological Society (RES) 54
Rubus fruticosus agg. 244
Ruffe *Gymnocephalus cernuus* 221
Ryland, J.S. 88
Rylands, Kev 312

saline lagoons 250
salps (pelagic tunicates) 218–19
San Diego Sea Squirt *Botrylloides diegensis* 219
Sand Lizard *Lacerta agilis* 226
Sand Smelt *Atherina presbyter* 222
Sandham, Robin 344
Sandwich Bay Bird Observatory 356
Saproxylic Quality Index 334
Sarcophagidae (flesh flies) 188
Sardet, Eric 150
sawflies (Symphyta) 168–9
scale insects 163
Scaly Cricket *Pseudomogoplistes vicentae* 15
Scarabaeidae 181–2
Scarce 7-Spot Ladybird *Coccinella magnifica* 174
Scarce Bordered Straw *Helicoverpa armigera* 283
Scarlet Darter *Crocothemis erythraea* 148
Scathophagidae (dung flies) 186
Scenopinidae (window flies) 187
Sciomyzidae (snail-killing flies) 188–9
Sciopteryx soror 169
scorpionflies (Mecoptera) 211–12

scorpions (Scorpiones)   119–28
Scott, Brad   313
Scutigerella palmonii   132
Scythris and *Ochsenheimeria* species   206
sea anemones   95
Sea Bass *Dicentrarchus labrax*   222
Sea Club-rush *Bolboschoenus maritimus*   250
sea cucumbers (Holothurioidea)   216
sea fans (Alcyonacea)   96
Sea Gherkin *Pawsonia saxicola*   216
Sea Hare *Aplysia punctata*   106
sea mats   108–9
Sea Mouse *Aphrodita aculeata*   112
sea pens (Pennatulacea)   96
sea slugs (Nudibranchia)   104–6
sea squirts (Ascidiacea)   218–19
sea stars (Asteroidea)   217
sea urchins (Echinodea)   216
search image   270–1
seaweeds   60
Secombe, Matthew   354
seed bugs (Lygaeoidea)   158
seed shrimps (Ostracoda)   139
Segestria florentina   124
Self, Matt   4
Shanny *Lipophrys pholis*   105, 222
shieldbugs and allies   156
shore bugs   159
Shore Lark *Eremophila alpestris*   1
Short-winged Earwig *Apterygida media*   153
shorthand, recording   265
Shotbolt, Richard   351
Shredded Carrot Sponge *Amphilectus fucorum*   87
Sialis lutaria   213
Signal Crayfish *Pacifastacus leniusculus*   134
Silver-washed Fritillary *Argynnis paphia*   198
Skevington, Mark   312–13, 348–50
Skinner, Bernard   2
Slender Sea Pen *Virgularia mirabilis*   96
Slow-worm *Anguis fragilis*   227
slugs   99
Small, Julian   35, 313
Small Marbled *Eublemma parva*   206
Small Scabious *Scabiosa columbaria*   244
Small-spotted Catshark *Scyliorhinus canicula*   355
Smith, C.W.   68

Smith, Robert   351
Smooth Newt *Lissotriton vulgaris*   304
Smooth Snake *Coronella austriaca*   226
snails   99–107
snakeflies (Raphidioptera)   212–13
Snakelocks Anemone *Anemonia viridis*   95
Snakeskin Grisette *Amanita ceciliae*   318
Snow Flea *Boreus hyemalis*   212
Snowdon Lily *Gagea serotina*   10, 258
social media and natural history   298–300
   Facebook group etiquette   298
   'post hijacking', avoiding   300
   PSL Facebook group   298
   using experts   299
sodium hypochlorite (bleach) (NaOCl)   68
soft corals   96
soil mites   127
Solar-powered Sea Slug *Elysia viridis*   105–6
soldier beetles (Cantharidae)   176
soldier flies and allies   187–8
Sombre Brocade *Dryobotodes tenebrosa*   283
South Downs effect   296
Southampton Flatworm *Caenoplana variegata*   115
Southern Emerald Damselfly *Lestes barbarus*   147–8
Southern Migrant Hawker *Aeshna affinis*   26, 147–8
Southwood, T.R.E.   155
Sparganophilidae   110
Spaull, Zak   344
species names   284–8
   common names
      capitalising   285–6
      on social media   288
      use of   284
   correct way to write   289
   genus or generic level   288
   hyphens   287
   new common names, creating   284–5
Sphagnum mosses   249, 266
Spider Recording Scheme (SRS)   121
spider year-listing challenge   322–4
spiders (Araneae)   119–23
Spiny Squat Lobster *Galathea strigosa*   256–7
Spiny Starfish *Marthasterias glacialis*   216
Spiny Toad *Bufo spinosus*   228

Spitting Spider *Scytodes thoracica* 262
Spotted Flycatcher *Muscicapa striata* 21, 318
Sprawler *Asteroscopus sphinx* 8
springtails 141–2
Spurn Bird Observatory 354
squashbugs and allies 156
squids 107
Squinancywort *Asperula cynanchica* 84
St Piran's Crab *Clibanarius erythropus* 136
Stace, Clive A. 42, 82
stalked jellyfish (Staurozoa) 94
Staphylinus dimidiaticornis 183
Starfish. *See* sea stars 217
Starry Smooth-hound *Mustelus asterias* 225
Steatoda albomaculata 350, 353
Stenosoma lancifer 137
Sterry, Paul 70–1
stick-insects 152
Stigmella plagicolella 35
stilt bugs (Berytidae) 158
Stinking Hellebore *Helleborus foetidus* 193
Stone Loach *Barbatula barbatula* 220
Stonechat *Saxicola rubicola* 263
stoneflies (Plecoptera) 209–10
stoneworts (Charophyceae) 59
Stratiomyidae (soldier flies) 187
Stratiomys longicornis 187
Strawberry Anemone *Actinia fragacea* 343
Strawberry Worm *Eupolymnia nebulosa* 113
Streeter, David 82
Stripe-winged Grasshopper *Stenobothrus lineatus* 292
Strumpshaw Fen 221
stylops (Strepsiptera) 213
Summerfield Books 53
Sun-fish *Mola mola* 224
surveying and monitoring 327–37
structured surveys 262
Sussex Biodiversity Record Centre 23
Sussex Tiger *Nephrotoma sullingtoniensis* 325–6
Sussex Wildlife Trust (SWT) 4, 23
Sutton, Peter 149
Sweet Chestnut *Castanea sativa* 74
Sweet Vernal-grass *Anthoxanthum odoratum* 265
Syagrius intrudens 346, 347
symphylans (Symphyla) 132

Synageles venator 354
Syrphidae (hoverflies) 189–91

Tabanidae (horseflies) 187, 194–5
Tachina grossa 194
Tachyporus hypnorum 182
Tall Sea Pen *Funiculina quadrangularis* 96
Tansy Beetle *Chrysolina graminis* 350
Tanymastix stagnalis 140
tapeworms 115–17
tardigrades or water bears (Tardigrada) 241–2
Tawny Cockroach *Ectobius pallidus* 152
Tawny Owl *Strix aluco* 306
taxonomic groups
  accessing 51–243. *See also individual entries*
  breakdown of 54
  collective effort and top lister 53
  ease 52–3
    difficult 52
    easy 52
    moderate 52
    unknown 53
    very difficult 53
    very easy 52
  essential texts 53–4
  Field Studies Council's ID Resource Finder 54
  scale 52
  vital equipment 54
Tebenna micalis 206
Telfer, Mark 4, 12, 22–3, 268, 302–3, 346, 348–50
Tenebrionid *Omophlus pubescens* 352
Terpios gelatinosa 87
Terrestrial flatworms 115–16
Terrestrial slugs 101–2
Terrestrial snails 99–100
Testate amoebae (shell-bearing rhizopod protozoa) 65–7
Tethya citrina 88
Theonoe minutissima 341
Therevidiae (stiletto flies) 187
Theridiosoma gemmosum 283
Thomas, Sam 375
threats to PSL 371–6
  apps use for identification with no alternative 374

consultancy trap   376
eDNA   375–6
long-term effects of regular use of apps   373–5
See also AI identification apps and PSL
Thrift *Armeria maritima*   352
thrips (Thysanoptera)   163
Thyme-leaved Speedwell *Veronica serpyllifolia*   84
ticks (Ixodida)   119–28
  and Lyme disease   272
TinyRecorder   343
Tipulidae   192–3
Tompot Blenny *Parablennius gattorugine*   222
Topknot *Zeugopterus punctatus*   248
tortrix moths (Tortricidae)   202–3
Transparent Burnet *Zygaena purpuralis*   379
Trapezium Shieldbug *Coptosoma scutellatum*   154
Traveller's-joy *Clematis vitalba*   153
Tree-cricket *Oecanthus pellucens*   150
Tree Sparrow *Passer montanus*   1
Trematoda (flukes)   115, 117
Trichoceridae   192–3
triggers   367
Tub Gurnard *Chelidonichthys lucerna*   223–4
tunicates (sea squirts and salps)   218–19
Turbellaria (flatworms)   115
turning over stones and logs   271–2
Turtle Dove *Streptopelia turtur*   1, 232
Twin Fan Worm *Bispira volutacornis*   112
twitching   19–21, 230
Two-spotted Goby *Gobiusculus flavescens*   222

Ueda, Ken-ichi   378
UK Species Inventory (UKSI)   25, 43, 51
  number of species available on   51
Uloborus plumipes   251–2
Undulate Ray *Raja undulata*   225
urban settings   255
Urwin, Bill   314, 350
USA, PSL in   377

Vagrant Emperor *Anax ephippiger*   148
Van Toller, Simon   314, 354
Variable Chafer *Gnorimus variabilis*   181

vascular plants   81–6
  microspecies and difficult families   86
  vegetative botany   85–6
Vaulted Earthstar *Geastrum britannicum*   74
Vestal *Rhodometra sacraria*   206
Vesterholt, Jan   71
vice-counties   265–6
Villa cingulata   187
Viper's Bugloss *Echium vulgare*   207
Viviparous (Common) Lizard *Zootoca vivipara*   226

Walckenaeria mitrata   119
Walker, Dave   150
Wall Lizard *Podarcis muralis*   227
Wallace, Ian   211
Walrus *Odobenus rosmarus*   20
Warlock's Butter *Exidia nigricans*   285
Wart-biter *Decticus verrucivorus*   292
Washington, Clive   347, 349
water bears (Tardigrada)   241–2
water beetles (several disparate families)   178–9
water fleas (Diplostraca)   140
water mites (Hydrachnidia)   128
Water Stick-insect *Ranatra linearis*   159
Wavy Hair-grass *Avenella flexuosa*   81
waxcaps   70, 72
waxflies   211
weather   270
web-spinners (Embioptera)   213
weevils (Curculionidae)   180–1, 244, 245, 347, 354
Western Green Lizard *Lacerta bilineata*   227
Whelan, Paul   68
Whild, Sarah   364
White Admiral *Limenitis camilla*   198
White-clawed Crayfish *Austropotamobius pallipes*   134
Whitethroat *Curruca communis*   232
Wild Cat *Felis silvestris*   237
Wild Marjoram *Origanum vulgare*   84
Wild Thyme *Thymus drucei*   84
Willow Tit *Poecile montanus*   1
wing-wave flies (Ulidiidae)   189
Wireweed *Sargassum muticum*   105
Witch *Glyptocephalus cynoglossus*   223
Witches' Butter *Exidia glandulosa*   284
Witts, Harry   353

women
  competitiveness not being appealing 357–9
  fewer number in PSL, reasons for 356–9
  gender pay gap 357
  lack of time due to work and childcare commitments 357
  personal safety 357
  reluctance to engage in a male-dominated hobby 357
  steered away from such hobbies in childhood 358
Wood Sage *Teucrium scorodonia* 84
Woodcock *Scolopax rusticola* 318
woodlice 133–4
woodlouse flies (Rhinophoridae) 186–7
Woodpigeon *Columba palumbus* 306

Wormwood (moth) *Cucullia absinthii* 205

Xylomyidae (wood soldier flies) 187
Xylophagidae (awl flies) 187

Yates, Barry 343
Yellow Horned-poppy *Glaucium flavum* 305
Yellow Loosestrife *Lysimachia vulgaris* 318
Yellow-plumed Sea Slug *Berthella plumula* 106
Yellow-rattle *Rhinanthus minor* 84
Yellow-tailed Scorpion *Euscorpius flavicaudis* 124
Young Ornithologists Club (YOC) 1

Zygentoma 144–5